Horses and Risk

Horses and Risk

Editor

Kirrilly Thompson

MDPI • Basel • Beijing • Wuhan • Barcelona • Belgrade • Manchester • Tokyo • Cluj • Tianjin

Editor
Kirrilly Thompson
College of Health, Medicine
and Wellbeing
University of Newcastle
Awabakal land
Callaghan, NSW
Australia

Editorial Office
MDPI
St. Alban-Anlage 66
4052 Basel, Switzerland

This is a reprint of articles from the Special Issue published online in the open access journal *Animals* (ISSN 2076-2615) (available at: www.mdpi.com/journal/animals/special_issues/horses-risk).

For citation purposes, cite each article independently as indicated on the article page online and as indicated below:

LastName, A.A.; LastName, B.B.; LastName, C.C. Article Title. *Journal Name* **Year**, *Volume Number*, Page Range.

ISBN 978-3-0365-5046-6 (Hbk)
ISBN 978-3-0365-5045-9 (PDF)

Contents

About the Editor . . . vii

Jasmine Tanner, Chris Rogers, Charlotte Bolwell, Naomi Cogger, Erica Gee and Wayne McIlwraith
Analysis of Failure to Finish a Race in a Cohort of Thoroughbred Racehorses in New Zealand
Reprinted from: *Animals* **2016**, *6*, 36, doi:10.3390/ani6060036 . . . 1

Meredith Chapman and Kirrilly Thompson
Preventing and Investigating Horse-Related Human Injury and Fatality in Work and Non-Work Equestrian Environments: A Consideration of the Workplace Health and Safety Framework
Reprinted from: *Animals* **2016**, *6*, 33, doi:10.3390/ani6050033 . . . 13

Gabriella Gronqvist, Chris Rogers and Erica Gee
The Management of Horses during Fireworks in New Zealand
Reprinted from: *Animals* **2016**, *6*, 20, doi:10.3390/ani6030020 . . . 27

Julie M. Fiedler and Paul D. McGreevy
Reconciling Horse Welfare, Worker Safety, and Public Expectations: Horse Event Incident Management Systems in Australia
Reprinted from: *Animals* **2016**, *6*, 16, doi:10.3390/ani6030016 . . . 35

Melissa Starling, Andrew McLean and Paul McGreevy
The Contribution of Equitation Science to Minimising Horse-Related Risks to Humans
Reprinted from: *Animals* **2016**, *6*, 15, doi:10.3390/ani6030015 . . . 47

Christina-Maria Ikinger, Jana Baldamus and Achim Spiller
Factors Influencing the Safety Behavior of German Equestrians: Attitudes towards Protective Equipment and Peer Behaviors
Reprinted from: *Animals* **2016**, *6*, 14, doi:10.3390/ani6020014 . . . 61

Denzil O'Brien
Look Before You Leap: What Are the Obstacles to Risk Calculation in the Equestrian Sport of Eventing?
Reprinted from: *Animals* **2016**, *6*, 13, doi:10.3390/ani6020013 . . . 81

Jodi DeAraugo, Suzanne McLaren, Phil McManus and Paul D. McGreevy
Improving the Understanding of Psychological Factors Contributing to Horse-Related Accident and Injury: Context, Loss of Focus, Cognitive Errors and Rigidity
Reprinted from: *Animals* **2016**, *6*, 12, doi:10.3390/ani6020012 . . . 105

Barbara Padalino, Evelyn Hall, Sharanne Raidal, Pietro Celi, Peter Knight and Leo Jeffcott et al.
Health Problems and Risk Factors Associated with Long Haul Transport of Horses in Australia
Reprinted from: *Animals* **2015**, *5*, 1296-1310, doi:10.3390/ani5040412 . . . 115

Jarkko Leppälä, Christina Lunner Kolstrup, Stefan Pinzke, Risto Rautiainen, Markku Saastamoinen and Susanna Särkijärvi
Development of a Safety Management Web Tool for Horse Stables
Reprinted from: *Animals* **2015**, *5*, 1136-1146, doi:10.3390/ani5040402 . . . 131

Karen Ruse, Aidan Davison and Kerry Bridle
Jump Horse Safety: Reconciling Public Debate and Australian Thoroughbred Jump Racing Data, 2012–2014
Reprinted from: *Animals* **2015**, *5*, 1072-1091, doi:10.3390/ani5040399 . . . . **143**

Markku Saastamoinen, Susanna Särkijärvi and Seppo Hyyppä
Reducing Respiratory Health Risks to Horses and Workers: A Comparison of Two Stall Bedding Materials
Reprinted from: *Animals* **2015**, *5*, 965-977, doi:10.3390/ani5040394 . . . . **163**

Christopher B. Riley, Jessica R. Liddiard and Kirrilly Thompson
A Cross-Sectional Study of Horse-Related Injuries in Veterinary and Animal Science Students at an Australian University
Reprinted from: *Animals* **2015**, *5*, 951-964, doi:10.3390/ani5040392 . . . . **177**

Beverley A. Curry, Peta L. Hitchens, Petr Otahal, Lei Si and Andrew J. Palmer
Workplace Injuries in Thoroughbred Racing: An Analysis of Insurance Payments and Injuries amongst Jockeys in Australia from 2002 to 2010
Reprinted from: *Animals* **2015**, *5*, 897-909, doi:10.3390/ani5030390 . . . . **191**

Kirrilly Thompson and Chelsea Matthews
Inroads into Equestrian Safety: Rider-Reported Factors Contributing to Horse-Related Accidents and Near Misses on Australian Roads
Reprinted from: *Animals* **2015**, *5*, 592-609, doi:10.3390/ani5030374 . . . . **205**

Laura Haigh and Kirrilly Thompson
Helmet Use Amongst Equestrians: Harnessing Social and Attitudinal Factors Revealed in Online Forums
Reprinted from: *Animals* **2015**, *5*, 576-591, doi:10.3390/ani5030373 . . . . **223**

Kirrilly Thompson, Paul McGreevy and Phil McManus
A Critical Review of Horse-Related Risk: A Research Agenda for Safer Mounts, Riders and Equestrian Cultures
Reprinted from: *Animals* **2015**, *5*, 561-575, doi:10.3390/ani5030372 . . . . **239**

About the Editor

Kirrilly Thompson

Dr. Kirrilly Thompson is a galloping cultural anthropologist who has researched the cultural and symbolic importance of horses in human lives, the psycho-social dimensions of equestrian sport and horse welfare, and the influence of human–horse relationships on human natural disaster preparedness and response. To this body of largely anthropocentric research, Kirrilly's 2018 book "(Un)stable relations: Horses, Humans and Social Agency" with Lynda Birke considers the philosophical, practical and ethical responsibilities of acknowledging horses as social actors in their relations with humans. Kirrilly is currently working in One Health, where she is concerned with the promotion of human behaviour for healthy people, animals and environments.

animals MDPI

Article

Analysis of Failure to Finish a Race in a Cohort of Thoroughbred Racehorses in New Zealand

Jasmine Tanner [1], Chris Rogers [1,*], Charlotte Bolwell [1], Naomi Cogger [2], Erica Gee [1] and Wayne McIlwraith [3]

[1] Equine Research Centre, Institute of Veterinary Animal and Biomedical Sciences, Massey University, Palmerston North 4474, New Zealand; jaztanner@gmail.com (J.T.); c.bolwell@massey.ac.nz (C.B.); e.k.gee@massey.ac.nz (E.G.)
[2] Epicentre, Institute of Veterinary Animal and Biomedical Sciences, Massey University, Palmerston North 4442, New Zealand; n.cogger@massey.ac.nz
[3] Equine Orthopaedic Research Center, Colorado State University, Fort Collins, CO 80523, USA; Wayne.Mcilwraith@ColoState.EDU
* Correspondence: c.w.rogers@massey.ac.nz; Tel.: +64-06-356-9099 (ext. 85162)

Academic Editor: Kirrilly Thompson
Received: 11 January 2016; Accepted: 23 May 2016; Published: 25 May 2016

Simple Summary: Overall, the failure to finish rate in New Zealand, 2.88 per 1000 horse starts (95% CI 2.64–3.12), was lower than international figures for race day catastrophic injury. Racing and environmental variables such as horse experience, race distance and season were associated with failure to finish a race. Catastrophic injury accounted for approximately half the failure to finish events. Jockey falls were positively associated with less experienced jockeys and horses.

Abstract: The objective was to describe the incidence of failure to finish a race in flat-racing Thoroughbreds in New Zealand as these are summary indicators of falls, injuries and poor performance. Retrospective data on six complete flat racing seasons (n = 188,615 race starts) of all Thoroughbred flat race starts from 1 August 2005 to 31 July 2011 were obtained. The incidence of failure to finish events and binomial exact 95% confidence intervals were calculated per 1000 horse starts. The association between horse-, rider- and race-level variables with the outcomes failure to finish, pulled up/fell and lost rider were examined with a mixed effects Poisson regression model. A total of 544 horses failed to finish in 188,615 race starts with an overall incidence of 2.88 per 1000 horse starts (95% CI 2.64–3.12). The incidence of failure to finish horses across each race year showed little variability. In the univariable analysis race distance, larger field size, season, and ratings bands showed association with failing to finish a race. The overall failure to finish outcome was associated with season, race distance and ratings bands (horse experience and success ranking criteria). In the multivariable analysis, race distance and ratings bands were associated with horses that pulled-up/fell; season, apprentice allowances and ratings bands were associated with the outcome lost rider. The failure to finish rate was lower than international figures for race day catastrophic injury. Racing and environmental variables were associated with failure to finish a race highlighting the multifactorial nature of race-day events. Further investigation of risk factors for failure to finish is required to better understand the reasons for a low failure to finish rate in Thoroughbred flat races in New Zealand.

Keywords: horse; thoroughbred; horse racing; injury; jockey; falls

1. Introduction

In recent years, there has been much attention focused on the quantification of catastrophic and musculoskeletal injury, and risk factors for these in Thoroughbred flat racing [1].

Epidemiological studies have emphasised the multifactorial nature of musculoskeletal events and the complexity of the issues in reducing the risk factors identified [2–5]. Furthermore, the pattern of training, track surfaces, racing conditions and regulation (e.g., medication use) vary between racing jurisdictions, resulting in different risk factors and rates of musculoskeletal injury worldwide [1,6].

Whilst variation between racing jurisdictions may prevent the application of a single global solution, it does provide the opportunity to examine the moderating effects of the racing industry structure on risk factors for, and rates of, injury. To fully interpret these moderating effects it is important to describe the complexity and structure of the racing system within each jurisdiction. Within New Zealand, the production process preceding race training (growth and development leading to yearling preparation) and the training milestones leading up to the first race start have been well described [7–10], including description of the tissue response leading to first trial start [11,12]. Annually 5562 horses contribute to the 31,488 starters during the season. Most horses enter race training as two-year-olds and will race until the end of their five years old season. During a season flat racing horses start a median of 5 (IQR 2–8) times over race distances of 1400 m (IQR 1200-1670, [12,13]. The temperate climate of New Zealand permits racing year round on approximately 50 different 1800 m turf tracks. Despite the number of different tracks there is consistency in the general dimensions of the tracks and the going of the racing surfaces throughout the season (few with the racing surface classified as "fast" and the majority described as "good" or "dead") [14].

The interaction of the production process with the progression through training [15], and the subsequent influence attainment of these early milestones has on racing success and longevity further emphasize the complexity of the racing system and the need to understand how variables limit progression of the horse through the racing production cycle [16,17].

It is important to consider each event reported (race day injury, fracture, catastrophic musculoskeletal injury) as part of an integrated continuum of the interaction of cyclic load (frequency of high speed/gallop strides) and environmental challenge, and not as discrete entities [18,19]. The complexity of the interaction of cyclic load and tissue response is highlighted by the non-linear and dynamic relationships of canter and gallop exercise accumulated during training with fracture risk [20].

Racing and racing injuries could therefore be regarded as part of a complex system. A complex system is a term used to describe how relationships between parts give rise to the collective behaviors of a system and how the system interacts and forms relationships with its environment [21]. An example of the complexity of the racing system was highlighted in a review article [22] with the example given being the apparent greater risk of fatal musculoskeletal injury on turf rather than dirt races in the United States, which may be compounded by the association of longer races on turf compared to dirt [23].

Many racing jurisdictions present detailed racing event data from which the horses failing to finish a race can be identified. As part of the rules of racing the Racing Integrity Unit are responsible for producing a stipendiary stewards' report for each race, detailing any events that occurred during the race. Failure to finish data incorporate a spectrum of events ranging from jockeys "pulling-up" a horse because it was failing to "run on its merits" and suspected injury through to catastrophic injury and jockey falls. As such, failure to finish data provides a holistic system based measure that includes components of racehorse welfare, jockey safety and confidence of the betting public. Thus failure to finish provides an opportunity to screen the performance of a racing jurisdiction for the reliability and consistency of racing.

There is little published information on the number of horses failing to finish races in New Zealand or possible associations relating to horses not finishing races. Additionally, there appears to be limited analysis of risk factors for failure to finish data in racehorse populations and there is an opportunity to gain greater understanding of the events that occur during a race that prevent horses from finishing. The aim of this study was to describe the incidence of failure to finish a race and investigate risk factors for failure to finish events pulled up and lost rider in flat-racing Thoroughbreds in New Zealand.

2. Materials and Methods

2.1. Data

A retrospective cohort study was used to investigate all Thoroughbred flat race starts in the six years from 1 August 2005 to 31 July 2011; data were obtained from New Zealand Thoroughbred Racing (NZTR). Data available included race date, race track, race number, race class, race distance, track condition (or 'going'), penetrometer reading, horse name, horse age, horse sex, trainer, trainer location (city), finishing position in race, barrier draw (position in the starting gates), carded weight (weight allocated by race handicapper), carried weight (carded weight less any apprentice weight allowance) and domestic rating (analogous to the British horse racing "official rating" system).

2.2. Case Definitions

A horse start occurred whenever a horse entered the starting gate for a race and the gate was released. Any horse involved in an incident that occurred prior to the release of the starting gate and was deemed a late scratching (non-starter) by race day officials was excluded from the study. A horse could contribute several starts and more than one failure to finish event over the study period. Failure to finish occurred whenever a horse started in a race (*i.e.*, the starting gate was released) and failed to cross the finish line. Failure to finish was classified as: pulled up (when the jockey pulled the horse out of the race), fell (when the horse fell during the race), lost rider (when the jockey was dislodged from the horse), brought down (when the horse fell due to collision with another fallen horse), and ran off (when the horse ran off to the outside of the racecourse).

2.3. Statistical Analysis

Data were structured for analysis in Microsoft Excel 2007 and Microsoft Access 2007 (Microsoft Corporation, Redmond, WA, USA) and screened for errors using exploratory data analysis. Continuous data were assessed for normality using the Shapiro-Wilks test. The continuous variables that were categorised into groups included: weight carried (quartiles), race distance (quartiles) and field size (number of starters in the race) (quartiles). New variables were created for race year, season (spring, summer, autumn, winter), field size and whether or not the jockey had an apprentice allowance. Apprentice jockeys in New Zealand claim a weight reduction (allowance) on the handicap weight of horses they ride depending on previous experience. Under current rules an apprentice that has 0–5 career wins claims 4 kg, 6–30 wins claims 3 kg, 31–60 wins claims 2 kg and 61–100 wins claims 1 kg [24]. Ratings were categorised based on the ratings bands recognised by the New Zealand handicapping system. Within the rating system a horses is allocated a numerical rating reflecting its relative performance and its eligibility to compete in differing classes/grades of races. The rating is a dynamic measure of performance that is recalculated within two days of a horse's most recent race start [24] and is analogous to the rating system used by the British Horse Racing Board [25]. Domestic ratings were categorised into ratings bands as recognised by the New Zealand handicapper. Ratings band 50–54 were maiden (non-race winning) horses, ratings band 55–65 included horses that have won one race and two or more race winners with an extended run of poor form, ratings band 66–75 included most two win horses and three win horses with a recent loss of form and four or more win horses with an extended loss of form. Ratings band 76–85 included most three and four win horses and some open class horses with recent poor form. Ratings band 86–115 were open class (elite) horses [24].

The incidence of failure to finish and corresponding binomial exact 95% confidence intervals were calculated and reported as events per 1000 horse starts for all variables. Failure to finish events were sub-categorised into a pulled-up outcome and a lost rider outcome. Poisson regression was used to estimate incidence rate ratios (IRR) with 95% confidence intervals (95% CI) for the outcomes failure to finish, pulled-up, and lost rider within the univariable analysis. Variables showing association ($p < 0.2$) with the outcomes were analysed in multivariable mixed effects Poisson regression models (for each outcome separately) fitted in a backwards step-wise fashion. A postestimation goodness-of-fit test was performed to test for overdispersion in each multivariable model using the Pearson chi-square and deviance chi-square test, then a random effect for horse was added to each model to adjust for clustering at the horse-level. Biologically plausible interaction terms were assessed in the final models. A Kaplan-Meier curve was used to graphically present the failure to finish by race distance. Statistical significance was set at $p < 0.05$ and analysis conducted in STATA 12 (StataCorp, College Station, TX, USA).

3. Results

There were 188,616 race starts for 16,646 individual horses during the study period. The data represented 6072 2-year-old starts, 43,228 3-year-old starts and 139,316 4-year-old and older starts. During the study period the horses contributed a median of 7 (IQR 3–16) race starts. There were 544 failure to finish events providing an overall incidence of 2.88 per 1000 horse starts (95% CI 2.64–3.12). There was little variation in the incidence of failure to finish between racing years. The lowest incidence rate was 2.66 per 1000 starts (95% CI 2.13–3.28) in the 2009/10 racing year and the highest was 3.10 per 1000 starts (95% CI 2.52–3.78) in the 2007/08 racing year. There was no significant effect of horse age associated failure to finish, and no significant difference between ages in the older horse category of 6 years old and older. Of the 544 failure to finish events, 507 (93.2%) horses had single events, 17 (6.2%) horses contributed two events, and one (0.6%) horse had three events. Overall there were 269 (49.4%) pulled-up, 72 (13.2%) fell and 179 (32.9%) lost rider events, other failure to finish events were brought down ($n = 17$) and ran off ($n = 7$).

Univariable Poisson regression analysis of the failure to finish, pulled-up and lost rider outcomes are presented in Table 1. Horses racing over a distance of 1671 m or greater were more likely to fail to finish a race or be pulled-up compared to horses racing ≤1200 m (Figure 1). Horses racing in fields of 12–13 runners and 14–18 runners had a higher rate of failure to finish compared to horses racing in fields of 3–9 runners (Table 1). Race year, sex of horse, age of horse, barrier draw and race number on card (order of race at the race meeting) were not significantly associated with failure to finish.

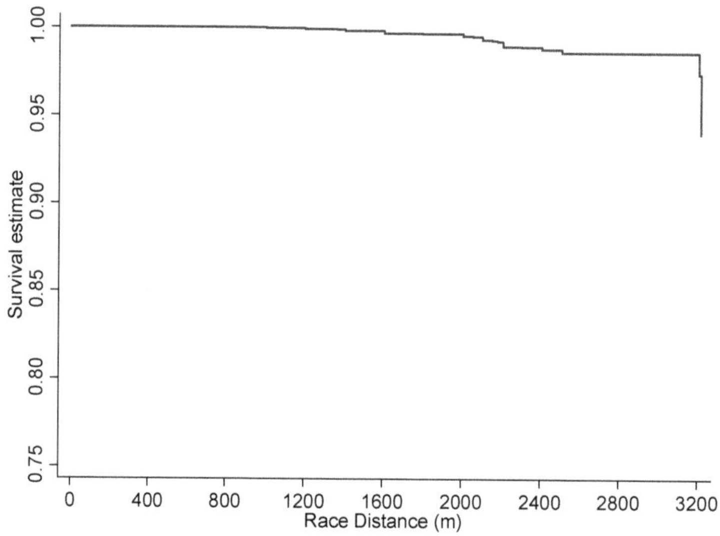

Figure 1. Kaplan-Meier survival curve for incidence of failure to finish and increasing race distance (m) for all Thoroughbred flat race starts in the 2005/06–2010/11 racing years in New Zealand (N = 188,615).

Table 1. Univariable Poisson regression for the outcomes failure to finish, pulled-up and lost rider showing incidence rate ratios (IRR) for all Thoroughbred flat race starts in the 2005/06–2010/11 racing years in New Zealand.

			Failure to Finish				Pulled Up				Lost Rider			
Variable	Level	No. of Starts	No. of Failure to Finish Events	Incidence Rate Ratio [a]	95% CI	*p*-Value [b]	No. of Pulled up Events	Incidence Rate Ratio [a]	95% CI	*p*-Value [b]	No. of Lost Rider Events	Incidence Rate Ratio [a]	95% CI	*p*-Value [b]
Race Year	2005/06	29,751	84	Ref		*0.921*	34	Ref		*0.448*	34	Ref		*0.608*
	2006/07	30,574	93	1.08	0.81–1.46	0.599	47	1.37	0.88–2.14	0.165	26	0.75	0.45–1.25	0.262
	2007/08	31,276	97	1.11	0.83–1.49	0.496	42	1.22	0.77–1.93	0.396	35	0.98	0.61–1.57	0.917
	2008/09	33,061	95	1.03	0.76–1.38	0.869	53	1.45	0.94–2.26	0.094	32	0.84	0.52–1.37	0.492
	2009/10	32,349	86	0.95	0.70–1.28	0.730	41	1.15	0.72–1.82	0.557	27	0.73	0.44–1.21	0.218
	2010/11	31,605	89	1.01	0.75–1.36	0.959	52	1.48	0.96–2.3	0.079	25	0.69	0.41–1.17	0.17
Season	Spring	49,620	155	Ref		*0.001*	79	Ref		*0.229*	58	Ref		*0.002*
	Summer	52,647	183	1.12	0.90–1.38	0.320	82	0.98	0.72–1.34	0.917	65	1.06	0.74–1.51	0.742
	Autumn	48,484	106	0.7	0.55–0.90	0.005	55	0.72	0.51–1.01	0.06	31	0.55	0.36–0.85	0.008
	Winter	37,865	100	0.84	0.65–1.08	0.183	53	0.88	0.62–1.25	0.468	25	0.56	0.35–0.9	0.016
Sex	Female	84,011	252	Ref		*0.447*	126	Ref		*0.522*	83	Ref		*0.620*
	Male	104,605	292	0.94	0.79–1.11	0.447	143	0.92	0.72–1.18	0.522	96	0.93	0.68–1.25	0.62
Track Condition	Fast	5,478	19	Ref		*0.134*	11	Ref		*0.462*	5	Ref		*0.820*
	Good	73,231	227	0.9	0.56–1.44	0.653	110	0.75	0.4–1.4	0.369	75	1.13	0.46–2.81	0.787
	Dead	44,481	125	0.81	0.50–1.32	0.398	59	0.66	0.35–1.26	0.208	43	1.07	0.42–2.7	0.889
	Slow	32,310	72	0.64	0.39–1.07	0.088	38	0.59	0.3–1.15	0.12	25	0.85	0.33–2.23	0.747
	Heavy	33,116	101	0.88	0.54–1.44	0.608	51	0.76	0.4–1.47	0.421	31	1.03	0.4–2.65	0.952
Age	2 yo	6,072	19	Ref		*0.936*	8	Ref		*0.098*	7	Ref		*0.420*
	3 yo	43,228	121	0.9	0.55–1.46	0.656	50	0.88	0.42–1.87	0.74	48	0.97	0.44–2.15	0.937
	4 yo	56,374	168	0.96	0.59–1.54	0.853	75	1.03	0.49–2.15	0.94	58	0.9	0.41–1.97	0.787
	5 yo	42,439	116	0.88	0.54–1.44	0.615	63	1.17	0.56–2.47	0.674	34	0.7	0.31–1.58	0.387
	6 yo+	40,503	120	0.96	0.59–1.56	0.864	73	1.45	0.69–3.05	0.324	32	0.68	0.3–1.55	0.361
Apprentice Allowance	No	144,005	398	Ref		*0.079*	213	Ref		*0.279*	116	Ref		*<0.001*
	Yes	44,611	146	1.19	0.98–1.44	0.079	56	0.85	0.63–1.14	0.279	63	1.76	1.3–2.4	<0.001
Race Distance	≤1200 m	49,554	120	Ref		*0.002*	58	Ref		*<0.001*	47	Ref		*0.357*
	1201–1400 m	47,914	129	1.12	0.87–1.43	0.387	47	0.84	0.57–1.24	0.394	55	1.22	0.83–1.81	0.314
	1401–1670 m	44,587	125	1.15	0.89–1.48	0.293	61	1.19	0.83–1.71	0.351	36	0.86	0.55–1.33	0.487
	1671 m+	46,561	172	1.54	1.22–1.95	<0.001	103	1.95	1.4–2.72	<0.001	41	0.93	0.61–1.42	0.739
Weight Carried	46–54.5 kg	55,382	180	Ref		*0.266*	86	Ref		*0.582*	68	Ref		*0.009*
	54.6–55.5 kg	40,677	116	0.87	0.69–1.10	0.261	50	0.78	0.55–1.11	0.17	46	0.92	0.63–1.34	0.678
	55.6–56.9 kg	38,547	103	0.82	0.64–1.05	0.111	54	0.9	0.64–1.26	0.538	29	0.61	0.4–0.95	0.027
	57–76 kg	54,010	145	0.82	0.66–1.03	0.086	79	0.94	0.69–1.28	0.686	36	0.54	0.36–0.82	0.003
Rating Bands	50–54	46,817	176	Ref		*<0.001*	78	Ref		*0.129*	68	Ref		*0.002*
	55–65	45,695	143	0.84	0.67–1.04	0.115	77	1.03	0.75–1.42	0.856	38	0.57	0.38–0.85	0.006
	66–75	57,524	130	0.61	0.48–0.76	<0.001	69	0.73	0.53–1.02	0.066	39	0.47	0.31–0.69	<0.001
	76–85	21,639	58	0.72	0.53–0.97	0.031	27	0.77	0.49–1.2	0.244	19	0.6	0.36–1.01	0.053
	86–115	16,941	37	0.59	0.41–0.85	0.004	18	0.66	0.39–1.12	0.122	15	0.62	0.35–1.09	0.096

Table 1. *Cont.*

Variable	Level	No. of Starts	Failure to Finish: No. of Failure to Finish Events	Incidence Rate Ratio [a]	95% CI	*p*-Value [b]	Pulled Up: No. of Pulled up Events	Incidence Rate Ratio [a]	95% CI	*p*-Value [b]	Lost Rider: No. of Lost Rider Events	Incidence Rate Ratio [a]	95% CI	*p*-Value [b]
Field Size	3–9	43,560	99	Ref		*0.040*	51	Ref		*0.362*	35	Ref		*0.498*
	10–11	42,003	121	1.26	0.97–1.65	0.086	61	1.23	0.85–1.79	0.269	42	1.24	0.79–1.94	0.348
	12–13	45,245	149	1.44	1.12–1.86	0.005	74	1.39	0.97–1.98	0.074	50	1.38	0.89–2.12	0.147
	14–18	57,808	175	1.33	1.04–1.70	0.025	83	1.22	0.86–1.73	0.268	52	1.12	0.73–1.72	0.604
Barrier	1–3	51,029	149	Ref		*0.772*	75	Ref		*0.692*	49	Ref		*0.888*
	4–6	50,625	152	1.03	0.82–1.29	0.806	75	1.01	0.73–1.39	0.957	49	1.01	0.68–1.5	0.956
	7–9	44,443	118	0.91	0.71–1.16	0.441	55	0.84	0.59–1.19	0.332	38	0.89	0.58–1.36	0.596
	10–21	42,519	125	1.01	0.79–1.28	0.960	64	1.02	0.73–1.43	0.904	43	1.06	0.7–1.6	0.784
Race Number	1	16,942	51	Ref		*0.266*	22	Ref		*0.955*	13	Ref		*0.450*
	2	18,389	55	1	0.68–1.46	0.998	25	1.06	0.6–1.89	0.837	22	1.56	0.79–3.1	0.203
	3	18,596	47	0.84	0.57–1.26	0.405	27	1.13	0.64–1.99	0.667	15	1.05	0.5–2.22	0.888
	4	18,840	55	0.98	0.67–1.43	0.897	23	0.95	0.53–1.71	0.87	20	1.38	0.69–2.78	0.364
	5	19,364	51	0.88	0.60–1.30	0.520	33	1.33	0.77–2.28	0.302	13	0.88	0.41–1.9	0.742
	6	19,732	64	1.09	0.75–1.58	0.647	32	1.28	0.74–2.2	0.379	21	1.4	0.7–2.8	0.343
	7	19,925	55	0.93	0.63–1.36	0.691	25	0.98	0.55–1.75	0.956	14	0.92	0.43–1.96	0.833
	8	20,507	59	0.97	0.66–1.41	0.859	30	1.15	0.66–2	0.616	18	1.15	0.56–2.36	0.698
	9	17,487	54	1.04	0.71–1.53	0.844	27	1.22	0.7–2.16	0.482	24	1.8	0.92–3.55	0.088
	10+	18,834	53	0.95	0.64–1.39	0.776	25	1.05	0.59–1.87	0.866	19	1.33	0.65–2.69	0.434

[a] Adjusted for potential clustering at horse-level; [b] Likelihood Ratio Test *p*-value *(italics)* reported as a test, Wald test *p*-value (non-italics) reported as a test for linear trend for continuous variables and ordered categories.

Multivariable mixed effects Poisson regression models of variables significantly associated with the outcomes failure to finish, pulled-up and lost rider are presented in Table 2. The Pearson goodness of fit statistic for the failure to finish model was $p = 0.86$, for the pulled up model $p = 0.09$, and the lost rider model $p = 0.60$, all indicating good model fit. Season, race distance and ratings band were significantly associated with failure to finish. The failure to finish rate was significantly greater in longer races compared to short races. There was a significant trend for with increasing race distance ($p < 0.001$) with both failure to finish and pulled-up. There was a greater failure to finish rate in autumn compared to spring, and at all rating bands 55 or greater compared to rating band 50–54. The rate of pulled-up was significantly lower in rating bands 66–75 and 86–115 compared to rating band 50–54.

Table 2. Results of multivariable mixed effects Poisson regression models of the variables significantly associated with the outcomes: failure to finish, pulled-up, and lost rider (adjusted for potential clustering at horse-level), for all Thoroughbred flat race starts in the 2005/06–2010/11 racing years in New Zealand (N = 188,615).

Variable	Level	Incidence Rate Ratio	95% CI	Wald Test *p*-Value [a]	LRT *p*-Value [b]
Outcome: Failure to Finish					
Season	Spring	Ref			*0.002*
	Summer	1.08	0.87–1.34	0.502	
	Autumn	0.7	0.55–0.89	0.005	
	Winter	0.82	0.63–1.05	0.117	
Race Distance	≤1200 m	Ref			*<0.001*
	1201–1400 m	1.18	0.92–1.51	0.198	
	1401–1670 m	1.24	0.96–1.60	0.102	
	1671 m+	1.73	1.36–2.20	<0.001	
Rating Bands	50–54	Ref			
	55–65	0.79	0.63–0.99	0.042	*<0.001*
	66–75	0.56	0.44–0.70	<0.001	
	76–85	0.65	0.48–0.88	0.005	
	86–115	0.53	0.37–0.76	0.001	
Horse [c]					*0.017*
Outcome: Pulled Up					
Race Distance	≤1200 m	Ref			*<0.001*
	1201–1400 m	0.88	0.60–1.29	0.517	
	1401–1670 m	1.26	0.87–1.81	0.218	
	1671 m+	2.14	1.53–3.00	<0.001	
Rating Bands	50–54	Ref			*0.019*
	55–65	0.94	0.69–1.31	0.722	
	66–75	0.64	0.46–0.90	0.010	
	76–85	0.65	0.42–1.03	0.064	
	86–115	0.56	0.33–0.95	0.033	
Horse [d]					*0.009*
Outcome: Lost Rider					
Season	Spring	Ref			*<0.001*
	Summer	1.08	0.76–1.54	0.678	
	Autumn	0.56	0.36–0.87	0.009	
	Winter	0.54	0.34–0.87	0.011	
Apprentice Allowance	No	Ref			*<0.001*
	Yes	1.78	1.30–2.43	<0.001	
Rating Bands	50–54	Ref			*0.004*
	55–65	0.59	0.39–0.87	0.009	
	66–75	0.48	0.32–0.71	<0.001	
	76–85	0.61	0.36–1.02	0.058	
	86–115	0.65	0.37–1.16	0.145	
Horse [e]					*0.005*

[a] Wald Test *p*-value reported as a test for linear trend for continuous variables and ordered categories; [b] LRT: Likelihood Ratio Test; [c] = horse level variance 0.54 (0.21–1.34); [d] = horse level variance 1.31 (0.59–2.92); [e] = horse level variance 1.35 (0.64–2.86).

Season, apprentice allowance and rating band were associated with lost rider in the multivariable model (Table 2). The rate of lost rider events was greater for apprentice allowances compared to no apprentice allowance. Autumn had a significantly lower rate of lost rider events compared to spring, as did rating bands 55–65 and 66–75 compared to rating band 50–54. None of the interaction terms were found to be statistically significant. No variables were found to be significantly associated with the outcome fell within the univariable analysis.

The addition of the horse as a random effect improved the goodness of fit of the model. The Poisson model was the best fit for the data rather than a negative binomial regression. The large number of horses included as a random effect (n = 16,646) prevented the use of Cooks Distance to test for model validity.

4. Discussion

Failure to finish data represent a broad category of events, including musculoskeletal injury, that prevent a horse from completing the race, which does not appear to have been previously reported within flat racing in New Zealand . The failure to finish rates reported in this paper would appear low, and this is reinforced by consideration of international data on race day musculoskeletal injury, which represented a component (~55%) of the New Zealand failure to finish records. Internationally, within the literature, musculoskeletal injury reported on race day ranges from 3.1 per 1000 starts [26] to 4.4 per 1000 starts [27], which is greater than the 2.88 per 1000 starts for failure to finish reported in this paper. As the records utilised in this study were the official racing records it is unlikely that errors in recording or failure to record events has contributed to this low rate. The restricted sampling frame in this study of the racing event may mean that some data outside the racing event, such as the loss of a rider prior to race start, provided some under reporting. It is reported that 47% of jockey falls in New Zealand occur prior to the race start [28]. The data reported in this paper also represents all flat horses racing within New Zealand across a number of years and thus should have minimal bias due to the effect of racing location or seasonal/annual variations in data.

Compared to international studies there is also a low incidence of race day falls by flat race jockeys in New Zealand (2.2/1000) [28] *vs.* 3.7–4.4/1000 for the United Kingdom/ Ireland and France [29] and Australia (4.2/1000) [30], which supports the low failure to finish data reported here. Within the lost rider multivariable model apprentice jockeys claiming a weight allowance were over represented. This observation is in support of data reported out of Australia [31], where the rate of falls by apprentice jockeys was inversely proportional to experience. The dataset did not contain jockey names only the weight a horse was carded and the weight the horse actually carried during the race. This prevented us from including jockey as a random effect within the model, and thus clustering of these events with certain jockeys.

Season (Spring) and race grade (maiden/lower rating horses) were positively associated with lost rider. The race grade effect was also supported by a study of predictors of jockey falls in flat racing in Australia and followed up with the study on early career jockeys [31,32]. The season effect may be due to the horses starting a new campaign in spring and the start of a racing career for young and less experienced horses. Lower grade horses also generally have less racing experience and thus are less predictable and tractable during race riding placing them at greater risk of interference. The lack of interaction between the three variables (rating, season and apprentice jockeys) indicates independent risk and not the presentation of the "perfect storm" of inexperienced jockeys riding "fresh" and "excited" inexperienced horses.

Contributing factors to the low failure to finish rate may relate to the structure and the type of racing within New Zealand. Racehorses within New Zealand typically run over 1400 m in races with 11 horse fields and have a median of 5 (IQR 2–8) starts per racing season [13]. These parameters *per se*/individually are not uniquely different from other major racing jurisdictions with similar range of field sizes and number race starts reported for flat racing horses in the UK [33] and USA [34]. Distributions for race distances appear sparsely reported within the literature. Most racing jurisdictions appear skewed towards sprint and the lower end of middle distance races, with a mean of approximately 1 mile (1600 m) [23,30,35]. Given the relative international uniformity in many of the racing production parameters reported the low expression of risk identified must relate to some subtle interaction or moderating effect in the production parameters within the complex system.

Greater detailed examination of the characteristics of the pattern of racing (changes in racing distance and the timing of the race starts relative to each other) and training (possibly use of the non-totalisator race starts (barrier trials) within the conditioning programme) may help elucidate why the lower than expected rate of risk.

The apparent low level of race starts, despite a relatively low cost racing structure, may be due to the use of trials (non-totalisator/qualifying races) by racehorse trainers in New Zealand for education and in the final stages of race preparation. A cross-sectional survey of 2-year-old training practices identified the use of trials for education and training milestones within this age group as well as a strong emphasis in 2-year-old training being for education and conditioning purposes rather than with the primary objective of obtaining a race start [10].

The presence of the rating band 50–54, which is associated with maiden and 2-year-old racehorses, within the multivariable model indicated that despite these potentially positive characteristics of New Zealand racing there is still an increased rate of failure to finish, and specifically lost rider, associated with younger and more inexperienced horses. This pattern of greater risk with the less experienced/lower grade horses has been reported with jockey falls in Australia [30]. Within the Australian data there was an interaction of jockey experience with horse experience and lower grade races which accentuated the risk, which was not apparent in our dataset.

Previous studies have shown that racehorse trainers in New Zealand provide horses with voluntary breaks from training (a "spell") in order to allow the horse time to "strengthen and develop" [10,36]. The willingness of New Zealand trainers and owners to spell a horse may relate to the structure of the racing calendar, which for most horses has a uniform pattern [37]. The consistency of the racing offered throughout the racing season means that provision of a spell may delay a race start but does not impact on the opportunity to have a race start in the appropriate grade/class of race.

Despite racing on turf tracks within a temperate climate, racetrack surfaces in New Zealand are consistently reported in the good to dead range of going (61% of races, median penetrometer reading 2.3–2.7). It is only in winter that the median going decreases to heavy (penetrometer reading 4.3) [14]. However, even with moderate between season differences in rate of going there was a consistent and limited within season variation in track condition. The consistency in racing surface may be due to focused management at the track level, and at national level the scheduling of the pattern of racing so that the free-draining tracks are used most in winter, which should provide a predictable and relative consistent racing environment for both horse and jockey. The limited variation in racing surface going within a race meeting and across the seasons may explain the lack of significant effect identified for season or going within the multivariable models. The absence of fast tracks (less than 3% of races run in a year) may also be a protective factor in relation to the reported failure to finish rate, as fast tracks have been reported to be associated with an increased risk for race day musculoskeletal injury and fracture [38,39] and race day falls [30].

5. Conclusions

Overall, the failure to finish rate was lower than international figures for race day catastrophic injury. Racing and environmental variables such as horse experience, race distance and season were associated with failure to finish a race, highlighting the multifactorial nature of race-day events. Investigation of the biological and industry based drivers of the risk factors, particularly season and horse experience are required to identify pragmatic management changes to reduce the risk of failure to finish. Further investigation of risk factors for failure to finish is required to better understand the reasons for a low failure to finish rate in Thoroughbred flat races in New Zealand.

Acknowledgments: The authors thank New Zealand Thoroughbred Racing for provision of the data.

Author Contributions: Chris Rogers, Wayne McIlwraith, Charlotte Bolwell and Erica Gee designed the study Jasmine Tanner collected the data and conducted the analysis in association with Chris Rogers, Charlotte Bolwell, Naomi Cogger and Erica Gee. Chris Rogers and Charlotte Bolwell drafted the manuscript, which was revised by all authors. All authors read and approved the final manuscript.

Conflicts of Interest: The authors declare no conflict of interest.

References

1. Parkin, T.D.H. Epidemiology of racetrack injuries in racehorses. *Vet. Clin. N. Am. Equine Pract.* **2008**, *24*, 1–19. [CrossRef] [PubMed]
2. Pinchbeck, G.L.; Clegg, P.D.; Boyde, A.; Barr, E.D.; Riggs, C.M. Horse-, training- and race-level risk factors for palmar/plantar osteochondral disease in the racing thoroughbred. *Equine Vet. J.* **2013**, *45*, 582–586. [CrossRef] [PubMed]
3. Reed, S.R.; Jackson, B.F.; Wood, J.L.N.; Price, J.S.; Verheyen, K.L.P. Exercise affects joint injury risk in young thoroughbreds in training. *Vet. J.* **2013**, *196*, 339–344. [CrossRef] [PubMed]
4. Parkin, T.D.H.; Clegg, P.D.; French, N.P.; Proudman, C.J.; Riggs, C.M.; Singer, E.R.; Webbon, P.M.; Morgan, K.L. Risk factors for fatal lateral condylar fracture of the third metacarpus/metatarsus in UK racing. *Equine Vet. J.* **2005**, *37*, 192–199. [CrossRef] [PubMed]
5. Perkins, N.R.; Reid, S.W.J.; Morris, R.S. Risk factors for injury to the superficial digital flexor tendon and suspensory apparatus in thoroughbred racehorses in New Zealand. *N. Z. Vet. J.* **2005**, *53*, 184–192. [CrossRef] [PubMed]
6. Verheyen, K.L.P. Reducing injuries in racehorses: Mission impossible? *Equine Vet. J.* **2013**, *45*, 6–7. [CrossRef] [PubMed]
7. Rogers, C.W.; Gee, E.K.; Firth, E.C. A cross-sectional survey of thoroughbred stud farm management in the north island of New Zealand. *N. Z. Vet. J.* **2007**, *55*, 302–307. [CrossRef] [PubMed]
8. Stowers, N.L.; Rogers, C.W.; Hoskin, S.O. Management of weanlings on commercial thoroughbred studs farms in the north island of New Zealand. *Proc. N. Z. Soc. Anim. Prod.* **2009**, *69*, 4–9.
9. Bolwell, C.F.; Rogers, C.W.; Firth, E.C.; French, N. Management and exercise of thoroughbred yearlings during preparation for yearling sales: A cross-sectional survey. *Proc. N. Z. Soc. Anim. Prod.* **2010**, *70*, 157–161.
10. Bolwell, C.F.; Russell, L.J.; Rogers, C.W.; Firth, E.C. A cross-sectional survey of training practices of 2-year-old racehorses in the north island of New Zealand. *Comp. Exerc. Phys.* **2010**, *7*, 37–42. [CrossRef]
11. Firth, E.C.; Rogers, C.W. Musculoskeletal responses of 2-year-old thoroughbred horses to early training. Conclusions. *N. Z. Vet. J.* **2005**, *53*, 377–383. [CrossRef] [PubMed]
12. Bogers, S.H.; Rogers, C.W.; Bolwell, C.F.; Roe, W.D.; Gee, E.K.; McIlwraith, C.W. The effect of 2-year-old race training on BMDV and its spatial distribution in the distal epiphysis of the third metatarsal bone. *Vet. J.* **2014**, *201*, 353–358. [CrossRef] [PubMed]
13. Bolwell, C.F.; Rogers, C.W.; Gee, E.K.; Rosanowski, S.M. Descriptive statistics and the pattern of horse racing in New Zealand: Part 1 thoroughbred racing. *Anim. Prod. Sci.* **2014**. [CrossRef]
14. Rogers, C.W.; Bolwell, C.F.; Gee, E.K.; Peterson, M.L.; McIlwraith, C.W. Profile and surface conditions of New Zealand thoroughbred racetracks. *J. Equine Vet. Sci.* **2014**, *34*, 1105–1109. [CrossRef]
15. Bolwell, C.F.; Rogers, C.W.; French, N.; Firth, E.C. Associations between yearling exercise and interruptions to race training in thoroughbred racehorses. *Am. J. Vet. Res.* **2012**, *73*, 1610–1616. [CrossRef] [PubMed]
16. Tanner, J.C.; Rogers, C.W.; Firth, E.C. The relationship of training milestones with racing success in a population of standardbred horses in New Zealand. *N. Z. Vet. J.* **2011**, *59*, 323–327. [CrossRef] [PubMed]
17. Tanner, J.C.; Rogers, C.W.; Firth, E.C. The association of 2-year-old training milestones with career length and racing success in a sample of thoroughbred horses in New Zealand. *Equine Vet. J.* **2013**, *45*, 20–24. [CrossRef] [PubMed]
18. Tull, T.M.; Bramlage, L.R. Racing prognosis after cumulative stress-induced injury of the distal portion of the third metacarpal and third metatarsal bones in thoroughbred racehorses: 55 cases (2000–2009). *J. Am. Vet. Med. Assoc.* **2011**, *238*, 1316–1322. [CrossRef] [PubMed]
19. Whitton, R.C.; Trope, G.D.; Ghasem-Zadeh, A.; Anderson, G.A.; Parkin, T.D.H.; Mackie, E.J.; Seeman, E. Third metacarpal condylar fatigue fractures in equine athletes occur within previously modelled subchondral bone. *Bone* **2010**, *47*, 826–831. [CrossRef] [PubMed]
20. Verheyen, K.; Price, J.; Lanyon, L.; Wood, J. Exercise distance and speed affect the risk of fracture in racehorses. *Bone* **2006**, *39*, 1322–1330. [CrossRef] [PubMed]

21. Motta, S. From immune system to semiconductors—What next? Comment on "thermostatted kinetic equations as models for complex systems in physics and life sciences" by Carlo Bianca. *Phys. Life Rev.* **2012**, *9*, 406–409. [CrossRef] [PubMed]
22. Stover, S.M. The epidemiology of thoroughbred racehorse injuries. *Clin. Tech. Equine Pract.* **2003**, *2*, 312–322. [CrossRef]
23. Hernandez, J.; Hawkins, D.L.; Scollay, M.C. Race-start characteristics and risk of catastrophic musculoskeletal injury in thoroughbred racehorses. *J. Am. Vet. Med. Assoc.* **2001**, *218*, 83–86. [CrossRef] [PubMed]
24. Anonymous. *New Zealand Thoroughbred Racing Handicaping Guide*; New Zealand Thoroughbred Racing Inc.: Wellington, New Zealand, 2013; p. 27.
25. Rutherford, D.J.; Bladon, B.; Rogers, C.W. Outcome of lag screw treatment of incomplete, frontal plane, third carpal bone radial facet fractures in horses. *N. Z. Vet. J.* **2007**, *55*, 94–99. [CrossRef] [PubMed]
26. Williams, R.B.; Harkins, L.S.; Hammond, C.J.; Wood, J.L.N. Racehorse injuries, clinical problems and fatalities recorded on British racecourses from flat racing and national hunt racing during 1996, 1997 and 1998. *Equine Vet. J.* **2001**, *33*, 478–486. [CrossRef] [PubMed]
27. Cohen, N.D.; Mundy, G.D.; Peloso, J.G.; Carey, V.J.; Amend, N.K. Results of physical inspection before races and race-related characteristics and their association with musculoskeletal injuries in thoroughbreds during races. *J. Am. Vet. Med. Assoc.* **1999**, *215*, 654–661. [PubMed]
28. Bolwell, C.; Rogers, C.; Gee, E. Descriptive epidemiology of race-day jockey falls and injuries in New Zealand. *Comp. Exerc. Phys.* **2014**, *10*, 49–55. [CrossRef]
29. McCrory, P.; Turner, M.; LeMasson, B.; Bodere, C.; Allemandou, A. An analysis of injuries resulting from professional horse racing in France during 1991–2001: A comparison with injuries resulting from professional horse racing in Great Britain during 1992–2001. *Br. J. Sports Med.* **2006**, *40*, 614–618. [CrossRef] [PubMed]
30. Hitchens, P.L.; Blizzard, C.L.; Jones, G.; Day, L.M.; Fell, J. The incidence of race-day jockey falls in Australia, 2002–2006. *Med. J. Aus.* **2009**, *190*, 83–86.
31. Hitchens, P.L.; Blizzard, C.L.; Jones, G.; Day, L.M.; Fell, J. The association between jockey experience and race-day falls in flat racing in Australia. *Inj. Prev.* **2012**, *18*, 385–391. [CrossRef] [PubMed]
32. Hitchens, P.L.; Blizzard, C.L.; Jones, G.; Day, L.; Fell, J. Predictors of race-day jockey falls in flat racing in Australia. *Occup. Environ. Med.* **2010**, *67*, 693–698. [CrossRef] [PubMed]
33. Anonymous. *British Racing Statistics 2012*; British Horse Racing Authority: London, UK, 2012; p. 73.
34. Anonymous. *The Jockey Club 2014 Fact Book*. Available online: http://www.jockeyclub.com/default.asp?section=Resources&area=1 (accessed on 7 January 2015).
35. Anonymous. *British Horse Racing Fact Book 2011/12*; British Horse Racing Authority: London, UK, 2012; p. 42.
36. Perkins, N.R.; Reid, S.W.J.; Morris, R.S. Profiling the New Zealand thoroughbred racing industry. 2. Conditions interfering with training and racing. *N. Z. Vet. J.* **2005**, *53*, 69–76. [CrossRef] [PubMed]
37. Rosanowski, S.M.; Rogers, C.W.; Cogger, N. The movement pattern of horses around race meetings in New Zealand. *Anim. Prod. Sci.* **2015**, *55*, 1075–1080. [CrossRef]
38. Bailey, C.J.; Reid, S.W.J.; Hodgson, D.R.; Bourke, J.M.; Rose, R.J. Flat, hurdle and steeple racing: Risk factors for musculoskeletal injury. *Equine Vet. J.* **1998**, *30*, 498–503. [CrossRef] [PubMed]
39. Oikawa, M.; Kusunose, R. Fractures sustained by racehorses in Japan during flat racing with special reference to track condition and racing time. *Vet. J.* **2005**, *170*, 369–374. [CrossRef] [PubMed]

animals

MDPI

Article

Preventing and Investigating Horse-Related Human Injury and Fatality in Work and Non-Work Equestrian Environments: A Consideration of the Workplace Health and Safety Framework

Meredith Chapman [1,2,*,†] and Kirrilly Thompson [1,†]

[1] The Appleton Institute, Central Queensland University, 44 Greenhill Road, Wayville, SA 5034, Australia; kirrilly.thompson@cqu.edu.au

[2] Safety in Focus, PO Box 711, Narrabri, New South Wales 2390, Australia

* Correspondence: meredith@safetyinfocus.com.au; Tel.: +61-2-6792-2342

† The authors contributed equally to this work.

Academic Editor: Clive Phillips

Received: 18 November 2015; Accepted: 28 April 2016; Published: 6 May 2016

Simple Summary: Attempts to reduce horse-related injuries and fatalities to humans have mostly focused on personal protective equipment like helmets. In organizational contexts, such technical interventions are considered secondary to reducing the frequency and severity of accidents. In this article, we describe the Workplace Health and Safety (WHS) framework that has been associated with reduced risks in industries and organisations. We consider how such a framework could be used to reduce horse-related risks in workplaces, as well as non-work equestrian competition and leisure environments. In this article, we propose that the simplicity and concepts of the WHS framework can provide risk mitigation benefits to both work and non-work equine identities.

Abstract: It has been suggested that one in five riders will be injured due to a fall from a horse, resulting in severe head or torso injuries. Attempts to reduce injury have primarily focussed on low level risk controls, such as helmets. In comparison, risk mitigation in high risk workplaces and sports is directed at more effective and preventative controls like training, consultation, safe work procedures, fit for purpose equipment and regular Workplace Health and Safety (WHS) monitoring. However, there has been no systematic consideration of the risk-reduction benefits of applying a WHS framework to reducing horse-related risks in workplaces, let alone competition or leisure contexts. In this article, we discuss the different dimensions of risk during human–horse interaction: the risk itself, animal, human and environmental factors and their combinations thereof. We consider the potential of the WHS framework as a tool for reducing (a) situation-specific hazards, and (b) the risks inherent in and arising from human–horse interactions. Whilst most—if not all—horses are unpredictable, the majority of horse-related injuries should be treated as preventable. The article concludes with a practical application of WHS to prevent horse-related injury by discussing effective evidence-based guidelines and regulatory monitoring for equestrian sectors. It suggests that the WHS framework has significant potential not only to reduce the occurrence and likelihood of horse-related human accident and injury, but to enable systematic accident analysis and investigation of horse-related adverse events.

Keywords: horses; people; risk; mitigation; safety; WHS; injury; deaths; workplace

1. Introduction

Despite workhorses being largely replaced by machinery in the industrial era, horses have continued to 'work' for humans. They are our athletes, entertainers and therapists. Horses aid humans in law enforcement and agriculture, and take centre stage in racing. They are supported by an equine industry serviced by human specialists: riders, handlers, trainers, coaches, farriers, veterinarians, transporters and more. These humans are exposed to horse-related risks on a regular —if not daily—basis. Although the risks to humans of interacting with horses are well known [1–5], there have been no significant reductions in rates of injury or death over the past decades [6]. In contrast, there have been significant decreases in injuries and fatalities in high risk workplace settings such as mining and construction [7]. One significant difference between injury and fatality rates in high-risk industry compared to high-risk interspecies interactions across the combined sport and leisure sectors is Workplace Health and Safety (WHS). There has been widespread adoption and implementation of WHS principles in industry [8], with a focus on improved risk management. Broadly speaking, the same cannot be said for Australia's horse industry where formal WHS application is inconsistent and restricted to select sectors like Thoroughbred Racing. This may result from diversity; with the equestrian sector including racing, sport, competition, recreation and leisure. Despite these activities making a significant economic contribution to Australia [9], they are not immediately—or equally—recognisable as workplaces. However, it seems prudent to consider if the application of a WHS framework to horse-related interactions across the broad horse industry could provide the same benefits in injury prevention and reduction that have been documented in other industries.

The aim of this article is to consider the potential benefits of applying a WHS framework to horse-related interactions to (a) reduce horse-related risks, and (b) enable investigation of adverse horse-related incidents. To determine the applicability of WHS to horse-related risks, we first discuss the multiple dimensions of risk in human–horse interactions: horse, human, environment and combinations thereof. Given its track record of reducing risks in workplaces, we consider how applicable WHS is to human–horse interactions in work and non-work contexts. We determine that the WHS framework can be easily translated to horse-related interactions. In fact, despite the WHS framework being developed without explicit reference to animals, we identify particular potential for WHS to address the inherent unpredictability commonly attributed to horses. Moreover, we identify an additional role for the WHS framework in guiding horse-related accident and injury investigations. To our knowledge, this is the first considered proposition for the usefulness of a WHS framework in reducing horse-related human injury and fatality. This article suggests a research agenda that outlines the empirical research necessary to evaluate our propositions and concludes with the challenge that other high risk industries have reduced workplace injuries and deaths with WHS, so what's preventing equine from achieving similar results.

2. Horse-Related Risks

2.1. The Horse

The horse is considered an animal with a fight/flight instinct that humans spend many hours trying to train, coerce and desensitize to any adverse external stimuli [10,11]. Horses intrinsically are herd animals, with 'leader' and 'follower' instincts, having a mind of their own, which leads to 'unpredictable' behaviours [12–14], although predictability is largely dependent on human–horse knowledge and capabilities [5]. When placed in a situation where they feel threatened or insecure, horses can display dangerous behaviours of running, biting, crushing or kicking [15]. Research has determined that horses have a 'fear memory' that can be 'turned on' by human interaction, or may be 'toned down' (directed by a human) to enable the development of 'trust' [16–18]. Whilst there is more to learn and understand about horses as a species that directly affects safe and successful human–horse interactions [19], the majority of horse behavioural 'problems' are thought to be caused by equine confusion or a lack of understanding between the human and horse [20,21].

2.2. The Human

The degree of 'risk' that extends to horse riders or handlers can be affected by their own subjective perception of their capability and 'horsemanship' skill. High perceptions of 'self-efficacy' [22,23] could increase the likelihood of an injury or even death. A human's level of horse experience and knowledge can be defined by their age, the number of horse interactive hours, along with their type of supervision and training [24,25]. When there is a deficit in one of these components, the likelihood of harm increases [24,26–28]. A particularly important determinant of horse-related risk is the suitability of horse and human combinations [29,30]. A mismatch at any stage during contact between the human and horse could potentiate a negative incident. Moreover, competitive drives amongst professional or ambitious equestrians can compromise their safety, in the form of 'goal seduction' [31,32].

Similarly, humans tend to devalue the importance of equine safety at point of sale [33], possibly where sellers can be seduced by financial return. For example, human desire for financial benefit might result in knowledge of undesirable horse traits and/or dangerous behaviours being withheld from a buyer. Alternatively, it may result from a buyer's desire to own a horse regardless of such concerns (perhaps due to high self-efficacy in addressing them). More naively, selling unsuitable horses to riders may be facilitated by a lack of buyer expectation and devaluation of safety.

2.3. The Environment

A key element in human–horse interaction may be the environment in which the connection occurs. The environment may include the physical location and terrain, whether the area is confined by barriers (fencing, yard or crush), weather conditions and the degree of visibility for the horse, rider or handler [34]. Any one of these occurrences may affect consequence. If there is a distraction, other animals, unfamiliar or loud noises or a change in routine, the environment is changed from its initial state. Horses are very visual and reactive to changes in the environment as well as changes in the appearance of other humans and animals.

The environment can be considered through the concept of an 'affective atmosphere' [35]. A poor safety culture or environment in a workplace, can be described as untidy, having unfinished jobs and where workers take risks or short cuts, with little or no communication, [36]. A positive safety culture evolves from the combination of both individual and group efforts towards values, attitudes, goals and proficiency of an organisation's WHS program [37–40]. The type of safety culture that exists at a workplace is determined by a broad, organisation-wide approach to safety management. A safety conscious manager empowers workers to prioritise safety, which translates into a safer work environment. Therefore, the feelings and behaviours of humans can determine the type of environment that is displayed and generated [41].

3. What is Workplace Health and Safety? WHS

All humans should have the right to be safe, maintain good health and enjoy life. This right extends to a workplace setting, with a belief that a worker can go to a work, perform their assigned duties and return home safely, injury free. This ideology will only be successful at work if all stakeholders (workers, employers, suppliers *etc.*) are thinking 'safety first', planning for unforeseeable events that cause harm and implement systems to manage harmful exposures. At its simplest, WHS is a set of 'processes and standards', mandated by legal obligations, that workers are expected to follow to promote and maintain their personal health, safety and welfare and that of others. WHS legislation defines the context of work for a person who conducts business or undertaking, whilst providing stakeholders with a clear understanding about their obligations (duty of care) and the consequences for neglecting them [42]. Furthermore, WHS legislation prescribes the need for those in control of a workplace to provide safe premises, safe machinery/materials, suitable training, supervision, work environment and facilities, supported by safe systems of work [43,44].

Many countries have adopted a regulatory WHS framework to assist workplaces in meeting their obligations and keeping workers healthy and safe. If safe work systems are adopted that are easily understood, workers and their families have more financial security, unencumbered by injury or death. By using a predetermined and industry specific WHS framework, employers benefit from uninjured workers through improved productivity and lost time at work. Some of the items in a WHS framework include the provision of defined policies, procedures and clear processes for worker communication and participation about their own work practices and safety. Providing workers with the opportunity for training and skills development, ensuring proactive risk management practices are implemented and monitoring, measuring and reviewing workplace activities. Items within the WHS framework require mandatory compliance. However, each workplace has the opportunity to develop and adopt their own set of processes to demonstrate 'due diligence' and compliance as reasonably practicable [45–49].

Horses as 'Working Animals': Working Horse Safety

In Australia, the majority of notifiable deaths and serious injuries each year occur in what Australia's WHS regulator classifies as the top five 'high risk' workplaces: Transport, Agriculture, Construction, Mining and Manufacturing. All of these industries have a legislated 'duty of care' with designated accountabilities and responsibilities for WHS, including mandatory compliance using a suitable risk management approach. In 2014, Australia reported 20 work-related traumatic injury fatalities due to falls from a height, with Agriculture being rated the second highest contributing industry with 'horse related' human deaths accounting for three (SWA, [50]). Many other potential accidents are prevented, due to suitable training in the use and maintenance of equipment, farm plant and machinery. Likewise, 'working horses' require maintenance of their hooves ('tyres'), general health and nutrition ('fuel'), with knowledge of their level of education and training ('fit for purpose use'). However, horses are not explicitly classified as a 'tool of trade' when used in Agriculture. It would appear that the 'risk' of human injury resulting from horses at work, the importance of record keeping and information seeking about this 'working animal', is given less attention when compared to other high risk workplaces that reply on suitable and safe work equipment. Recognising this oversight provides a significant opportunity to improve how horse-related risks are managed in work and non-work situations. In particular, there is a need to explicate how horses and horse-related injuries 'fit' within industry and understand the legal obligations that result from any classification.

In general, prosecutions for any offence against WHS legislation have identified that a 'duty of care' is owed to a person due to the nature of the worker and employer relationship. Examples of human and horse work arrangements include cowgirls/jillaroos and cowboys/jackaroos working on cattle stations and feedlots, as well as stable hands, polo grooms, track work riders, animal transporters, coaches and instructors. When engaging a 'worker' either paid or unpaid, WHS law usually prescribes compliance with a set of risk management responsibilities, such as suitable training, supervision, consultation, monitoring and the implementation of hazard controls. In the case of human–horse interactions, all horses can be considered as workers, attended to by small to large hosts of human workers.

To date, attempts to reduce injuries amongst this human workforce have focussed on technical interventions such as back protectors, inflatable vests and frangible pins that reduce the risk of rotational falls at fixed obstacles [26,27,51,52]. However, technical intervention is not considered the most effective means of reducing injury [53]. Wearing a helmet is the 'lowest' or 'least' effective form of risk control within the WHS 'Hierarchy of Controls' [54–57]. Moreover, helmets generally only apply to riders (not handlers), who may only wear them when they are compulsory, such as during competition [58].

Many researchers have identified a need for attention to shift from a preoccupation with incident data to risk management and injury prevention [59–62]. More information is needed to identify the variances in human 'risk perception' [63] and the 'beliefs' that shape human behaviour and the environment in a variety of equine activities. This has commenced with preliminary studies on risk perception and the socio-cultural dimensions of risk amongst equestrians [5,58,64]. However, there has been no evaluation of the potential benefits of applying WHS principles to prevent and reduce horse-related injury and fatality. This is surprising, given that coronial findings and case law have identified the exact kinds of horse-related risk contributors that a generic WHS framework could easily prevent; an inability to recognise the level of risk; an unsuitable match of horse, rider or handler; and a lack of appropriate levels of supervision and training [65–67]. Whilst the ability for a WHS framework to introduce regulatory compliance is clear, how else might it systematically reduce horse-related risks? In the following sections, we consider how a WHS framework can be applied to human and horse interactions. We note its simplicity of documentation; procedures and reporting; risk assessment; skills assessment; training and supervision; and structured communication.

4. Applying the WHS Framework to Horse-Related Human Injury and Fatality

4.1. Documentation, Procedures and Reporting

To assist in reducing workplace injuries, standardization of safety processes is necessary. Standardization promotes organizational consistency and details requirements for best practice performance, whereby it then becomes the 'backbone' to continuous systems improvement. Moreover, a standardized WHS system assists an organization or industry to identify what information needs recording and monitoring to reduce human exposure to potential hazards. Clear, concise documentation, procedures and reporting processes support humans to rationalize critical safety decisions. A WHS system also provides 'evidence' for those making safety judgments in the event of an incident, whereby demonstrating 'due diligence' [68,69].

For human–horse interactions, 'due diligence' can be achieved by implementing inspection checklists of rider/handler equipment; noting and reporting environmental conditions; formulating safe work procedures; assessing equine level of 'risk'; determining rider/handler capabilities and maintaining training records. All of these processes provide consistency and communicate any changes that may occur [70,71]. Having a structure for equine workers to follow promotes group cohesiveness. Furthermore, it identifies who is responsible and accountable for what; enabling a reporting mechanism for near miss events, injuries, hazards and highlights areas for WHS systems improvement [72]. In directing and documenting delegated worker duties and activities, a WHS system evolves, forming a framework for the production of safer human–horse relations.

4.2. Risk Assessment

In a generic WHS framework, risk assessment refers to identifying hazards such as equipment or plant. In the case of horse-related activities, horses are also a hazard, albeit one with significantly more autonomy and capable of exercising their own will in unpredictable ways [73]. The Victorian Injury Surveillance Unit at Monash University in Australia classified horses for the purpose of data collection as being a form of farm transport [74], Also, some WHS prosecutions suggest horses may 'fit' within the workplace during human–horse related undertakings.

From this perspective, proactive equine Risk Management (RM) [75] is about identifying, assessing and managing relevant risks prior to and during human and horse contact. A horse assessment may capture dangerous behaviours elicited during an exposure to various stimuli and situational circumstances, with the assessment occurring during a variety of conditions [76]. Earlier studies with similar assessment criteria are those measuring a horses response to 'human approach' [77], testing 'social separation' responses in a horse [78–80] and 'bridge testing' when a horse and handler cross over a novel surface [81–83]. Other scientific horse studies have measured physiological heart rate (HR) responses to various stimuli and challenging situations [84,85], however these are not easily measured and present some ambiguity given an increase in HR generally links to an increase in physical activity. A more complex horse assessment similar to a 'novel object test' [78,82,86] with a combination of assessments is needed for consistency.

Risk assessment should commence when a human first approaches and makes contact with a horse. The aim is to identify what positive and negative behaviours a horse displays during ground work activities such as catching, tying up, picking up feet, lunging, and during preparation for riding. To identify the suitability of a horse, an experienced handler/rider may perform a 'test ride' [87,88]. This allows for the observation and anticipation of undesirable behaviours and changes in the horse's response whilst exposed to a variety of stimuli and obstacles. The degree of change in the horse's response can give a subjective indication as to its level of sensitivity or de-sensitivity. Measuring a horse's response, whilst it is in contact with various obstacles and activities could give the rider an indication of how risky this horse may be during future human contact, and for what kind of rider it is most suitable. This process provides an RM platform, to identify a risk rating for a horse e.g., low, medium, high or unacceptable, thus providing an indication of how risky this horse may be [89]. Based on these practical horse assessment guidelines, WHS may provide 'reasonably practicable' insight into a foreseeable or unplanned event during human and horse interaction, whereby limiting harmful exposures that may not otherwise be accounted for. However, further scientific research is needed to identify validated and reliable tools that can be used in the field along with an audit of pre-existing tools.

4.3. Skills Assessment

Skill matching to assigned activities can save lives, but in equestrian environments, making appropriate decisions about horse/rider combinations requires more than just an assessment of the horse. A similar assessment of the level of skill and 'horsemanship' of riders and handlers should also be undertaken [87]. This type of assessment would be based on determining a human's level of competency in performing a required task or activity. This process is used in other workplaces where assessing a worker's ability to perform the inherent demands of a job promotes a suitable match [90,91]. Furthermore, it identifies a workers capability to perform their duties safety, to determine if they require further education, training, supervision or a change in job task [41]. In equine workplaces, a handler/rider could demonstrate their practical skills in handling and riding a horse in a controlled environment.

Similarly to a horse assessment, the human would show their level of skill in approaching a horse safely, tacking-up a horse in preparation for riding, riding the horse through a series of activities and obstacles in various environments, before ceasing their interaction with the horse [92]. A set of parameters defining beginner skill level to an advanced rider would be tabled using similar activities to support more reliable outcomes. The assessor can determine a handler/rider level of competency when they either start to show unsafe horsemanship and the assessment is ceased or they reach a pre-determined level successfully without signs of incompetence.

4.4. Training and Supervision

The WHS framework supports competency based training regimes as the most effective method to produce safe outcomes in work performance [93–96]. It is only through training, supervision and experience that humans learn the skills necessary to make safe decisions, whereby protecting them from unnecessary work and non-work related injuries and deaths. Sharing experiences and relevant information maintains open channels for workers and others to address issues as they arise. Through demonstration of job activities, assessment and adequate supervision, a worker can improve their level of skill which results in better performance and reduced risk taking behaviors [97]. Training regimes approved by accredited training organizations such as Australian Skills Quality Authority (ASQA) are rarely disputed. Due to regulatory monitoring of training delivery and certification requirements, there is some level of reassurance that when a worker obtains a license, skill or qualification there is both visual and documented evidence that this is so.

In contrast, equestrian activities have a long history and are subject to many different philosophies, applications and cultures [98–103], as well as different types of horses with their own dispositions [29].

There is no single, agreed upon approach to interacting with horses, let alone a shared understanding of the 'safest' ways to interact. Indeed, it is easier to talk about equestrian sub-cultures and their conflict—such as classical equitation [104–106], modern equitation [29,64,107–117], natural equitation [99,101,102,118] and scientific equitation [20,119–122]—rather than an unchallenged equestrian culture. Current practices differ, are resistant to change, are antiquated or may have become dislocated from their original conditions. Some serve no other purpose than aesthetic pleasure, such as high school dressage, the purpose of which is frequently justified in relation to obsolete practices of military riding and clearing the ground of foot soldiers. Whilst the burgeoning field of equitation science is making significant progress in providing an evidence base to understanding equine ethology and identifying its implications for safe human–horse interaction, the existing WHS framework could be readily implemented to ensure that horse-related training and supervision is established as commonplace and subject to regular review.

To ensure effective training programs for safe horse handling skills, an individual training plan [123] needs to be developed, especially with workers engaging horses. This would include recognition of demonstrated prior knowledge and skill, forming a current assessment of handler/rider ability, and developing a plan to increase skills to a more advanced and unsupervised level. The process of learning would commence with a foundation in safe approach and handling of a horse (demonstrated on horses with a variety of handling experience, in confined and open spaces, over a designated duration), and discussion to identify underpinning knowledge relative to various horse-related activities. This would be followed by a rider validating their perceived level of riding skill in a controlled environment, progressing as deemed safe and competent to higher levels of demonstrated rider ability involving precision, control and completion of assigned activities [124]. Where a rider exhibits unsafe and/or inadequate skill, further training and supervision would be required to promote a riders ability to achieve an assigned task. For example, in the job of mustering cattle, if the skill level of a rider was rated at a lower level, they would 'tail a herd of cattle' where they are positioned riding at the rear and under supervision. As the riders' skills increased they could 'turn back cattle' (retrieve those that escaped the herd) and eventually, as an advanced rider, 'lead the herd of cattle', being positioned up front. In any workplace where a duty of care exists between an employer and its workers (paid or unpaid), WHS legislation applies. The equine industry is no exception. The duty of keeping workers 'healthy and safe' (as reasonably practical) is clearly tabled by defined statutory duties (accountabilities and responsibilities) embedded within the WHS Acts and Regulations [125–127]. Where this mandatory obligation applies, every horse and human activity deemed as a work undertaking or workplace would benefit from the legal defense of using a WHS framework with a documented safety management system.

4.5. Structured Communication

Consultation and communication is a legal requirement and an essential part of managing safety risks. In order to achieve a safe workplace, everyone involved needs to communicate with each other to identify hazards and risks, talk about safety concerns and work together to find solutions [128,129]. Communication assists an organisation by drawing on the knowledge and experience of its workers, enabling more informed decisions to be made about how work should be carried out safety [130]. For communication to be successful, it needs to be easily understood, relevant and 'effective'.

Developing 'safe' human–horse relationships requires the views of others being heard and greater co-operation and trust. Communicating important safety information such as poor horse behaviours, risks when handling or riding horses, rider capabilities, and/or emergency stop preparedness, prior to and during horse and human interaction will promote the delivery of a 'best practice' model. It provides feedback to support value added training programs, enhances safety awareness and promotes a 'safety first' culture. For example, making time to discuss horse health gives an opportunity to understand adverse horse reactions to pain or discomfort [131]. Prior to any horse and human interaction, communication may be relayed by a daily meeting, to generate discussions on pre-existing horse conditions, horse performance or health care needs (teeth, farrier, worming) and future management, similar to the patient handovers undertaken in hospital and care settings [132–134].

Communication can take the form of verbal informal chats; record keeping of horse health and treatments; evaluating a horse's response to different activities; group discussions about exposures and handler/rider activities. Noting or reporting any subtle changes in a horse's usual behaviour and herd responses are keys to managing safe horse and human interactions where this dynamic relationship is ever changing.

5. What Can We Do about What is Missing in the Equine Industry?

Every human has a 'duty of care' to one another, be that directive under Common, Civil or WHS law. The obligation is to demonstrate 'due diligence' and work to improve horse-related human injury or death by identifying suitable and plausible risk mitigation strategies. In Australia, ASQA has audited all horse related courses delivered nationally to ensure course competencies and training providers are maintaining 'due diligence' [135] (unpublished work). This is one step in the right direction, but more can be done to deliver better and safer equine safety and training standards, especially in those human–horse interactions that are not delivered by an ASQA accredited Registered Training Provider (RTO).

This article proposes that a WHS framework has significant potential to reduce horse-related injury and fatality in work as well as non-work contexts. One novel proof of concept is to use the WHS framework to structure a systematic review of horse-related human deaths where there are multiple sources of information that make it possible to: (a) reconstruct the event; (b) conduct a root cause analysis [136,137]; (c) identify key points at which a WHS framework was absent; (d) retrospectively reconstruct the event with the recommended WHS framework in place; (e) compare the hypothetical outcome with the actual event; and (f) identify improvements to prevent future incidents. By using investigative tools like root cause analysis and pre-event incident exploration, it should be possible to identify predisposing factors to the incident and theoretically consider what preventative safety measures could have been adopted, and should be routinely maintained. Having access to coroners' reports, interviewing consenting injured parties and analyzing accident information against a safety standard (e.g., AS/NZ: 4801, ISO1400 or WHS Legislation) will provide valuable information about imminent 'risks' during horse and human interaction.

Further empirical research is required to determine how translatable a WHS framework and its benefits are in reducing horse-related human injuries and deaths in non-work environments. Such research will be essential to identifying areas of necessary adaptation or extension, barriers for implementation and resistance from the target audience, and existing fora to promote safety awareness education and initiatives. There is also a particular need to identify best practice industry role models [58], as future practical initiatives may benefit from involving those already implementing WHS successfully in work environments (e.g., Thoroughbred Racing) with participants in the non-work equine sector (e.g., as Pony Clubs).

6. Conclusions

The above discussion suggests that a WHS framework can support a reduction in horse-related human injuries and deaths through documented procedures and reporting; risk assessment; skills assessment; training and supervision and improve safety culture in the equestrian industry. By being 'risk aware' of the inherent dangers with horses and taking conscious steps to reduce harm, horse handlers and riders can become insightful and responsible, therefore endorsing safe behaviors that stimulate a proactive equine safety culture [40,138]. As such, a WHS framework has significant potential to reduce the risks inherent in, and arising from horse and human interaction, regardless of whether they occur in work, private, public, amateur or professional contexts. It could also pave the way for improved education and behavior change strategies, especially to overcome any fatalistic acceptance that horse-related human injuries and deaths are inevitable [40].

Given that there is a pre-existing framework of WHS that has had demonstrated success in other work contexts, any form of complacency around horse-related human injury and fatality is inexcusable. But how do we anticipate and ever hope to reduce horse-related human injury and fatality when equine associated legal obligations are still un-prescribed, with no best practice guidance or enforcement of safety principles? Moreover the problem remains; where do horses fit into in current WHS law amongst specific definitions such as 'plant', 'structures' and 'substances'. This omission and failure of a clear definition of a 'working animal' (e.g., horses) used at work or during human–horse interaction in a work setting leaves this high risk activity open to interpretation.

Therefore, there is a pressing need for the adoption of minimal and consistent standards for qualifications, training, supervision, consultation, monitoring and the implementation of hazard controls in accordance with a defined hierarchy. Given the evidence that WHS has assisted in the reduction of work-related injuries and deaths, it seems sensible to take advantage of an established framework to guard the lives and livelihoods of those working, playing and competing with horses.

Acknowledgments: The authors would like to thank Laura Haigh for her assistance in preparing the manuscript.

Author Contributions: The article is based on research undertaken primarily by Meredith Chapman with supervision, advice, conceptualization and writing by Kirrilly Thompson.

Conflicts of Interest: The authors declare no conflict of interest.

References

1. Chambless, D. Traumatic brain injury in equestrian sport. In Proceedings of the 2nd Helmet Safety Symposium, Lexington, KY, USA, 23 July 2011.
2. Firth, J. *Equestrian Injuries*; Williams & Wilkins: Baltimore, MD, USA, 1985.
3. Åman, M.; Henriksson Larsén, K. Acute injuries in 35 sports; incidences and severity. *Br. J. Sports Med.* **2014**, *48*, 674.
4. Schneider, S.; Seither, B.; Tönges, S.; Schmitt, H. Sports injuries: Population based representative data on incidence, diagnosis, sequelae, and high risk groups. *Br. J. Sports Med.* **2006**, *40*, 334–339. [CrossRef] [PubMed]
5. Carmichael Ii, S.P.; Davenport, D.L.; Kearney, P.A.; Bernard, A.C. On and off the horse: Mechanisms and patterns of injury in mounted and unmounted equestrians. *Injury* **2014**, *45*, 1479–1483. [CrossRef] [PubMed]
6. National Coronial Information System (NCIS). *Animal-Related Deaths*; NCIS: Melbourne, Australia, 2011; pp. 1–8.
7. Arocena, P.; Núñez, I. An empirical analysis of the effectiveness of occupational health and safety management systems in smes. *Int. Small Bus. J.* **2010**, *28*, 398–419. [CrossRef]
8. Coulter, S. *Advancing Safety around the World. History, Geography, Religion, and Culture are Major Influences on Safe Work Applications in Coutries across the Globe*; Occupational Health & Safety: Dallas, TX, USA, 2009.
9. Gordon, J. *The Horse Industry—Contributing to the Australian Economy*; Centre for International Economics: Canbera, Austrilian, 2001.
10. McGreevy, P. The advent of equitation science. *Vet. J.* **2006**, *174*, 492–500. [CrossRef] [PubMed]
11. McCall, C.; Hall, S.; McElhenney, W.; Cummins, K. Evaluation and comparison of four methods of ranking horses based on reactivity. *Appl. Anim. Behav. Sci.* **2006**, *96*, 115–127. [CrossRef]
12. Sighieri, C.; Tedeschi, D.; de Andreis, C.; Petri, L.; Baragli, P. Behaviour patterns of horses can be used to establish a dominant-subordinate relationship between man and horse. *Anim. Welf.* **2003**, *12*, 705–708.
13. Grandin, T. Safe handling of large animals. *Occup. Med.* **1999**, *14*, 195–212. [PubMed]
14. Budiansky, S. *The Nature of Horses*; Phoenix Illustrated: London, UK, 1998.
15. Visser, E.K.; van Reenen, C.G.; Blokhuis, M.Z.; Morgan, E.K.M.; Hassmén, P.; Rundgren, T.M.M.; Blokhuis, H.J. Does horse temperament influence horse–rider cooperation? *J. Appl. Anim. Welf. Sci.* **2008**, *11*, 267–284. [CrossRef] [PubMed]
16. Grandin, T. Assessment of stress during handling and transport. *J. Anim. Sci.* **1997**, *75*, 249–257. [PubMed]
17. Grandin, T.; Johnson, C. *Animals in Translation: Using the Mysteries of Autism to Decode Animal Behaviour*; Sunny Press: Albany, NY, USA, 2009.

18. Houpt, K.; Mills, D. Wht horse behaviour is important to the equine clinician. *Equine Vet. J.* **2006**, *38*, 386–387. [CrossRef] [PubMed]
19. Vandierendonck, M.; Goodwin, D. The human-animal relationship. Forever and a day. In *Social Contact in Horses: Implcations for Human-Horse Interaction*; de Jonge, F.H., van den Bos, R., Eds.; Royal van Gorcum: Assen, The Netherlands, 2005; pp. 65–81.
20. McGreevy, P.D. The advent of equitation science. *Vet. J.* **2007**, *174*, 492–500. [CrossRef] [PubMed]
21. McGreevy, P.; McLean, A. The domestic horse, the origins, development and management of its behaviour. In *Behavioural Problems with the Ridden Horse*; Cambridge University Press: Cambridge, UK, 2005; pp. 196–211.
22. Bandura, A. Percieved self-efficacy in cognitive development and functioning. *Educ. Psychol.* **1993**, *28*, 117–148. [CrossRef]
23. Zimmerman, B. Self-efficacy: An essential motive to learn. *Contemp. Educ. Psychol.* **2000**, *25*, 82–91. [CrossRef] [PubMed]
24. Hawson, L.A.; McGreevy, P.D.; McLean, A.N. Roles of equine ethology and applied learning theory in horse-related human injuries. *J. Vet. Behav.* **2010**, *5*, 324–338. [CrossRef]
25. Keeling, L.; Blomberg, A.; Ladewig, J. Horse-riding accidents: When the human-animal relationship goes wrong! In Proceedings of the 33rd International Congress of the International Society for Applied Ethology, Lillehammer, Norway, 17–21 August 1999.
26. Havlick, H. *Equestrian Sport-Related Injuries: A Review of Current Literature*; Charlottee Medical Clinic: Charlotte, NC, USA, 2010; pp. 299–302.
27. Ball, C.; Ball, J.; Kirkpatrick, A.; Mulloy, R. Equestrian injuries: Incidence, injury patterns, and risk factors for 10 years of major traumatic injuries. *Am. J. Surg.* **2007**, *193*, 636–640. [CrossRef] [PubMed]
28. Griffen, M.; Boulanger, B.R.; Kearney, P.A.; Tsuei, B.J.; Ochoa, J. Injury during contact with horses: Recent experience with 75 patients at a level 1 trauma center. *South. Med. J.* **2002**, *95*, 441–445. [CrossRef] [PubMed]
29. Thompson, K.; Birke, L. The horse has got to want to help: Human-animal habituses and networks in amateur show jumping. In *Sport, Animals, and Society*; Gillett, J., Gilbert, M., Eds.; Routledge: New York, NY, USA, 2014; pp. 69–84.
30. Morgan, K.; Zetterqvist, M.; Hassmen, P.; Visser, E.; Rundgren, M.; Blokhuis, H. Riders personality and the perception of the co-operation between rider and horse. In Proceedings of 51st Annual Meeting of the Eurpoean Association for Animal Production, The Hague, The Netherlans, 21–24 August 2000; p. 396.
31. Owen, C. *Human Factors Challenges in Emergency Management: Enhancing Individual and Team Performance in Fire and Emergency Services*; Ashgate Publishing Ltd.: Farnham, UK, 2014.
32. Bearman, C.; Paletz, S.; Orasanu, J. Situational pressures on aviation decision making: Goal seduction and situation aversion. *Aviat. Space Environ. Med.* **2009**, *80*, 556–560. [CrossRef] [PubMed]
33. Hawson, L.A.; Oddie, C.; McLean, A.N.; McGreevy, P.D. Is safety valued in the australian pony market? *J. Vet. Behav.* **2011**, *6*, 254–260. [CrossRef]
34. Leppälä, J.; Kolstrup, C.; Pinzke, S.; Rautiainen, R.; Saastamoinen, M.; Särkijärvi, S. Development of a safety management web tool for horse stables. *Animals* **2015**, *5*, 0402. [CrossRef] [PubMed]
35. Bissell, D. Passenger mobilities: Affective atmospheres and the sociality of public transport. *Environ. Plan. D Soc. Space* **2010**, *28*, 270–289. [CrossRef]
36. Cooke, D.; Rohleder, T. Learning from incidents: From normal accidents to high reliability. *Syst. Dyn. Rev.* **2006**, *22*, 213–239. [CrossRef]
37. Toole, M. The relationship between employees' perceptions of safety and organizational culture. *J. Saf. Res.* **2002**, *33*, 231–243. [CrossRef]
38. Clarke, S. Perception of organizational safety: Implications for the development of safety culture. *J. Organ. Behav.* **1999**, *20*, 185–198. [CrossRef]
39. DeJoy, D. Behaviour change *versus* culture change: Divergent approached to manaing workplace safety. *Saf. Sci.* **2005**, *43*, 105–129. [CrossRef]
40. Thompson, K.; McGreevy, P.; McManus, P. A critical review of horse-related risk: A research agenda for safer mounts, riders and equestrian cultures. *Animals* **2015**, *5*, 561–575. [CrossRef] [PubMed]
41. Tracey, J.B.; Tannenbaum, S.I.; Kavanagh, M.J. Applying trained skills on the job: The importance of the work environment. *J. Appl. Psychol.* **1995**, *80*, 239. [CrossRef]
42. Barrett, R.; Mayson, S.; Bahn, S. Small firms and health and safety harmonisation: Potential regulatory effects of a dominant narrative. *J. Ind. Relat.* **2014**, *56*, 62–80. [CrossRef]

43. Breslin, P. What are the benefits and costs of developing, implementingand reviewing safe work method statementsfor high-risk works in the australian construction industry? *J. Health saf. Environ.* **2013**, *5*, 269–283.
44. Deighan, C.; Lansdown, T.; Brotherton, C. Using 'stage of change' and 'business activity models' to assess and improve health and safety behaviours in smes. *Policy Pract. Health Saf.* **2009**, *15*, 69–83.
45. Tooma, M.; Titterton, A.; Carnell, M. The End of Work Health and Safety Harmonisation? Available online: https://sia.org.au/downloads/publications/ohs_professional_magazine/ohs-march-2012.pdf (accessed on 30 September 2015).
46. Windholz, E. Ohs harmonisation: Delivering a more consistent (and effective and efficient) regulatory experience. *Employ. Law Bull.* **2012**, *18*, 1–7.
47. Johnstone, R. Harmonising Occupational Health and Safety Regulation in Australia: The First Report of the National OHS Review. 2008. Available online: http://search.informit.com.au/documentSummary;dn=656390346316766;res=IELHSS (accessed on 30 September 2015).
48. Bahn, S. *Workplace Hazard Identification: What do People Know and How is it Done?*; Griffith University: Nathan, Australia, 2012.
49. Bluff, L.; Johnstone, R. *The Relationship between'Reasonably Practicable'and Risk Management Regulation*; National Research Centre for Occupational Health and Safety Regulation: Canberra, Australia, 2004.
50. Safe Work Australia. Worker Fatalities. Available online: http://www.safeworkaustralia.gov.au (accessed on 5 November 2015).
51. Buckley, S.; Chalmers, D.; Langley, J. Injuries due to falls from horses. *Aust. J. Public Health* **1993**, *17*, 269–271. [CrossRef] [PubMed]
52. Kiss, K.; Swatek, P.; Lenart, I.; Mayr, J.; Schmidt, B.; Pinter, A.; Hollwarth, M. Analysis of horse-related injuries in children. *Pediatr. Surg. Int.* **2008**, *24*, 1165–1169. [CrossRef] [PubMed]
53. Vredenburgh, A.G. Organizational safety: Which management practices are most effective in reducing employee injury rates? *J. Saf. Res.* **2002**, *33*, 259–276. [CrossRef]
54. Viner, D. *Accident Analysis and Risk Control*; Sonali Publishing House: New Delhi, India, 1996.
55. Holmes, N.; Lingard, H.; Yesilyurt, Z.; de Munk, F. An exploratory study of meanings of risk control for long term and acute effect occupational health and safety risks in small business construction firms. *J. Saf. Res.* **1999**, *30*, 251–261. [CrossRef]
56. Hopkins, A. What are we to make of safe behaviour programs? *Saf. Sci.* **2006**, *44*, 583–597. [CrossRef]
57. Makin, A.M.; Winder, C. Managing hazards in the workplace using organisational safety management systems: A safe place, safe person, safe systems approach. *J. Risk Res.* **2009**, *12*, 329–343. [CrossRef]
58. Haigh, L.; Thompson, K. Helmet use amongst equestrians: Harnessing social and attitudinal factors revealed in online forums. *Animals* **2015**, *5*, 576–591. [CrossRef] [PubMed]
59. Smartt, P.; Chalmers, D. A new look at horse-related sport and recreational injury in new zeland. *J. Sci. Med. Sport* **2009**, *12*, 376–382. [CrossRef] [PubMed]
60. Thomas, K.; Annest, J.; Gilchrist, J.; Bixby-Hammett, D. Non-fatal horse related injuries treated in emergency departments in the united states, 2001–2003. *Br. J. Sports Med.* **2006**, *40*, 619–626. [CrossRef] [PubMed]
61. Mayberry, J.; Pearson, T.; Wiger, K.; Diggs, B.; Mullins, R. Equestrian injury prevention efforts need more attention to novice riders. *J. Trauma* **2007**, *62*, 735–739. [CrossRef] [PubMed]
62. Wolframm, I. *The Science of Equestrian Sports: Theory, Practice and Performance of the Equestrian Rider*; Routledge: Abingdon, UK, 2013; p. 208.
63. Kouabenan, D.R. Role of beliefs in accident and risk analysis and prevention. *Saf. Sci.* **2009**, *47*, 767–776. [CrossRef]
64. Thompson, K.; Nesci, C. Over-riding concerns: Developing safe relations in the high-risk interspecies sport of eventing. *Int. Rev. Sociol. Sport* **2013**. [CrossRef]
65. Workcover (Inspector Calvez) v Tafe Commission (2014) NSWDC: 2014. Available online: www.austlii.edu.au/databases.html (accessed on 15 July 2015).
66. Inspector Chadwick v B H Maclachlan (NSW) Pty Ltd. NSWIR Comm: 2004. Available online: www.austlii.edu.au/databases.html (accessed on 1 July 2015).
67. Whitaker v Delmina Pty Ltd. Supreme Court of Victoria: 1998. Available online: http://www.magistratescases.com.au/case/whittaker-v-delmina-pty-ltd/ (accessed on 28 August 2015).
68. Ryan, P. How the model whs laws will affect you. 2012. Available online: http://search.informit.com.au/documentSummary;dn=075378610944429;res=IELENG (accessed on 28 August 2015).

69. Tooma, M.; Carnell, M. Modernisation gets compliant. 2013. Available online: http://search.informit.com.au/documentSummary;dn=142663844558645;res=IELHSS (accessed on 28 August 2015).
70. O'Neill, S.; Wolfe, K. Officers' due diligence: Making work health and safety an accounting problem? 2012. Available online: http://ro.uow.edu.au/acsear2012/2012/papers/29/ (accessed on 1 March 2013).
71. Bentton, E. Demonstrating due Diligence. 2014. Available online: http://search.informit.com.au/documentSummary;dn=482696937049536;res=IELENG (accessed on 28 August 2015).
72. Makin, A.; Winder, C. A new conceptual framework to improve the application of occupational health and safety management systems. *Saf. Sci.* **2008**, *46*, 935–948. [CrossRef]
73. Thompson, K.; Matthews, C. Inroads into equestrian safety: Rider-reported factors contributing to horse-related accidents and near misses on australian roads. *Animals* **2015**, *5*, 18. [CrossRef] [PubMed]
74. Cassell, E.; Clapperton, A.; Reid, N. *Preventing Unintentional Farm Injury*; Monash University: Melbourne, Australia, 2008; pp. 1–12.
75. Rasmussen, J.; Suedung, I. *Proactive Risk Management in a Dynamic Society*; Swedish Rescue Services Agency: Karlstad, Sweden, 2000.
76. Carol, H.; Huws, N.; White, C.; Taylor, E.; Owen, C.; McGreevy, P. Assessment of ridden horse behavior. *J. Vet. Behav.* **2013**, *8*, 62–73.
77. Hausberger, A.; Muller, C. A brief note on some possible factors in the reactions of horses to humans. *Appl. Anim. Behav. Sci.* **2002**, *76*, 339–344. [CrossRef]
78. Lansade, L.; Bouissou, M.; Erhard, H. Reactivity to isolation and association with conspecifices: A temperament trait stable accross time and situations. *Appl. Anim. Behav. Sci.* **2008**, *109*, 355–373. [CrossRef]
79. Momozawa, Y.; Ono, T.; Sato, F.; Takeuchi, Y.; Mori, Y.; Kusunose, R. Assessment of equine temperament by a questionnaire survey to caretakers and evaluation of its relability by simultaneous behaviour test. *Appl. Anim. Behav. Sci.* **2003**, *84*, 127–138. [CrossRef]
80. Wolff, A.; Hausberger, M.; LeScolan, N. Experimental tests to assess emotionality in horses. *Behav. Proc.* **1997**, *40*, 209–221. [CrossRef]
81. Lesimple, C.; Fureix, C.; LeScolan, N.; Richard-Yris, M.; Hausberger, M. Housing conditions and breed are associated with emotionality and cognitive abilities in riding school horses. *Appl. Anim. Behav. Sci.* **2011**, *129*, 92–99. [CrossRef]
82. Visser, E.; van Reenen, C.; Hopster, H.; Schilder, M.; Knaap, J.; Barneveld, A.; Blokhuis, H. Quantifying aspects of young horses' temperament: Consistency of behaviourable variables. *Appl. Anim. Behav. Sci.* **2001**, *74*, 241–258. [CrossRef]
83. Von Borstel, U.; Pasing, S.; Gauly, M. Towards a more objective assessment of eqione personality using behavioural and physiological observations from performance test training. *Appl. Anim. Behav. Sci.* **2011**, *135*, 277–285. [CrossRef]
84. Paul, E.; Harding, E.; Mendl, M. Measuring emotional processesin animals: The utility of a cognitive approach. *Neurosci. Biobehav. Rev.* **2005**, *29*, 469–491. [CrossRef] [PubMed]
85. Munsters, C.; Visser, K.; van der Broek, J.; van Oldruitenborgh-Oosterbaan, M. The influence of challenging objects and horse-rider matching on heart rate, geart rate variability and behavioural score in riding horses. *Vet. J.* **2012**, *192*, 75–80. [CrossRef] [PubMed]
86. LeScolan, N.; Hausberger, A.; Wolff, A. Stability over situations in temperamental traits of horses revelaed by experimental and scoring approaches. *Behav. Proc.* **1997**, *41*, 257–266. [CrossRef]
87. Axel-Nilsson, M. The Match between Horse and Rider. Ph.D. Thesis, Acta Universitatis Agriculturae Suexiae, Uppsala, Sweden, 2015.
88. Greene, E.A.; Trott, J.F. The self-guided horse facility analysis: A proactive safety education tool for equine facilities. *J. Extens.* **2004**, *42*, 6.
89. Slovic, P.E. *The Perception of Risk*; Earthscan publications: New York, NY, USA, 2000.
90. Legge, J. Pre-employment functional assessment as an effective tool for controlling work-related musculoskeletal disorders: A review. *Ergonom. Aust.* **2014**, *18*, 27–30.
91. Legge, J.; Burgess-Limerick, R. Reliability of the jobfit system pre-employment functional assessment tool. *Work* **2007**, *28*, 299–312. [PubMed]
92. Schollhorn, W.; Peham, C.; Licka, T.; Scheidl, M. A pattern recognition approach for the quantification of horse and rider interactions. *Equine Vet. J. Suppl.* **2006**, *36*, 400. [CrossRef] [PubMed]
93. Burke, J.W. *Competency Based Education and Training*; Psychology Press: Hove, UK, 1989.

94. Dingsdag, D.P.; Biggs, H.C.; Sheahan, V.L.; Cipolla, D.J. *A Construction Safety Competency Framework: Improving OHS Performance by Creating and Maintaining a Safety Culture*; CRC for Construction Innovation: Brisbane, Australia, 2006.
95. O'Connor, P. Workplaces as sites of learning. *Think. work* **1994**, *1*, 257–295.
96. Bowden, J.; Masters, G.N. *Implications for Higher Education of a Competency-Based Approach to Education and Training*; Australian Government Publishing Service: Canberra, Australia, 1993.
97. Noe, R.A.; Schmitt, N. The influence of trainee attitudes on training effectiveness: Test of a model. *Pers. Psychol.* **1986**, *39*, 497–523. [CrossRef]
98. Boot, M.; McGreevy, P.D. The x files: Xenophon re examined through the lens of equitation science. *J. Vet. Behav. Clin. Appl. Res.* **2015**, *8*, 367–375. [CrossRef]
99. Latimer, J.; Birke, L. Natural relations: Horses, knowledge and technology. *Sociol. Rev.* **2009**, *57*, 2–27. [CrossRef]
100. Patton, P. Language, power and the training of horses. In *Zoontologies: The Question of the Animal*; Wolfe, C., Ed.; University of Minnesota Press: Minneapolis, MN, USA, 2003; pp. 83–99.
101. Birke, L. "Learning to speak horse": The culture of "natural horsemanship". *Soc. Animal.* **2007**, *15*, 217–239. [CrossRef]
102. Birke, L. Talking about horses: Control and freedom in the world of "natural horsemanship". *Soc. Animal.* **2008**, *16*, 107–126. [CrossRef]
103. DeAraugo, J.; McLean, A.; McLaren, S.; Caspar, G.; McLean, M.; McGreevy, P. Training methodologies differ with the attachment of humans to horses. *J. Vet. Behav. Clin. Appl. Res.* **2014**, *9*, 235–241. [CrossRef]
104. Raber, K.; Tucker, T.J. *The Culture of the Horse: Status, Discipline, and Identity in the Early Modern World*; Palgrave Macmillan: New York, NY, USA, 2005.
105. Raber, K.L. "Reasonable creatures" william cavendish and the art of dressage. In *Renaisance Culture and the Everyday*; Fumerton, P., Hunt, S., Eds.; University of Pennsylvania Press: Philadelphia, PA, USA, 1999; pp. 42–66.
106. van Orden, K. From *gens d'armes* to *gentilshommes*: Dressage, civility, and the *ballet à cheval*. In *The Culture of the Horse: Status, Discipline, and Identity in the Early Modern World*; Raber, K., Tucker, T.J., Eds.; Palgrave Macmillan: New York, NY, USA, 2005; pp. 197–222.
107. Cassidy, R. *The Sport of Kings: Kinship, Class and Thoroughbred Breeding in Newmarket*; Cambridge University Press: Cambridge, UK, 2002.
108. Cassidy, R. Falling in love with horses: The international thoroughbred auction. *Soc. Animal.* **2004**, *13*, 51–68. [CrossRef]
109. Cassidy, R.L. *Horse People: Thoroughbred Culture in Lexington and Newmarket*; Johns Hopkins University Press: Baltimore, MD, USA, 2010.
110. Wipper, A. The partnership: The horse-rider relationship in eventing. *Symb. Interact.* **2000**, *23*, 47–72. [CrossRef]
111. Game, A. Riding: Embodying the centaur. *Body Soc.* **2001**, *7*, 1–12. [CrossRef]
112. Adelman, M. *The Work of Amazons: The Construction of Women'S Identity as Professional Athletes in Brazilian Showjumping*; American Sociological Association (ASA): Washington, DC, USA, 1998.
113. Adelman, M. Mulheres atletas: Re-significacoes da corporalidade feminina. *Estudos Feministas* **2003**, *11*, 445–465. [CrossRef]
114. Adelman, M. O desafio das amazonas: A construção da identidade de mulheres como atletas e amazonas do hipismo clássico (salto) brasileiro' [the amazons' challenge: Women building their identities as athletes and horsewomen]. In *O mundo psicossocial da mulher no esporte: Comportamento, gênero, desempenho [the psychosocial world of women in sport: Behavior, gender and performance]*; Simões, A.C., Knijik, J.D., Eds.; Aleph: São Paulo, Brazil, 2004; pp. 277–303.
115. Adelman, M. Women who ride: Constructing identities and corporalities in equestrian sports in Brazil. In *L'identité genrée au coeur des transformations : Du corps sexué au corps genré*; Grenier-Torres, C., Ed.; L'Harmattan: Paris, France, 2010; pp. 105–125.
116. Adelman, M.; Moraes, F.A. Breaking their way in: Women jockeys at the racetrack in Brazil. *Adv. Gender Res.* **2008**, *12*, 99–123.
117. Coulter, K. Horse power: Gender, work and wealth in Canadian show jumping. In *Gender and Equestrian Sports*; Adelman, M., Knijnik, J., Eds.; Springer: New York, NY, USA, 2013; pp. 165–181.

118. Birke, L.; Brandt, K. Mutual corporeality: Gender and human/horse relationships. *Women's Stud. Int. Forum* **2009**, *32*, 189–197. [CrossRef]
119. Goodwin, D.; McGreevy, P.; Waran, N.; McLean, A. How equitation science can elucidate and refine horsemanship techniques. *Vet. J.* **2009**, *181*, 5–11. [CrossRef] [PubMed]
120. McGreevy, P.D.; McLean, A.N. *Equitation Science*; John Wiley & Sons: New York, NY, USA, 2011.
121. McGreevy, P.D.; Murphy, J. Equitation science offers new horizons in the understanding of equine performance and horse-human relationships. *Vet. J.* **2009**, *181*, 1–5. [CrossRef] [PubMed]
122. McLean, A.N.; McLean, M.; Stuart, N. *Academic Horse Training: Equitation Science in Practice*; Australian Equine Behaviour Centre: Clonbinane, Australia, 2008.
123. Leung, W.-C. Competency based medical training: Review. *BMJ* **2002**, *325*, 693. [CrossRef] [PubMed]
124. Rutherford, P.D. Competency based assessment: A guide to implementation. *J. Manag. Organ.* **1995**, *2*, 49.
125. Cooper, R.A.; Briggs, T.; Tooma, M.; Garlapati, A. *International Leadership Forum: An Insight into Most Recent International SHE Regulations*; American Society of Safety Engineers: Park Ridge, IL, USA, 2014.
126. Sherriff, B.N.; Tooma, M. *Understanding the Model Work Health and Safety Act*; CCH Australia Limited: Sydney, Australia, 2010.
127. Foster, N.J. Directors and due diligence in workplace safety. In Proceedings of Safety in Action Conference 2012, Sydney, Australia, 17–19 April 2012.
128. Leonard, M.; Graham, S.; Bonacum, D. The human factor: The critical importance of effective teamwork and communication in providing safe care. *Qual. Saf. Health Care* **2004**, *13*, i85–i90. [CrossRef] [PubMed]
129. Campion, M.A.; Medsker, G.J.; Higgs, A.C. Relations between work group characteristics and effectiveness: Implications for designing effective work groups. *Pers. Psychol.* **1993**, *46*, 823–850. [CrossRef]
130. Safe Work Australia (SWA). *Work Health and Safety Consultation, Co-Operation and Co-Ordination*; SWA: Canberra, Australia, 2011; pp. 1–27.
131. McGreevy, P.; McLean, A. Roles of learning theory amd ethology in equitation. *J. Vet. Behav.* **2007**, *2*, 108–118. [CrossRef]
132. Lamond, D. The information content of the nurse change of shift report: A comparative study. *J. Adv. Nurs.* **2000**, *31*, 794–804. [CrossRef] [PubMed]
133. Brown, M. Communicating information about workplace hazards: Effects on worker attutudes towards risks. In *The Social and Cultural Construction of Risk*; Springer: Dordrecht, The Netherlands, 1987; Volume 3, pp. 251–274.
134. Alessandra, T. *Communicating at Work*; Simon and Schuster: New York, NY, USA, 1993.
135. Australian Skills Quality Authority (ASQA). *Training in Equine Programs Australia*; ASQA: Melbourne, Australia, 2015; p. 120.
136. Goh, Y.; Brown, H.; Spickett, J. Applying systems thinking concepts in the analysis of major incidents and safety culture. *Saf. Sci.* **2010**, *48*, 302–309. [CrossRef]
137. Rooney, J.; Vanden Heuvel, L. Root cause analysis for beginners. Available online: http://asq.org/quality-progress/2004/07/quality-tools/root-cause-analysis-for-beginners.html (accessed on 12 May 2014).
138. Woolley, J.; Bailey, T.; Raftery, S. *What can Work Health and Safety Learn from Road Safety?*; The University of Adelaide: South Australia, Australia, 2013; p. 68.

animals

MDPI

Article

The Management of Horses during Fireworks in New Zealand

Gabriella Gronqvist *, Chris Rogers and Erica Gee

Massey Equine, Institute of Veterinary, Animal and Biomedical Sciences, Massey University, Private Bag 11-222, Palmerston North 4442, New Zealand; c.w.rogers@massey.ac.nz (C.R.); e.k.gee@massey.ac.nz (E.G.)

* Correspondence: G.Gronqvist@massey.ac.nz; Tel.: +64-6-350-5701; Fax: +64 6-350-5618

Academic Editor: Kirrilly Thompson

Received: 5 January 2016; Accepted: 23 February 2016; Published: 9 March 2016

Simple Summary: The negative effects of fireworks on companion animals have been reported, but little has been documented on the impact on horses. Horse anxiety was commonly associated with fireworks, and 26% of owners reported horse injuries as a result of fireworks. Many management strategies were seen as ineffective. The majority of horse owners were in favour of a ban on the sale of fireworks for private use.

Abstract: Within popular press there has been much coverage of the negative effects associated with firework and horses. The effect of fireworks has been documented in companion animals, yet no studies have investigated the negative effects, or otherwise, of fireworks on horses. This study aims to document horse responses and current management strategies to fireworks via an online survey. Of the total number of horses, 39% (1987/4765) were rated as "anxious", 40% (1816/4765) "very anxious" and only 21% (965/4765) rated as "not anxious" around fireworks. Running (82%, 912/1107) was the most common behaviour reported, with no difference between property type ($p > 0.05$) or location ($p > 0.05$). Possibly as a consequence of the high frequency of running, 35% (384/1107) of respondents reported having horses break through fences in response to fireworks and a quarter (26%, 289/1099) reported that their horse(s) had received injuries associated with fireworks. The most common management strategy was moving their horse(s) to a paddock away from the fireworks (77%) and to stable/yard them (55%). However, approximately 30% reported these management strategies to be ineffective. Of the survey participants, 90% (996/1104) were against the sale of fireworks for private use.

Keywords: fireworks; horses; anxiety; behaviour; fear

1. Introduction

In New Zealand, both public and private firework displays are common, especially during Guy Fawkes Day on 5 November. The Hazardous Substances and New Organisms Act 1996 govern the sale of fireworks for private use and limits sales to three days (2–5 November). However, while the sale of fireworks is restricted to only three days, there are no restrictions on the when fireworks can be used. Many counties, such as Canada, South Africa, Australia and Finland, have strict limitations or bans on private firework displays. A recent Parliamentary select committee rejected a ban on the private use of fireworks in New Zealand citing any changes as unnecessary and unenforceable [1]. The lack of change to the Hazardous Substances and New Organisms Act 1996 relating to the use of fireworks has prompted initiation of a subsequent petition calling for the ban of private use of fireworks [2]. The petition has the support of the New Zealand Police, Fire services and New Zealand Society for the Prevention of Cruelty to Animals (NZSPCA).

Previous studies have reported on fear behaviours in response to fireworks in companion animals both in New Zealand [3] and internationally [4]. Owners have reported negative effects of fireworks including escaping, vocalisation, urination or defecation, trembling and destructive behaviour.

Within the popular press and social media there has been much coverage of the negative effects associated with firework displays and horses [5]. Nevertheless, no studies appear to have investigated the possible problems, or otherwise, of fireworks on horses. Despite this lack of reporting in the peer reviewed literature, there are a number of publications from equestrian organisations and within the equestrian press providing guidelines on the management of horses and the negative effects of fireworks displays [6].

This lack of scientific literature on the topic may relate to the intensive management of horses within stables and therefore reduction in the exposure to the potentially noxious stimuli of fireworks. In November, in the United Kingdom approximately 70% of horses are stabled and 30% live outside. Of those horses that are stabled, almost 50% spend between 9 and 16 h inside daily [7]. The management of horses, even high level competition horses, in New Zealand is unique in that the temperate climate permits management of the horse at pasture year round [8]. This pastoral based management system may facilitate exposure to the visual, acoustic and olfactory stimuli of fireworks and be a reason for the perception that the private use of fireworks represents a hazard for horses at pasture.

In animals, fear responses to fireworks are believed to occur due to the intermittent and unpredictable high-intensity noise [9]. Cracknell and Mills (2008) report that the effects of secondary stimuli such as odours, light flashes and changes in barometric pressure on animals still remain largely unknown.

Horses are generally considered to be highly unpredictable flight animals [10] shown to be reactive to loud noises and flashing lights [11]. Fear is a reaction to perceived danger and is characterized by physiological and behavioural changes that heighten the individual's ability to deal with that danger [12,13]. Fear based behaviours in horses are numerous and include running, sweating and trembling [14]. Flight responses are particularly dangerous, with the potential to result in severe accidents of the horse and rider/handler [15].

At present, the lack of data on management strategies employed by horse owners, the perceived effectiveness of such changes and injuries encountered limits debate on the private use of fireworks and the consequences to horses. The aim of this study was to document horse responses and current management strategies to fireworks via an online survey.

2. Experimental Section

2.1. Questionnaire

Data were collected via an online survey using commercial survey software SurveyMonkey Audience (SurveyMonkey Inc.) (see supplementary file). The survey was initially distributed and "seeded" via six national and regional equestrian sport social media sites. The survey was open for 19 days from 14 October to 1 November 2015, prior to the first official day of the sale of fireworks for private use (2 November) in New Zealand.

The questionnaire could only be completed once per computer and all applicants remained anonymous. The survey was deemed to be low risk by the Massey University Human ethic committee and was registered as a low risk notification project.

The questionnaire consisted of 15 multi-choice and open ended questions in four categories covering property location and size, number of horses and primary use, the reaction of horses to fireworks in the previous year, preventative management and the occurrence of any injuries. Lastly, the participants were asked whether they were in favour of sale of firework for personal use.

2.2. Statistical Analysis

Data were described using simple descriptive statistics. In some instances the respondent may not have completed all questions and so the denominator for some questions may vary. For the data on anxiety and horses, the percentages reported here were based on the number of horses for which an anxiety score was given by the respondents. The distribution of property type, horse ownership, behaviour of horses during firework displays and owner support of the sale of fireworks for personal use were examined using a Chi squared test. The differences in behaviours reported between property types were tested using the Mann-Whitney test. Multivariate logistic regression was used to calculate the Odds Ratio (95% confidence intervals) of horse injury with property type and area of where the horse was kept. All statistical analyses were completed using the statistical software STATA 12 (StataCorp, TX, USA) and R 3.2.2 (Foundations for Statistical Computing, Vienna, Austria) with $p < 0.05$ set for significance.

3. Results

3.1. Demographics

Data were collected from 1111 respondents responsible for 6431 horses. It is estimated there are 110,000 horses in New Zealand and the AgriBase database identifies 13,072 properties with horses not identified as racing or commercial breeding properties [16]. Using these 13,072 properties as an approximate sampling frame, this represents a return rate of 9% (1111/13,072) of (non-racing) horse owners/properties in New Zealand. The majority of the respondents were from the North Island (89% 918/1111) and the greatest number of respondents were from the Auckland region (27%, 295/1111), followed by the Manawatu-Wanganui region (14%, 150/1111, Table 1).

Table 1. The total number and percentage (%) of respondents from the various regions in New Zealand and the number of property type (agistment [1], farm [2] or lifestyle [3]) in the various regions.

Area	N Respondents	% NZ	N Agistment	N Farm	N Lifestyle
Auckland	295	26.6	55	27	213
Bay of Plenty	45	4.1	2	2	41
Canterbury	91	8.2	11	12	68
Gisborne	7	0.6	-	-	7
Hawke's Bay	38	3.4	6	3	29
Manawatu	150	13.5	15	32	102
Marlborough	17	1.5	-	3	14
Nelson	3	0.3	1	-	2
Northland	61	5.5	3	11	46
Otago	56	5.1	2	20	34
Southland	13	1.2	-	3	10
Taranaki	72	6.5	8	12	52
Tasman	4	0.4	-	-	4
Waikato	125	11.3	12	22	91
Wellington	125	11.3	14	20	89
West Coast	8	0.7	1	2	5

[1] Livery service; [2] Commercial farming enterprise; [3] Small farm <4 ha in total area.

Approximately half the respondents identified themselves as living in an area consisting of predominantly lifestyle blocks (small farms < 10 ha, Table 2). Irrespective of the urban/rural classification, most respondents identified they lived and kept horses on lifestyle blocks (71%, 807/1107) with only 13% (130/1107) of respondents keeping horses on agistment properties (livery service). There were greater numbers of horses kept on an agistment property (15 (interquartile range IQR 7–26)) than on farms (commercial) or lifestyle blocks (5 IQR 3–10 *vs.* 4 IQR 2–6, $p < 0.001$). The number of horses owned by respondents also differed across property type, farm 4 (IQR 3–9), Lifestyle 4 (IQR 2–6), and agistment 4 (IQR 1–12), ($p < 0.001$). The most common use of horses were for sport (show jumping, dressage and eventing) (43%, 1979/4575) followed by trekking (25%, 1137/4573) Pony Club (17%, 774/4575) and racing (7%, 321/4575). The remaining categories (hunting, endurance and western) accounted for approximately 3% each.

Of the total number of horses, 39% (1987/4765) were rated as "anxious", 40% (1816/4765) "very anxious" and only 21% (965/4765) were rated as "not anxious" around fireworks or the Guy Fawkes period (Table 2). The levels of anxiety reported did not differ between property type (agistment, farm or lifestyle block, $p > 0.05$).

Table 2. The total number of respondents and horses according to location, property type and the behaviours exhibited during fireworks.

Categories	Total	Urban[1]		Semi-Rural[2]		Lifestyle[3]		Rural[4]	
	n	n	%	n	%	n	%	n	%
Respondents	1111	80	7.2	271	24.4	517	46.5	242	21.8
Property type									
Farm		7	8.8	34	12.5	23	9.5	106	20.5
Lifestyle block		28	35.0	193	15.9	457	188.8	112	21.7
Agistment		45	56.3	43	71.2	36	14.9	23	4.4
Horses/respondents (median and IQR)		3 (2–8)		4 (2–7)		4 (2–5)		4 (2–7)	
Behaviours									
Fence walking		39	48.8	136	50.2	237	45.8	94	38.8
Running		67	83.8	223	82.3	444	85.9	179	74.0
Decreased appetite		17	21.3	62	22.9	86	16.6	30	12.4
Breaking through fences		37	46.3	108	39.9	170	32.9	69	28.5
Weaving		9	11.3	19	7.0	14	2.7	34	14.0
Bucking/rearing		29	36.3	100	36.9	157	30.4	72	29.8
Sweating		46	57.5	156	57.6	316	61.1	114	47.1
Trembling		39	48.8	137	50.6	266	51.5	92	38.0
Injuries									
YES		25	33.3	84	33.3	138	28.9	42	20.4
Anxiety (n horses)	4765	338		1226		2113		1088	
not anxious		44	13.0	219	17.9	382	18.1	320	29.4
anxious		166	49.1	513	41.8	892	42.2	355	32.6
very anxious		128	37.9	494	40.3	839	39.7	413	38.0
Against the sale of fireworks for personal use		78	97.5	243	89.7	473	91.5	202	83.5

[1] within a town/urban environment; [2] adjacent to an urban area; [3] surrounded by lifestyle blocks; [4] surrounded by other large farms.

3.2. *Adverse Horse Behaviour and Anxiety*

The majority of respondents (1104/1111) reported that their horse(s) had previously exhibited at least one of the behaviours, listed in Figure 1, in response to fireworks. Running (82%, 912/1107) was the most common behaviour reported, with no significant difference between property type ($p = 0.412$) or location ($p = 0.068$). There were a group of behaviours with similar frequencies reported (trembling, sweating and fence walking). Similar frequencies of these behaviours were reported across property location identifiers (urban, semi-rural, lifestyle or rural). Possibly as consequence of the high frequency of running, 35% (384/1107) of respondents reported having horse(s) break through fences in response to fireworks.

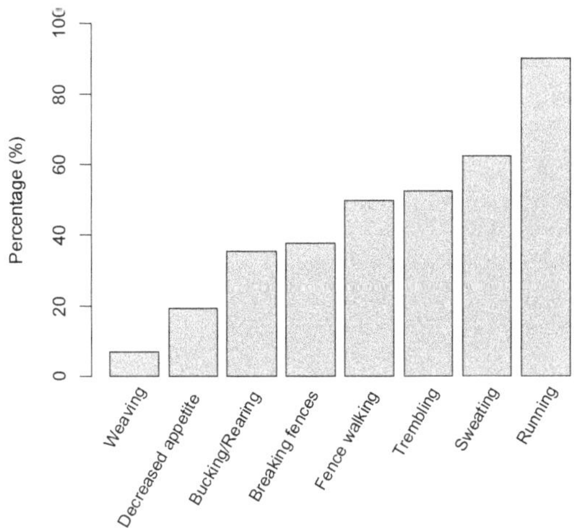

Figure 1. The percentage (%) of participants reporting that their horse(s) had exhibited the behaviors in association with fireworks.

A quarter (26%, 289/1099) of participants reported that their horse(s) had received injuries associated with fireworks. Multiple different injuries were reported; the most common were lacerations (40%, 194/289), strains/sprains (10%, 33/289) and broken limbs (7%, 11/289). The property type did not affect the odds of the horse(s) receiving injuries ($p > 0.05$, agistment = reference level, farm 1.5 (0.85–2.48), lifestyle 1.1 (0.71–1.7)). However, respondents that kept horses in rural areas were 0.6 (0.33–0.49) times as likely to report that their horses had received injuries due to fireworks, than respondents that kept horses in an area surrounded by lifestyle blocks ($p < 0.05$, reference level). The odds of horses receiving injuries did not differ between horses kept in semi-rural areas or urban areas compared with lifestyle blocks.

3.3. Duration of Firework Displays

Off the survey participants, 6% (63/1108) reported that their horse(s) had not been exposed to fireworks in the previous year. The remaining 94% (1045/1108) of survey respondents were asked the duration of time their horse(s) were exposed to intermittent fireworks. Thirty-three per cent (371/1108) reported fireworks continued for one or two weeks after Guy Fawkes, 26% (288/1108) for two or more months after Guy Fawkes, and 19% (209/1108) for up to a month after Guy Fawkes, while only 16% (177/1108) identified that their horse were exposed to fireworks only on Guy Fawkes Day.

3.4. Owner Management Strategies

The most common management strategy was the movement of the horse(s) to a paddock away from the fireworks (77% (779/1006)). However, 37% (374/779) reported this management strategy to be ineffective in reducing anxiety. Horse(s) had previously been stabled or yarded during fireworks by 55% (461/925) of respondents, but 30% (277/461) reported this to be ineffective. Only 30% (254/845) and 19% (152/808) of participants, respectively, had previously either moved their horse(s) off the property or sedated their horse(s) during fireworks. In both instances, 9% (73/254 and 66/152, respectively) of respondents deemed these approaches to be ineffective.

When asked about future management strategies, 20% (189/987) reported that they had no strategy planned. Of the participants, 55% (570/987) planned to move their horse(s) to a paddock further away from the fireworks, 24% (241/987) planned to stable or yard their horse(s), 12% (114/987) planned to sedate them and 10% (95/987) move them off the property. Participants were able to report on several management strategies.

3.5. *The Sale of Fireworks for Private Use*

Of the survey participants 90% (996/1104) reported that they did not support the sale of fireworks for private use, while 10% (108/1104) supported it. The majority of participants did not support the sale of fireworks even if their horse(s) had not previously been injured due to fireworks ($\chi^2 = 17.917$, df = 1, $p < 0.05$). A greater proportion of participants that kept their horses on lifestyle blocks did not support the sale of fireworks for private use ($\chi^2 = 16.799$, df = 2, $p < 0.05$), compared to those who kept horses on farms or agistment.

4. Discussion

The distribution of responses to the online survey was similar to that obtained by Rosanowski, Cogger, Rogers, Benschop and Stevenson [16] using a generalised random-tesselation stratified sampling design and indicates good agreement and reflection of the distribution of horse ownership location within New Zealand. The overrepresentation of the North Island may reflect that the surveys were seeded initially from social media sources based in the Manawatu and the North Island of New Zealand. The high level of response from respondents on lifestyle blocks was reflected in the bulk of respondents keeping their horses on their own property, rather than using agistment/livery yards, as is an option in Western Europe. Lifestyle blocks are typically less than 4 ha and so within a geographical proximity to neighbours where fireworks displays would provide obvious visual and auditory stimuli. The majority of the respondents kept horses for sport, rather than racing, which reflects the initial sampling frame of the survey and the pattern of horse ownership previously reported in New Zealand [16].

The timing of the survey was intentional to provide an overview of what was planned as a course of action during the "fireworks season". Sampling at this time provided minimisation of temporal bias, which is often a limitation in survey data the greater the duration between the event and the collection of the data. The survey was closed prior to the first official sale of fireworks and thus avoided bias in responses, or type of respondent, if adverse fireworks events were reported within the press. Motivation to complete the survey may have been greater in participants that had previously experienced an adverse event associated with fireworks. However, the large number of respondents should have attenuated this bias and implies that, within a pastoral management system, negative experiences with horses and fireworks are the norm rather than an exception. The inability to provide a tight definition around the term anxious and very anxious means some caution should be used when differentiating between these behaviour categories. Within the literature, fear and an ethogram for fear is well described. The use of a grading scale for stress could have been used to provide a tighter definition of the level of anxiety (stress) the horse experienced during fireworks displays [17]. However, in an attempt to increase opportunity for initiation and completion of the survey, the complexity of the anxiety was kept to simple low resolution descriptors. The objective of the survey was to obtain data on owners' perceptions, and general management strategies of their horses in relation to fireworks and not precisely quantify the level of anxiety/arousal to fireworks. Within the literature there are data on between breed and between individual levels of responsiveness to stimuli. These can also be tempered by changes in management. This is an area of behaviour research that requires investigation and possibly translation/dissemination to provide pragmatic management strategies for New Zealand horse owners during fireworks season.

Almost 80% of survey participants reported that their horses became anxious or very anxious during firework displays, with the remaining rating their horses as not anxious. These results support research by Young, Creighton, Smith and Hosie [17] who reported that the sound of fireworks played from compact disk caused higher cortisol levels in horses than the sound of coat clippers or social isolation. The present results are also in agreement with the fact that the majority of respondents described their horse(s) as presenting with, at least one, anxiety related behaviour during fireworks. The most common observation of "running" reflects the use of the flight response to escape noxious stimuli and the pasture based management system of horses in New Zealand. These factors are also reflected in the fact that almost 40% of respondents reported that their horse(s) had broken through fences.

A quarter of respondents reported that their horse(s) had received injuries they believed to be a result of the firework display. The injuries ranged from minor cuts and sprains to broken limbs resulting in death. The most prevalent injuries reported were lacerations varying from mild to severe. These injuries are possibly a reflection of the high percentage of horses reported to break through fences.

Respondents were asked whether, in previous years, they had moved their horse(s) to a paddock away from the fireworks displays, stable/yarded their horse(s), moved their horse(s) away from the property or sedated their horses during Guy Fawkes Day. Most owners had previously trialled a number of management strategies. The most prevalent management approaches were moving the horse(s) to a paddock further away from the fireworks or to stable/yard them. However, almost 40% deemed these methods unsuccessful in reducing anxiety. Nevertheless, these were also reported to be the two most common future management strategies, possibly as they are easier alternatives to relocating the horse off the property or sedation. It has previously been reported that the most common management methods of companion animals owners during fireworks are keeping the animals inside, comforting them, keeping blinds shut and distracting them with music [3]. The majority of these distraction strategies are not likely to be suitable for horses, especially in New Zealand where pastoral management systems are the norm. Moreover, trying to comfort or move a panicked horse can be dangerous for both the handler and the horse as they can charge blindly into humans, fences or other structures when highly aroused [15].

Habituation to repeatable stimuli often occurs with horses, as long as the behaviour is not reinforced with an adverse event [18,19]. The keeping of horses at pasture should permit exposure to fireworks and the opportunity for habituation. This may not, however, be occurring due to the generally focused exposure around the date of Guy Fawkes (November 5th) and then often intermittent exposure. A third of participants reported that they were exposed to fireworks for one or two weeks after Guy Fawkes. Another quarter of participants reported on exposure of two or more months after Guy Fawkes. However, the intermittent and possibly short bouts of exposure may not be enough to habituate horses. Furthermore, the lack of ability to plan and manage horses safely around fireworks has been cited repeatedly in social media and popular press articles [20].

The majority of participants reported that they did not support the sale of fireworks for personal use. This reflects the large number of participants who reported that their horse(s) have displayed anxiety and anxiety related behaviours during fireworks. It is possible that horse owners who have previously had negative experiences, such as injuries, were more motivated to respond to this survey. However, when asked whether the participants supported the sale of fireworks for private use, the majority answered "no", regardless of whether their horse(s) had previously been injured due to fireworks or not. Moreover, the large sampling size should have damped a potential bias. A greater proportion of participants whose horses were kept at lifestyle blocks (as compared to farms and agistment) were against the sale of fireworks for personal use. This may be a reflection of the relative close proximity to neighbouring properties and perhaps a greater number of neighbouring properties.

5. Conclusions

This study is the first to address the issue of horses and fireworks in New Zealand. The article provides a framework for discussing and reviewing legislation in relation to firework use and the risks posed or perceived by horse owners during Guy Fawkes Day firework displays.

Supplementary Materials: The following are available online at www.mdpi.com/2076-2615/6/3/20/, Questionnaire: Owner management of horses during Guy Fawke's Day.

Author Contributions: Gabriella Gronqvist, Chris Rogers and Erica Gee had the original idea for the study and, with all co-authors carried out the design. Gabriella Gronqvist was responsible for recruitment and follow-up of study participants. Gabriella Gronqvist was responsible for data cleaning and Gabriella Gronqvist and Chris Rogers carried out the analyses. Gabriella Gronqvist drafted the manuscript, which was revised by all authors. All authors read and approved the final manuscript.

Conflicts of Interest: The authors declare no conflict of interest.

References

1. Davison, I. Bid to Ban Private Use of Fireworks Rejected. Available online: http://www.nzherald.co.nz/nz/news/article.cfm?c_id=1&objectid=11498496 (accessed on 1 December 2015).
2. New Zealand Herald. Should We Ban Backyard Fireworks? Available online: http://www.nzherald.co.nz/opinion/news/article.cfm?c_id=466&objectid=11355743 (accessed on 1 December 2015).
3. Dale, A.R.; Walker, J.K.; Farnworth, M.J.; Morrissey, S.V.; Waran, N.K. A survey of owners' perceptions of fear of fireworks in a sample of dogs and cats in New Zealand. *N. Z. Vet. J.* **2010**, *58*, 286–291. [CrossRef] [PubMed]
4. Overall, K.L.; Dunham, A.E.; Frank, D. Frequency of nonspecific clinical signs in dogs with separation anxiety, thunderstorm phobia, and noise phobia, alone or in combination. *J. Am. Vet. Med. Assoc.* **2001**, *219*, 467–473. [CrossRef] [PubMed]
5. Peacock, A. Horse Injuries Prompt Renewed Calls for Firework Bans. Available online: http://www.stuff.co.nz/auckland/73704842/horse-injuries-prompt-renewed-calls-for-firework-bans (accessed on 7 December 2015).
6. British Horse Societ. *Advice on Horses and Fireworks*; The British Horse Society: Abbey Park, UK, 2015; p. 3.
7. Hockenhull, J.; Creighton, E. The day-to-day management of uk leisure horses and the prevalence of owner-reported stable-related and handling behaviour problems. *Animal Welfare* **2015**, *24*, 29–36. [CrossRef]
8. Verhaar, N.; Rogers, C.W.; Gee, E.; Bolwell, C.; Rosanowski, S.M. The feeding practices and estimated workload in a cohort of New Zealand competition horses. *J. Equine Vet. Sci.* **2014**, *34*, 79–84. [CrossRef]
9. Cracknell, N.R.; Mills, D.S. A double-blind placebo-controlled study into the efficacy of a homeopathic remedy for fear of firework noises in the dog (canis familiaris). *Vet. J.* **2008**, *177*, 80–88. [CrossRef] [PubMed]
10. LeGuin, E.; Raber, K.; Tucker, T.J. *Man and Horse in Harmony. In the Culture of the Horse: Status, Discipline, and Identity in the Early Modern World*; Palgrave Macmillan US: New York, NY, USA, 2005.
11. Landsberg, G.M. Behavioral Problems of Horses. Available online: http://www.merckvetmanual.com/mvm/behavior/normal_social_behavior_and_behavioral_problems_of_domestic_animals/behavioral_problems_of_horses.html (accessed on 1 December 2015).
12. Burggren, W.W.; French, K.; Randall, D.J. Eckert animal physiology: Mechanisms and adaptations. WH Freeman and Company, New York, NY, USA, 2002.
13. Boissy, A. Fear and fearfulness in animals. *Q. Rev. Biol.* **1995**, *70*, 165–191. [CrossRef] [PubMed]
14. Hahn, C. *Blackwell's Five-Minute Veterinary Consult: Equine*; Wiley: Fillmore, CA, USA, 2011.
15. McGreevy, P.D.; Henshall, C.; Starling, M.J.; McLean, A.N.; Boakes, R.A. The importance of safety signals in animal handling and training. *J. Vet. Behav. Clin. Appl. Res.* **2014**, *9*, 382–387. [CrossRef]
16. Rosanowski, S.M.; Cogger, N.; Rogers, C.W.; Benschop, J.; Stevenson, M.A. A description of the demographic characteristics of the New Zealand non-commercial horse population with data collected using a generalised random-tessellation stratified sampling design. *Prev. Vet. Med.* **2012**, *107*, 242–252. [CrossRef] [PubMed]
17. Young, T.; Creighton, E.; Smith, T.; Hosie, C. A novel scale of behavioural indicators of stress for use with domestic horses. *Appl. Anim. Behav. Sci.* **2012**, *140*, 33–43. [CrossRef]
18. Christensen, J.W.; Rundgren, M.; Olsson, K. Training methods for horses: Habituation to a frightening stimulus. *Equine Vet. J.* **2006**, *38*, 439–443. [CrossRef] [PubMed]
19. McGreevy, P.D.; McLean, A.N. Punishment in horse-training and the concept of ethical equitation. *J. Vet. Behav. Clin. Appl. Res.* **2009**, *4*, 193–197. [CrossRef]
20. Anonymous. Fireworks Heard Report. Available online: https://docs.google.com/forms/d/1XIBz7DhSDJWmFxSij5kVCzx3_DP7D6r466DOvj2ovgk/viewanalytics?usp=form_confirm (accessed on 24 November 2015).

animals

MDPI

Discussion

Reconciling Horse Welfare, Worker Safety, and Public Expectations: Horse Event Incident Management Systems in Australia

Julie M. Fiedler [1,*] **and Paul D. McGreevy** [2]

1 Faculty of Law, Education, Business and Arts, Charles Darwin University, Casuarina NT 0810, Australia
2 Faculty of Veterinary Science, School of Life and Environmental Sciences, University of Sydney, Sydney NSW 2006, Australia; paul.mcgreevy@sydney.edu.au
* Correspondence: julie.fiedler@students.cdu.edu.au; Tel.: +61-840-248-8306

Academic Editor: Kirrilly Thompson
Received: 4 November 2015; Accepted: 16 February 2016; Published: 24 February 2016

Simple Summary: Although often highly rewarding, human-horse interactions can also be dangerous. Using examples from equine and other contexts, this article acknowledges the growing public awareness of animal welfare, work underway towards safer equestrian workplaces, and the potential for adapting large animal rescue skills for the purposes of horse event incident management. Additionally, we identity the need for further research into communication strategies that address animal welfare and safety issues that arise when humans and horses interact in the workplace.

Abstract: Human-horse interactions have a rich tradition and can be highly rewarding, particularly within sport and recreation pursuits, but they can also be dangerous or even life-threatening. In parallel, sport and recreation pursuits involving animals, including horses, are facing an increased level of public scrutiny in relation to the use of animals for these purposes. However, the challenge lies with event organisers to reconcile the expectations of the public, the need to meet legal requirements to reduce or eliminate risks to paid and volunteer workers, and address horse welfare. In this article we explore incident management at horse events as an example of a situation where volunteers and horses can be placed at risk during a rescue. We introduce large animal rescue skills as a solution to improving worker safety and improving horse welfare outcomes. Whilst there are government and horse industry initiatives to improve safety and address animal welfare, there remains a pressing need to invest in a strong communication plan which will improve the safety of workplaces in which humans and horses interact.

Keywords: horse; risk; safety; injury; accident; management; mitigation; behaviour change

1. Introduction

Sports that use animals can operate only with a social licence [1]. Public expectations surrounding the use of animals in sporting and recreational contexts are rapidly evolving and are being shaped, in part at least, by media stories of animals at risk, coupled with the use of graphic imagery. In parallel, there are strong public expectations of safe working environments for people, enforced by law, including those workplaces where there are human-horse interactions. The current article focuses on three areas in this domain: first, the increasing public awareness of animal welfare; second, the requirement for safer workplaces where both humans and horses are present; and third, the need for an industry-led communication plan that will address animal welfare issues and safer workplaces involving horses.

Horse event incident management provides a context to a feasible solution to all of these emergent issues. Large animal rescue skills provide a framework to manage a potentially hazardous equine patient-centred incident which, in turn, provides a safer workplace for people. The incident will be better managed if responders are trained and safe systems of work are adopted, leading to improved welfare outcomes for the horse.

In this article, animal welfare relates to the state of an animal in its attempts to cope with its environment [2] (p. 524). While our discussion is based on experience working with Australian horse-related sports and recreation activities, the issues discussed are relevant to other sport and recreation contexts involving animals, disparate organizations, and sub-groups, such as, cattle, sheep, and pigs used for exhibition or competition.

2. Public Awareness and Animal Welfare

Traditional newspapers, radio, and television, coupled with social media, offer an unprecedented opportunity to shape public consciousness on a wide range of issues. Social media platforms act as a carriageway for calls to action which, in turn, mobilise the online community to participate in targeted grassroots activism. One example of grassroots campaigning is the Australian platform "*Get Up!*" [3] and its project "*Community Run*" [4]. The GetUp! website claims that GetUp! has over one million members and lists a high court win among its achievements. Issues, including animal cruelty, are able to gain an increased profile through mediating technologies, particularly where graphic images boost a story's impact.

In Australia, a recent newsfeed has featured animals-at-risk stories coupled with graphic images from the live cattle export trade following the ABC Four Corners program "*A Bloody Business*" [5] and the greyhound racing industry story "*Making a Killing*" [6]. Both documentaries have resulted in political and industry actions to improve animal welfare outcomes and manage threats to the long-term reputation of the organisations involved. Public outrage following exposure of animal cruelty in the live cattle export story resulted in actions which included the GetUp! 250,000 signature petition on live cattle export being presented to Prime Minister Julia Gillard [7], which contributed to the Australian Government suspending trade with Indonesia [8]. Similarly, public outrage following the broadcast of live-baiting practices in the greyound racing industry, has resulted in the removal of the Racing Queeensland board [9], the NSW Government establishing a Special Commissionof Inquiry [10], and the (custodial) sentencing of three greyhound trainers [1].

Citizens in many countries set expectations for animal welfare through elected government representatives and the making of laws. Examples of such law reform include the Swiss Government's requirements for meeting the social needs of horses [11] and, more recently, the New Zealand government's legal recognition of animal sentience [12]. In Australia, where animal law is not as advanced as in Europe or the UK, there is a boom in the tertiary study of animal law, perhaps due to the media profile given to animal welfare issues [13].

Modern technology provides easy access to information for individuals to become informed about animal welfare issues and provides the platform for animal welfare messages to be shared thousands of times, further contributing to the rapid shaping of public opinion. There are numerous examples of the media reporting animal welfare concerns and developments in public outrage and policy change. Indicative examples of such scenarios that have been reported in the media are outlined in Table 1.

Table 1. Examples of public awareness and animal welfare media stories.

Media	Animal Welfare Aspect	Storyline
It was Spain's "national fiesta". Now bullfighting divides its people [14]	Bulls are killed for public sport and entertainment	Next generation of youth does not see a role for bullfighting in Spain in an increasingly globalized world
Zimbabwe bans lion hunting after international outcry [15]	Lions, tigers and other exotic wildlife hunted for sport, offered as packaged tourism experiences	An international public outcry arising from the killing of a favourite lion by a U.S. citizen results in changes to hunting laws
Pigs to use Twitter and Facebook to challenge animal welfare criticism [16]	The housing conditions and care of animals raised for human consumption	Farmers use social media to inform public on how farm animals are raised
Starbucks to switch to 100 percent cage-free eggs by 2020 [17]	Ethical choices when sourcing ingredients	A commercial decision by Starbucks to remain competitive

3. Public Opinion and Horses at Risk

Media channels dedicated to horse-themed journalism provide a focus for the public's increasing concern for animal welfare. An example is *Epona.TV's* blog page [18] that, *inter alia*, publishes articles related to horses at risk of having their welfare compromised in sporting contexts. One blog example, "Akeem Foldager timeline" provides a chronicle of public, organizational, and industry participant actions following the publication of photographic evidence of a ridden horse being constricted by the two bits associated with a double bridle, resulting in the animal's tongue turning blue [19]. The science underpinning cardiovascular changes that occur when restrictive gear is used in equitation is well established [20] and the principles of ethical equitation are reasonably clear [21]. However, there still remain considerable gaps between equine welfare science and mainstream horse use. To build an understanding of how attitude, beliefs and values differ between horse welfare advocates and professional horse industry participants, and how incremental improvements in horse welfare can be achieved, there is a need for further investment in research from the social science fields.

For the purposes of this article, a sport or recreation horse activity is defined as a structured, managed environment where the public is invited to view proceedings. In this discussion, we focus on the type of competitions conducted under the rules of an incorporated association; for example, dressage, horse-racing, and endurance trials. Such competitive events attract competitors and spectators, with the latter including online or offline support crews, organisational members, officials, and the general public. Throughout the events, participants may use social media to share opinions in real-time and worldwide. As an example, if a horse's safety appears to have been compromised, opinions are formed and shared by the public without consent from the event organisers, horse owner, or competitor. The online dialogue may attract particular interest and following if the participants are injured, a horse is trapped, or a poor response is mounted by officials. Some examples of such scenarios are outlined in Table 2.

Table 2. Examples of public awareness and horses at risk media stories.

Media	Animal Welfare Issue	Storyline
Pictures of lame horse a PR disaster on the magic day Black Caviar came back to Sydney [22]	A lame horse was ridden by a reporter during the broadcast of an interview with the jockey of the winning horse, *Black Caviar*	The commercial decision by Channel 7 to continue with broadcasting the interview. The television station accused of a lapse in a duty of care
Clydesdale slips on Granite Island causeway [23]	Horse slips over on wet rubber matting, remaining recumbent for a period	Spectator reports the incident to RSPCA
Swiss Federation bans use of draw reins in 2016 [24]	Horses ridden with hyper flexed necks are under stress	Swiss Equestrian Federation moves to act on negative social media and calls for support of improved horse welfare by all equestrians

The shaping of public opinion has broadened to include the management of deceased animals, with a 2014 billboard featuring a deceased racehorse erected in Melbourne prior to the spring racing carnival [25], and the 2015 billboards on busses [26], displaying an image of a deceased cow, broadcasting a message against live export. The growing public scrutiny surrounding the euthanasia of horses is providing a rapidly evolving communication challenge for event organisers and veterinarians, charged with ensuring the welfare of participating horses is paramount. Traditional values, practices, and policies may be well accepted by participants but often do not have the same meaning for the general public. For community-level events, the handling of deceased horses with dignity is introduced in the Australian Horse Welfare and Well-being Toolkit [27] (p. 25). Covering planning, logistics, veterinary support and data collection, the toolkit sets out a checklist for organisers. However, there is no independent online portal of evidence-based information for the public and journalists seeking information on the welfare of sport and recreation horses, including those times when euthanasia is recommended.

4. Government Efforts to Manage Risks Posed by Horses in Workplace

Workplace safety laws, workers compensation claims and fines following the death of Sarah Waugh [28] are driving government-led initiatives targeting safety in workplaces involving horses. *Worksafe Australia* statistics reveals that, on average, one horse industry worker is hospitalised per day in Australia from a fall, kick, strike, or bite [29] (p. 2).

Workplaces involving horses have inherent risks for workers [30] (p. 324). Under Australian law, any horse organisation that engages a paid staff member or contractor is considered a "Person or Organisation Conducting a Business Undertaking (PCBU)" and Workplace Health and Safety laws apply. In Australia, competitive horse events, other than racing, are almost exclusively organised and managed by a volunteer workforce under the auspices of a peak body that employs staff in a national, regional, or state office. Therefore, the PCBU status applies to many horse events and the requirement for volunteer workplace inductions, defined job roles, and the use of personal protective equipment (PPE) applies.

Australian government agencies have taken action to manage some horse-related risks. These include development Worksafe Australia's "*Guidelines for reducing risk when new and inexperienced people interact with horses*" [29], the Australian Government Australian Skills and Quality Authority's report "*Training in equine programs in Australia*" [31] and implementation of Technical and Further Education TAFE New South Wales "*Procedures for delivery of horse industry training*" [32]. Media stories providing examples of government efforts to manage risks appear in Table 3.

Table 3. Examples of media stories of government efforts to manage risks.

Story	Safety Issue	Storyline
State to regulate equine industry after death of Hunter rider Sarah Waugh [28]	Injuries and death to people who work with horses	Public awareness raised by parents of Sarah Waugh, resulting in development of an enforceable Code of Practice by Workcover NSW
Three Queensland vets face prosecution over how they managed Hendra cases [33]	Laws require that workers need to take reasonable care of themselves and others in the workplace	Regulators act to enforce Workplace Health and Safety laws
Improving safety in horse racing: it's all in the data [34]	Research on the costs of Workers Compensation for racing industry riders	Development of a tool for comparing costs and risks associated with introducing strategies to improve safety

The level of risk increases for emergency service workers when attending an incident scene involving a horse. In research undertaken by Smith *et al.* [35] (p. 9), fire and rescue volunteers reported a concern about the physical management of large animals, inter-agency coordination, and dealing with owners to the extent that they seek further training opportunities in this domain. It is acknowledged that large animal rescues expose humans to a range of risks, including an unpredictable working environment attributed to horse behavioural characteristics, biosecurity, heavy manual handling, and injuries from being kicked.

Actions to manage risks undertaken by the South Australian and New South Wales State Emergency Services have resulted in the preparation of large animal rescue technical manuals, delivery of standardised training and purchase of fit-for-purpose equipment. The capabilities of the emergency response agencies are enhanced through partnerships with organisations specialising in emergency animal patient care, including Equine Veterinarians Australia's Large Animal Rescue Registry [36], RSPCA state bodies, and the non-profit South Australian Veterinary Emergency Management [37].

To date, there has been only limited investment by government or the horse industry in safety research or data collection systems that have the potential to result in safer human-horse interactions in the workplace.

5. Horse Industry Efforts to Manage Risks

In 2014, the Australian Horse Industry Council (AHIC) conducted a national survey of horse owners and industry workers [38]. Of the 3017 responses analysed, 38% of respondents had received a horse-related injury serious enough to require hospitalisation. After horse-racing, sports including polo, polocrosse, and events involving cattle (e.g., camp drafting), recorded the highest reported incidences of personal injury during training and competition. Furthermore, from the 2083 responses analysed, eight out of ten people at horse events are volunteers, with 31% volunteering at least monthly. The number of participants in the Australian horse industry is not known, so an incidence rate cannot be calculated.

The AHIC has addressed risks to horses and humans through the *HorseSafe Code of Practice* [39]. The voluntary *Code* provides a minimum standard for assessment and control of risks associated with people working around horses while horse welfare is the focus of the *Australian Horse Welfare Protocol* [40].

Horse industry participants have been further supported by evidence-based tools developed by the *International Society for Equitation Science* (ISES), including the ISES *Code of Conduct* that offers guidelines to ensure optimal horse and rider welfare and safety at competitive events. This code refers to other ISES position statements and embraces the *ISES's First Principles of Horse Training* [41] which, in turn, informs the selection and application of handling techniques.

Targeting horse event organisers, the Australian Horse Industry Council has developed the *Australian Horse Welfare and Well-being Toolkit* [27]. This resource provides an introduction to horse event incident management and recommends the appointment of a Horse Welfare Officer. The role of the officer is to work across organisational management structures, with oversight of the welfare of horses, including when an incident occurs [27].

6. Horse Event Incident Management

In this article, we have so far acknowledged the increased public awareness of animal welfare issues, the risks in workplaces where humans and horses interact, and government and horse industry efforts to promote safer work practices. We now showcase one solution for providing a safer working environment when humans and horses are involved in an event incident.

Technical rescue knowledge and skills drawn from the emergency services sector provide a potential model for adaptation to horse incident management. In particular, there are procedures that emerge from the emergency services sector. This includes preparation of an incident management plan, use of established communication systems, and selection of trained personnel for a safer work environment [42] (p. 15). Similar to human-centred incident responses, an assessment may determine that an equine trauma care patient can be transferred to a place of safety, triaged away from public scrutiny and, on occasion, prepared for transit to an equine hospital [43] (p. 80).

Large animal rescue (LAR) adopts an even-handed approach, balancing high-risk hazard management with the welfare of the horse. To avoid further injury to humans or horses, the management of an incident scene requires the assignment of roles to individual responders, the use of personal protective equipment and the establishment of three working zones (Figure 1), based on the level of risk [44] (pp. 6–10). The first, a "hot zone" nearest to the equine patient, has the highest risk, where only essential personnel are positioned. The second, a "warm zone", is outside of the kicking or head-tossing range of the horse, where the incident controller, safety officer, veterinarian, horse welfare officer, owner, and tool dump are positioned. The third is the "cold zone", a low-risk area where the media and spectators are positioned [27] (p. 24). In most situations, An introductory level LAR kit consists of strops (or straps), pole hooks to use as extensions of responders arms, a tool to thread the strops under the horse, rope, and a rescue glide [44] (pp. 111–113).

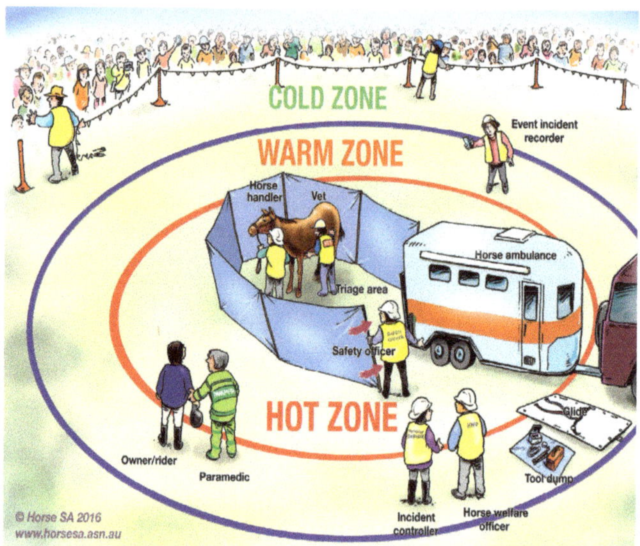

Figure 1. Large animal rescue three working zones.

Basic LAR manipulation techniques, aimed at manoeuvring the horse to a safer place, include the forward drag (Figure 2), backward drag (Figure 3), sideways drag (Figure 4) and the so-called Hampshire skid (Figure 5) [45] (pp. 21–25). The torso of the horse is used for manoeuvring, as use of the head, neck, legs or tail may result in further injury [44] (pp. 37–38). With a recumbent horse, the dependent eye needs to be protected [42] (p. 277). The techniques may also be used to place a live or deceased horse onto a rescue glide (Figure 6) which, in turn, is winched into the float [44] (p. 116), and transferred to a place of safety.

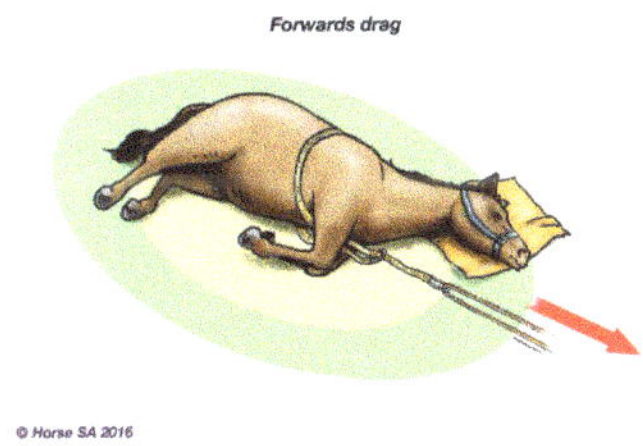

Figure 2. Forwards drag.

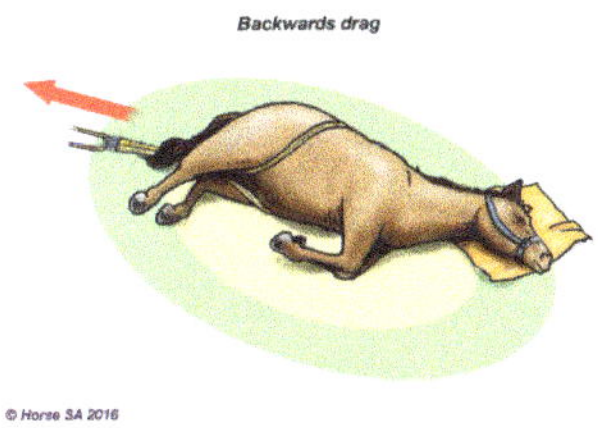

Figure 3. Backwards drag.

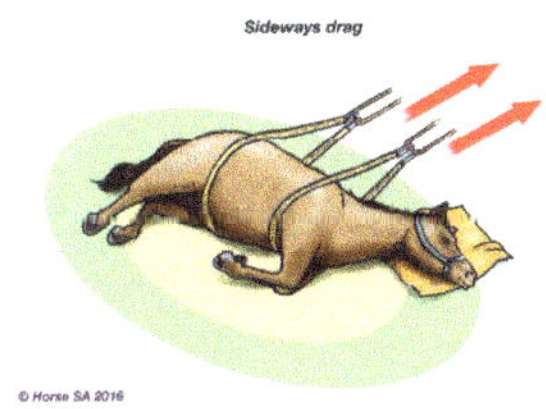

Figure 4. Sideways drag.

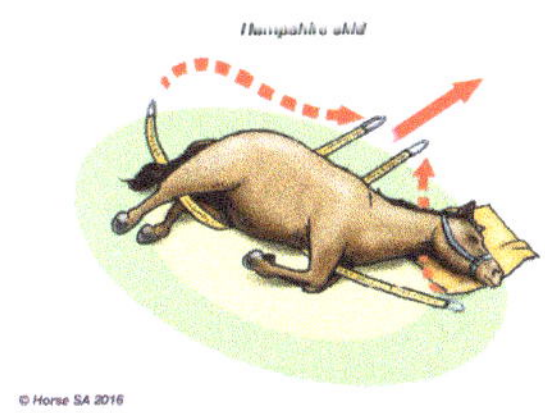

Figure 5. Hampshire Skid.

Figure 6. Horse on a rescue glide.

In these scenarios, a veterinarian works as a part of the horse event incident response crew in the same way a paramedic integrates into a human rescue scenario. Large animal rescue provides a casualty-centred approach, incorporating triage, immediate care and, if required, euthanasia, while all the time keeping people safer in what is often a dynamic rescue environment. These characteristics of a horse-related emergency mean that optimal handling of such incidents merits careful planning lest it falls under the glare of intense public scrutiny [46]. Horses that have had their innate sense of safety compromised, such as when an incident occurs, may react unpredictably and cause severe injury to themselves or people. Planned and managed incident responses provide an opportunity to improve horse welfare outcomes through the benchmarking of key outcomes, including response time, survival rate of equine patients and, for human responders, injury statistics. Through the provision of rescue and trauma care training, risks relating to human safety and horse welfare are mitigated [43] (p. 80).

Although many participants in horse incident management are volunteers, it should be noted that the Australian workplace safety laws requiring PCBUs to minimize or eliminate risk in the workplace [47] (p. 1) may still apply. Therefore, horse events, if categorized as a PCBU, have a legal requirement for a safe workplace. At events, volunteer workers may lack experience with horses, but nevertheless supervise subordinates and others [29] (p. 19). Volunteers may also be expected to deal with horse-related incidents, in addition to their regular role at the event, even though they may not have been provided with specialist training. This approach aligns with the observations of Thompson *et al.* [48] (p. 565), who note that horse riders and handlers often undertake activities a certain way simply "because they have always been done that way", and are rarely provided with advice on how to reduce or remove risks. A risk assessment by event organisers may determine that little capacity exists amongst the volunteers to respond effectively to an incident and, therefore, pre-event planning will need to address this gap.

Public scruity of horses involved in incidents will attract media attention, as will the manner in which any response is undertaken. For examples of media stories refer to Table 4.

Table 4. Examples of media stories relating to horse incidents.

Story	Management	Storyline
Shocking picture shows racehorse champion *Wigmore Hall* destroyed at packed course [49]	Horse racing incident response	Incomplete screening of a horse racing incident resulted in the witnessing of the euthanasia of *Wigmore Hall*. The photograph was subsequently published on the front page of the *Daily Mirror*
Why the long face motorists? Farce as M6 is shut in both directions after horse gets stuck in its horsebox [7]	Emergency management procedures followed for a patient-centred rescue	Traffic delays as a trapped horse was extricated from a horse box on a busy motorway

7. Communication Strategy Design

Our discussion leads us to propose the need for an industry-led communication strategy that addresses the dual messages about the growing public awareness of animal welfare issues and safer workplaces where interactions with horses take place. A communication plan should set a range of measurable targets, including the training of horse event volunteers in large animal rescue skills and educating the public on expectations relating to horse welfare in the settings of sport and recreation.

Few education or training resources are available to support horse event organising committees or volunteers. For emergency service agencies and horse owners, educational resources set in the context of the natural environment, farms, road transport and equestrian enterprises or workplaces include the books *Technical Large Animal Emergency Rescue* [42] and *Equine Emergency Rescue* [44].

However, there is a gap in resources to support volunteers responsible for managing horse event incidents where the additional elements of crowding by spectators and public scrutiny of animal welfare outcomes are factors contributing to a successful resolution. Horse event volunteers are unlikely to be experienced in working safely as part of an incident management team or be proficient at handling horses in stressful situations.

Forming a core element of a communication strategy, the development of targeted educational resources can provide guidance for safer practices in the workplace, and influence the adoption of new techniques. To ensure relevance for event volunteers, resources and training should be culturally appropriate, customised to address recognisable incidents, and easily accessible for "just in time" reminders. Communication prepared by organisers prior to the event, targeting participants and spectators, may help to manage expectations of incident responses and horse welfare outcomes.

Barriers to an effective change in practices may include horse industry participants' attitudes to injury, with many participants currently accepting injury as part of the job [50] (p. 897). Similarly, Thompson *et al.* [48] observes that throughout history, horses and riding have been described as forms of "art" that conflict with the practical application of risk mitigation. These authors go on to argue that the need for the horse to be safe could be reconfigured in a way to keep riders safe, too.

We hypothesize that if horses and people are kept safer in the workplace, public expectations relating to animal welfare are more likely to be met. An example of using the needs of an animal to promote safer decision-making can be drawn from the disaster management sector, notably Australian bushfire evacuation planning [51]. The Australian *National Planning Principles for Animals in Disasters* states that human welfare and safety will be improved if emergency management planning processes include animals [52]. Furthermore, the presence of animals in emergency situations impact human behaviour and safety; therefore, emergency service organisations need to work with communities on animal emergency management above standard preparedness (p. 7).

The Royal Commission into the 2009 Victorian Bushfires found that people died as they chose to stay with their pets or, because of their pets, they left too late [53] (p. ii). Through recognising the human-animal relationship [54], public messaging can encourage owners to take their pets to a safer place, which increases the liklihood that the humans will also stay safe. When the LAR ethos is applied to a horse event incidents, horse-centred messaging results in caring for the horse as a casualty that, in turn, creates a safer workplace [43] (pp. 77–81).

Development and implementation of an industry-led communication plan is likely to be more effective than if initiated by a non-sectorial industry body, for example, the Australian Horse Industry Council [55], through a participatory model that involves horse sport and recreation participants, veterinarians, and large animal rescue experts. We acknowledge that communication strategies will need to be designed to suit different horse sport and recreation activities, recognising the different drivers for participation and that the contexts in which horses perform have a wide range of variability.

8. Conclusions

In this article, we have discussed the increased public awareness of animal welfare issues, the risks in workplaces where humans and horses interact and government and horse industry efforts to promote safer work practices. We have emphasised the need for a communication plan that addresses the paired messages of growing public awareness of animal welfare issues and safer workplaces where humans and horses interact. We recommend further research into factors that will reduce risks in the workplace involving human-horse interactions. Furthermore, we recommend development of an industry-led communication plan which is undertaken in partnership with animal welfare advocacy organisations and experts from the fields of media communication and social sciences. The plan would set out measurable targets, including the training of horse event volunteers in large animal rescue skills and educating the public on expectations relating to horse welfare in the settings of sport and recreation. The plan will need to clearly identify how it will be implemented, maintained, and undergo evaluation by an independent organisation. A well-designed communication plan will, in turn, lead to a safer workplace for people interacting with horses.

Acknowledgments: The Horse Federation of South Australia Inc. (Horse SA) is thanked for its support of the discussions and workshops that have given rise to this article, and for permission to use illustrations. The authors would like to thank the reviewers for their insightful comments, which led to improvements in this article.

Author Contributions: Each author contributed equally to the content of the article. Julie M. Fiedler conceived the article and both authors developed the ideas that underpin it. The first draft was written by Julie M. Fiedler.

Conflicts of Interest: The authors declare no conflict of interest.

References

1. Spicer, D. Live baiting scandal: Three greyhound trainers sentenced for animal cruelty. Available online: http://www.abc.net.au/news/2015-12-16/three-greyhound-trainers-sentenced-for-animal-cruelty/7033436 (accessed on 26 December 2015).
2. Broom, D. Indicators of poor welfare. *Brit. Vet. J.* **1986**, *142*, 524–526. [CrossRef]
3. Lesté-Lasserre, C. Switzerland Bans Draw Reins from Jumping Competitions. Available online: http://www.thehorse.com/articles/36626/switzerland-bans-draw-reins-from-jumping-competitions (accessed on 16 January 2016).
4. CommunityRun. Available online: https://www.communityrun.org/categories (accessed on 26 December 2015).
5. Ferguson, S. A Bloody Business (ABC Four Corners). Available online: http://www.abc.net.au/4corners/special_eds/20110530/cattle/SRC (accessed on 28 December 2015).
6. Meldrum-Hanna, C.; Clark, S. Making a Killing. Available online: http://www.abc.net.au/4corners/stories/2015/02/16/4178920.htmSRC (accessed on 28 December 2015).
7. Spillett, R. Why the Long Faces, Motorists? Farce as M6 is Shut in BOTH Directions after Horse Gets Stuck in Its Horsebox. Available online: http://www.dailymail.co.uk/news/article-3275992/M6-grinds-standstill-stuck-horse-rescued-horsebox.htmlSRC (accessed on 28 December 2015).
8. Woodley, N. Government to Suspend Live Cattle Exports. Available online: http://www.abc.net.au/news/2011-06-08/government-to-suspend-live-cattle-exports/2750312 (accessed on 26 December 2015).
9. Vogler, S.; Guppy, D.; Snowdon, T. All Racing Queensland Boards to be Abolished as Live-Baiting and Animal Cruelty Crisis Cripples Greyhound Industry. Available online: http://www.couriermail.com.au/news/queensland/all-racing-queensland-boards-to-be-abolished-as-livebaiting-and-animal-cruelty-crisis-cripples-greyhound-industry/news-story/5a3fcf842dc04ac1820966f7d8150dfa (accessed on 28 December 2015).
10. Special Commission of Inquiry into the Greyhound Racing Industry in New South Wales. Available online: http://www.greyhoundracinginquiry.justice.nsw.gov.au/ (accessed on 26 December 2015).
11. Stamatelakys, I. Swiss Animal Protection Law Includes Horse Management Guidelines. Available online: http://www.thehorse.com/articles/22200/swiss-animal-protection-law-includes-horse-management-guidelines (accessed on 28 December 2015).
12. Landmark Animal Welfare Legislation Welcomed by Veterinarians. Available online: http://www.nzva.org.nz/newsstory/landmark-animal-welfare-legislation-welcomed-veterinarians?destination=node%2F4825 (accessed on 28 December 2015).
13. Hare, J. Animal Law Courses on the Rise in Australia. Available online: http://www.theaustralian.com.au/higher-education/animal-law-courses-on-the-rise-in-australia/news-story/8fcd348e16d6233e147b06933502029f (accessed on 28 December 2015).
14. Wheeler, D. It Was Spain's "National Fiesta". Now Bullfighting Divides Its People. Available online: http://www.theguardian.com/world/2015/oct/25/bullfighting-spain-national-fiesta-now-divides-its-peopleSRC (accessed on 1 November 2015).
15. Zimbabwe Bans Lion Hunting after International Outcry. Available online: http://www.9news.com.au/world/2015/08/03/12/51/zimbabwe-bans-lion-hunting-amid-international-outcrySRC (accessed on 1 November 2015).
16. McCarthy, M. Pigs to Use Twitter and Facebook to Challenge Animal Welfare Criticism. Available online: http://www.abc.net.au/news/2015-10-01/tweeting-pigs-animal-welfare/6820612SRC (accessed on 1 November 2015).

17. Huffstutter, P. Starbucks to Switch to 100 Percent Cage-Free Eggs by 2020. Available online: http://www.reuters.com/article/2015/10/01/us-starbucks-eggs-idUSKCN0RV5HM20151001SRC (accessed on 1 November 2015).
18. Taylor, J.; Thomsen, L. Epona.TV. Available online: http://epona.tv/blog (accessed on 26 December 2015).
19. Taylor, J. Akeem Foldager Time Line. Epona.TV: Volume 2015. Available online: http://epona.tv/blog/2014/december/akeem-foldager-time-line (accessed on 26 December 2015).
20. McGreevy, P.; Warren-Smith, A.; Guisard, Y. The effect of double bridles and jaw-clamping crank nosebands on facial cutaneous and ocular temperature in horses. *J. Vet. Behav. Clin. Appl. Res.* **2012**, *7*, 142–148. [CrossRef]
21. Jones, B.; McGreevy, P.D. Ethical equitation: Applying a cost-benefit approach. *J. Vet. Behav. Clin. Appl. Res.* **2010**, *5*, 196–202. [CrossRef]
22. Zachariah, R. Pictures of Lame Horse a PR Disaster on the Magic Day Black Caviar Came Back to Sydney. Available online: http://www.dailytelegraph.com.au/sport/superracing/pictures-of-lame-horse-a-pr-disaster-on-the-magic-day-black-caviar-came-back-to-sydney/story-fn67rc85–1226620255280SRC (accessed on 20 September 2015).
23. Clydesdale Slips on Granite Island Causeway. Available online: http://www.victorharbortimes. com.au/story/3324949/clydesdale-falls-on-causeway/SRC (accessed on 3 September 2015).
24. Sparks, C. Swiss Federation Bans Use of Draw Reins in 2016. Available online: http://www.horsecollaborative.com/swiss-federation-bans-use-of-draws-reins-in-2016/SRC (accessed on 24 December 2015).
25. Hutak, M. Flogging a billboard of a dead horse. Available online: http://www.abc.net.au/ news/2014-10-09/hutak-flogging-a-billboard-of-a-dead-horse/5801974 (accessed on 28 December 2015).
26. Wagstaff, J. Animals Australia Launches New Campaign against Live Exports. Available online: http://www.weeklytimesnow.com.au/agribusiness/animals-australia-launches-new-campaign-against-live-exports/story-fnkeqfxg-1227206295015 (accessed on 28 December 2015).
27. Australian Horse Industry Council. Australian Horse Welfare and Well-being Toolkit. Available online: http://www.horsecouncil.org.au/wp-content/uploads/pdfdocs/HOR0288_Horse_welfare_and_well-being_toolkit_WEB.pdfSRC (accessed on 19 September 2015).
28. Harris, M. Sarah Waugh's Horse-Riding Death: TAFE NSW Fined $300,000. Available online: http://www.smh.com.au/nsw/sarah-waughs-horseriding-death-tafe-nsw-fined-300000-20140715-zt7st.html (accessed on 26 December 2015).
29. Safe Work Australia. Guide to Managing Risks When New or Inexperienced People Interact with Horses. Available online: http://www.safeworkaustralia.gov.au/sites/swa/about/publications/pages/guide-managing-risks-interact-horsesSRC (accessed on 19 September 2015).
30. Hawson, L.A.; McLean, A.N.; McGreevy, P.D. The Roles of Equine Ethology and Applied Learning Theory in Horse-Related Human Injuries. *J. Vet. Behav. Clin. Appl. Res.* **2010**, *5*, 324–338. [CrossRef]
31. Australian Skills and Quality Authority. *Training in Equine Programs in Australia*; Australian Skills and Quality Authority: Melbourne, Australia, 2015.
32. NSW, TAFE. Implementation of the TAFE NSW Procedures for the Delivery of Equine Training. 2013. Available online: http://www.horsesa.asn.au/wp-content/uploads/2014/11/TAFE-NSW-Procedures-for-the-Delivery-of-Equine-Training-Implementation.pdf (accessed on 26 December 2015).
33. Buchanan, K. Three Queensland Vets Face Prosecution over How They Managed Hendra Cases. Available online: http://www.abc.net.au/news/2015-09-25/vets-prosecuted-over-hendra-cases/6801422 (accessed on 26 December 2015).
34. Hitchins, P. Improving Safety in Horse Racing: It's All in the Data. Available online: http://theconversation.com/improving-safety-in-horse-racing-its-all-in-the-data-48896 (accessed on 26 December 2015).
35. Smith, B.; Thompson, K.; Taylor, M. What's the big deal? Responder experiences of large animal rescue in Australia. *PLoS Curr.* **2015**. [CrossRef]
36. Equine Veterinarians Australia. Large Animal Rescue Registery. Available online: http://www.ava.com.au/equine/rescueregistry (accessed on 26 December 2015).
37. South Australian Veterinary Emergency Management. Available online: http://www.savem.org.au/ (accessed on 26 December 2015).
38. ASDF Research. *Horse Industry Research*; ASDF Research: Australia, 2015.

39. Australian Horse Industry Council. Horse Safe Code of Practice. Available online: http://www.horsecouncil.org.au/wp-content/uploads/pdfdocs/HorseSafe%20Code%20Of%20Practice%20Revised%202009.pdfSRC (accessed on 19 September 2015).
40. Australian Horse Industry Council. Australian Horse Welfare Protocol. Available online: http://www.australiananimalwelfare.com.au/content/horse-welfare-protocol (accessed on 26 December 2015).
41. International Society for Equitation Science. Principles of Learning Theory in Equitation. Available online: http://www.equitationscience.com/learning-theory-in-equitationSRC (accessed on 19 September 2015).
42. Gimenez, R.; Gimenez, T.; May, K.A. *Technical Large Animal Emergency Rescue*; Wiley-Blackwell: Singapore, 2008.
43. Green, J. Equine Incident Management: From Rescue Site to Racecourse. Available online: http://issuu.com/anderson-co/docs/european_trainer_issue_49_internet__f42c84ddb0dfc8?e=5870891/13860434SRC (accessed on 14 October 2015).
44. Leighton, M.; Staples, M. *Equine Emergency Rescue*, 3rd ed.; Velvet Vision Press: Queensland, Australia, 2015.
45. South Australian State Emergency Service. *Large Animal Rescue Learner Guide*; Government of South Australia: Adelaide, South Australia, 2015.
46. McGreevy, P. Firm but gentle: Learning to handle with care. *J. Vet. Med. Educ.* **2007**, *34*, 539–541. [CrossRef] [PubMed]
47. Work Health and Safety Act. Available online: http://www.legislation.sa.gov.au/LZ/C/A/WORK%20HEALTH%20AND%20SAFETY%20ACT%202012/CURRENT/2012.40.UN.PDFSRC (accessed on 31 October 2015).
48. Thompson, K.; McGreevy, P.; McManus, P. A critical review of horse-related risk: Agenda for safer mounts, riders and equestrian cultures. *Animals* **2015**, *5*, 561–575. [CrossRef] [PubMed]
49. Sommerlad, N. Shocking Picture Shows Racehorse Champion Wigmore Hall Destroyed at Packed Course. Available online: http://www.mirror.co.uk/news/uk-news/shocking-picture-shows-racehorse-champion-4290791SRC (accessed on 20 September 2015).
50. Curry, B.A.; Hitchins, P.L.; Si, L.; Palmer, A.J. Workplace injuries in thoroughbred racing: An analysis of insurance payments and injuries amongst jockeys in Australia from 2002 to 2010. *Animals* **2015**, *5*, 897–909. [CrossRef] [PubMed]
51. South Australian Country Fire Service. Pets and Livestock. Available online: http://www.cfs.sa.gov.au/site/prepare_for_bushfire/know_your_area/prepare_yourself/pets_and_livestock.jsp (accessed on 24 December 2015).
52. National Advisory Committee for Animals in Emergencies. National Planning Principles for Animals in Disasters. Available online: http://www.australiananimalwelfare.com.au/app/webroot/files/upload/files/PDF/FINAL%20National%20Planning%20Principles%20for%20Animals%20in%20Disasters.pdf (accessed on 24 December 2015).
53. Victorian Emergency Animal Welfare Plan. Victorian Government. 2011. Available online: http://agriculture.vic.gov.au/__data/assets/pdf_file/0006/257730/Victorian-Emergency-Animal-Welfare-Plan.pdf (accessed on 26 December 2015).
54. Thompson, K. Save me, save my dog: Increasing natural disaster preparedness and survival by addressing human-animal relationships. *Aust. J. Commun.* **2013**, *40*, 123–136.
55. Australian Horse Industry Council. Available online: http://www.horsecouncil.org.au/ (accessed on 26 December 2015).

animals

MDPI

Review

The Contribution of Equitation Science to Minimising Horse-Related Risks to Humans

Melissa Starling [1,*], Andrew McLean [2,†] and Paul McGreevy [1,†]

1 Faculty of Veterinary Science, University of Sydney, Sydney NSW 2006, Australia; paul.mcgreevy@sydney.edu.au
2 Australian Equine Behaviour Centre, Broadford VIC 3658, Australia; andrewmclean@esi-education.com
* Correspondence: mjstarling@fastmail.com.au; Tel.: +61-2-9351-5827 (ext. 5827); Fax: +61-2-9351-3957
† These authors contributed equally to this work.

Academic Editor: Kirrilly Thompson
Received: 28 October 2015; Accepted: 2 February 2016; Published: 23 February 2016

Simple Summary: Equitation science describes an approach to horse training and riding that focuses on embracing the cognitive abilities of horses, their natural behaviour, and how human riders can use signalling and rewards to best effect. This approach is concerned with both horse welfare and rider safety, and this review discusses how equitation science can minimise risk to humans around horses and enhance horse welfare.

Abstract: Equitation science is an evidence-based approach to horse training and riding that focuses on a thorough understanding of both equine ethology and learning theory. This combination leads to more effective horse training, but also plays a role in keeping horse riders and trainers safe around horses. Equitation science underpins ethical equitation, and recognises the limits of the horse's cognitive and physical abilities. Equitation is an ancient practice that has benefited from a rich tradition that sees it flourishing in contemporary sporting pursuits. Despite its history, horse-riding is an activity for which neither horses nor humans evolved, and it brings with it significant risks to the safety of both species. This review outlines the reasons horses may behave in ways that endanger humans and how training choices can exacerbate this. It then discusses the recently introduced 10 Principles of Equitation Science and explains how following these principles can minimise horse-related risk to humans and enhance horse welfare.

Keywords: horse-riding; ethology; safety; equitation science; learning theory

---

1. Introduction

Equitation science is defined as the art and practice of horsemanship and horse-riding [1]. It is based on the founding principle of promoting the use of an evidence-based approach to explain and emphasise best practice in horse training and riding [2]. It is an inclusive system that aims to embrace all forms of training that are effective and ethical [3]. Equitation embraces a cost-benefit approach to ethics. The greater the impact to horses of a practice, the stronger the justification for that practice needs to be [4]. Ideally, all impacts could be considered moderate or less [4,5]. It also offers guidelines about what may be considered ethical. For example, positive punishment is where a noxious stimulus is used to suppress behaviour. This is discouraged in ethical equitation, as will be discussed later in this article. The uptake of ethical equitation has been broad and encouraging, but there has been some resistance for various reasons, such as the attitude of riders towards science, a preference for instruction from those successful at elite levels, and an emotional appreciation for finding harmony with horses organically [6]. Clearly, traditional approaches to riding instruction have been recorded over centuries in the teachings of the so-called ancient masters [7,8] as well as more contemporary horse trainers [7,9]. Many of these approaches, but not all, withstand scientific scrutiny [10].

Equitation science has a strong focus on horse ethology, acknowledging the way horses learn and their adaptive behavioural tendencies [11]. In this way, riders and trainers can be guided to work with horses in ways that are within the species' cognitive limitations [12]. This is particularly important because horses and humans have different cognitive abilities [13]. Humans tend to attribute their own cognitive abilities to non-human animals [14–16], which can lead to unrealistic expectations about how quickly a horse can learn. The use of a concept such as "respect" may be important to the human but have little meaning for the horse. When training animals, trainers may become frustrated or assign human values and intentions to what are natural horse behaviours or cognitive limitations [17]. This in turn could lead to horse behaviour being misinterpreted as being deliberately challenging or wilfully disobedient, and the trainer adopting punishment to correct it [18]. Punishment is likely to impact negatively on equine welfare, but may also shift the focus of the horse to finding safety and relief from conflict rather than performing cued behaviours that are not relevant to their current goals [19]. This could lead to defensive behaviours or flight responses that threaten the safety of horse riders and trainers. Indeed, horse-riding is known to be a more dangerous activity than motorcycling, with most injuries occurring to the head, trunk, vertebrae, and wrists [20].

2. Causes of Dangerous Behaviour in Horses

2.1. Flight Responses

Like many animals, horses are often at their most dangerous both to themselves and to humans when highly aroused and attempting to escape perceived danger [21]. Horses are a large prey species reliant chiefly on flight at speed to keep themselves safe. This poses a problem for them in many human–horse interactions, as it is common for horses to be confined or restrained to varying degrees when around humans [22]. Being contained where the horse is worked with or housed reduces a horse's options for gaining safety. If sufficiently aroused, it may run blindly into fences or other structures, or into humans [23]. More commonly, inability to escape may pressure the animal into experimenting with confrontational behaviours such as charging, striking with hooves, and biting [24]. These behaviours are likely to meet with some success, since humans confronted by a large, aggressive animal will naturally prioritise their own safety and attempt to escape by retreating [23]. The alternative is for the human to meet confrontation with escalated aggression and intimidate the horse into withdrawal. The consequences of these actions are potentially serious. In the former scenario, the horse's aggressive behaviour is negatively reinforced by the human retreating, (*i.e.*, this has proved a successful way for it to create the space it needs to feel safer), or to avoid an event it had anticipated and found threatening or noxious [25]. Not only is this dangerous for the human, but it probably also perpetuates heightened arousal and negative emotional states in the horse [3]. Emotional conditioning may result in humans and/or training scenarios acting as signals for horses to become aroused and avoidant, or vigilant in anticipation of negative outcomes. Such anticipation primes animals to respond more strongly to aversive stimuli [26] and to interpret ambiguous signals as more likely to be potentially threatening [23]. This is unlikely to be comfortable for the horse, and is likely to trigger further defensive, high-arousal behaviours [12]. Furthermore, behaviours that have resulted in successful avoidance of an anticipated aversive experience can quickly become entrenched and very difficult to eliminate [27]. Deviating from avoidance behaviours known to be successful increases risk, so persistent avoidance behaviours are adaptive [28]. Re-training a horse that has learned undesirable avoidance or distance-increasing behaviours may prove challenging [29].

Punishing avoidance or distance-increasing behaviours is a strategy that comes with its own significant risks. If it is accepted that horses are at their most dangerous when trying to escape, then introducing further threats to a horse's sense of safety in the form of pain, or eliciting flight responses with predatory behaviour, such as chasing, stalking, and sudden movements, could potentially contribute to the underlying problem rather than improving it [19]. While punishment may successfully suppress some dangerous behaviours, it may serve to heighten the horse's arousal and negative emotional state while around humans, which could lead to the expression of other unsafe behaviours that have not been specifically punished [30]. Unwelcome behaviours may represent manifestations of the combined effects of arousal and emotional state.

2.2. *Confusion and Conflict*

A horse cannot necessarily read the intentions of its rider or trainer. There are many ways a rider or trainer's behaviour can induce confusion. For example, signals may be unclear or inconsistent, or the horse may be unaware of how to reduce pressure imposed on it, or what the consequences may be [31]. This confusion can lead to a horse becoming conflicted, with opposing motivations, such as both approach and avoidance competing to manifest as its behavioural response. Such a horse may become uncertain of which behavioural response to adopt and, if arousal is elevated, this could prompt it to experiment with inappropriate locomotory responses such as bolting, bucking, rearing or shying [32]. Such loss of predictability and controllability may render the horse increasingly insecure, repeating the cycle of conflict and further endangering the rider.

2.3. *Frustration*

Aggression has often been shown to stem from frustration [33]. In applied ethology, the term frustration is used to describe thwarted motivation [34–36]. In horses, frustration may be caused by them being unable to perform natural behaviours due to restraint [37]. It is believed that frustration may be responsible for some stereotypies and self-injurious behaviours [38,39], but it may also negatively affect a horse's behavioural inhibition and acceptance of handling procedures that might normally be tolerated [40]. Frustration can also arise during training in-hand and under-saddle, when reinforcement is inconsistent or non-existent and, as a result, horses are confused about how to escape pressure applied to them, or access other reinforcers, or avoid punishment [31]. This may have the effect of reducing self-regulation and increasing ambivalence so that horses may attempt to escape, becoming aggressive towards objects around them, including humans, or, alternatively, they may resort to apathy.

3. The 10 Principles of Ethical Equitation

Equitation science focuses on interacting with and managing horses in ways that avoid provoking dangerous behaviour in the first place [20]. This circumvents the aforementioned problems that can arise when addressing dangerous horse behaviours. Avoiding flight responses and minimising confusion and frustration in horses are the cornerstones of promoting safe behaviours and avoiding dangerous behaviours in equitation. Equitation science seeks to apply scientifically obtained data to training and riding horses to improve the safety and wellbeing of both horse and rider. It is not a single system or method, but it allows all methods of horse handling, training and riding to be assessed on the basis of a cost-benefit analysis that embraces their effectiveness and humaneness. There are significant challenges in empirically assessing the relative merits of approaches that are underpinned by the International Society for Equitation Science (ISES) principles because many other approaches to training include elements that align with these principles. If equitation does indeed guide horse riders and trainers towards safer interactions with horses, it is difficult to justify encouraging a deviation from this path simply to test its validity. Seeking horse riders who embrace either all of the principles of equitation science or a consistent subset of them to comparable degrees and skill in application remains problematic, so there are few data to support the effectiveness of the following ISES principles in isolation, (but any data that are available are cited). The cornerstones of promoting safety in equitation manifest in ISES's recently released 10 Principles of Ethical [41]. They are as follows:

1. Train according to the horse's ethology and cognition.
2. Use learning theory appropriately.

3. Train easy-to-discriminate signals.
4. Shape responses and movements.
5. Elicit responses one-at-a-time.
6. Train only one response per signal.
7. Form consistent habits.
8. Train persistence of responses (self-carriage).
9. Avoid and dissociate flight responses (because they resist extinction and trigger fear problems).
10. Demonstrate minimum levels of arousal sufficient for training (to ensure absence of conflict).

This article explains how these principles can minimise horse-related risks to humans as well as the welfare implications for horses.

3.1. Train According to the Horse's Ethology and Cognition

***Ethology** is the study of animal behaviour in light of how a species has evolved to live.* A horse's ethology informs what is known of horse social structures, including complex, dynamic social organisation with social rank determining which individuals receive priority access to resources. Equine ethology also shows that horses readily form attachment bonds and need the company of their own species, so isolation is detrimental [42]. A horse that is in an inappropriate social group may be less responsive during training and, equally, a horse that has encountered inconsistent training may be more likely to be distressed by marginally frustrating aspects of its world when not being ridden [3]. Horses have evolved to graze for about 16 h a day, which means they are moving for much of the day [29,43]. This has implications for horse management, as restricting physical movement ignores the motivation horses have for locomotion, and may result in frustration, post-inhibitory rebound [44] and subsequent behaviour problems [45].

***Cognition** refers to the ways animals process information about the world.* The equine prefrontal cortex is comparatively small to that of a human [46], so horses and humans may recall events differently from the way humans do [47]. Horses have evolved as prey animals that must be constantly aware of potential dangers, so they have developed excellent abilities to recognise stimuli that trigger responses such as the flight response [13]. Equine and human intelligence are qualitatively different, so care must be taken not to overestimate what they can perceive when it comes to which behaviours are "right" or "wrong" [12]. It is likely that incorrect behaviours are a product of training rather than a horse being wilfully disobedient. Equally, horses, like other animals, can develop emotional responses to stimuli that motivate their behavioural responses [48], so it is important not to underestimate their ability to react in a highly emotional way, which can make them unpredictable and dangerous, and have a significant impact on their welfare and willingness to work with humans [49,50].

Understanding equine ethology and cognition can help keep riders and trainers safe by helping them to appreciate what is most important to horses from moment to moment [51]. Understanding that horses must be vigilant and react very quickly to potential threats explains why they may be highly attentive to stimuli that signal threat, or even stimuli unfamiliar to them [29]. It is safer to assume that there is danger and act in the interests of self-protection than to assume that there is no danger and risk being injured or killed [52]. While a horse's safety is dependent on speed, their agility is limited in small spaces by their size. This should inform those working with horses that, in the face of threat, horses will have an urge to run, and the more startled or frightened they are, the more powerful this urge will be [20]. They often run without any apparent regard for their own safety, seemingly when above certain arousal thresholds to be unable to notice or process how to navigate obstacles. If they are not so aroused as to flee in this way, they may be able to trial alternative behaviours to reduce the perceived threat. However, the more aroused they become, the more they will default to natural, energetically and pathologically costly behavioural solutions.

The ability to predict how horses may respond to threatening stimuli is a feature of what is known as horsemanship [53] and, as such, it arms riders and trainers with valuable information against provoking dangerous horse behaviour. Understanding ethology can also help predict the kinds of stimuli that may provoke extreme responses in horses. Horses are neophobic, so any unfamiliar stimuli may be perceived as potentially threatening, and young and/or inexperienced horses are likely to be triggered to escape by stimuli to which more-experienced horses may have habituated [20]. Furthermore, a horse that has been triggered to escape is likely to recognise the stimuli that started this response very well, and may also recall other contextual stimuli, and associate them with the flight response [31]. This could result in a horse being triggered to escape by stimuli to which it had been previously habituated. Addressing these strong and problematic associations takes patience and care to slowly build up a horse's tolerance to stimuli that have previously set off a flight response [31].

Appreciating the social and locomotive needs of horses and being sure to meet them aids in avoiding frustration. Frustration can be a serious problem, as it can foster aggression [33].

3.2. Use Learning Theory Appropriately

Learning theory informs the ways horses learn that are common to all animals, and includes habituation, sensitisation, operant conditioning, shaping, and classical conditioning [25].

Habituation *is recognised when animals stop responding to events and stimuli as they become accustomed to them.* As discussed above, horses are innately neophobic and often find characteristics such as the size, novelty, proximity and sudden appearance of stimuli frightening or startling [54]. Movement may mimic more threatening stimuli, such as stalking or rushing predators or conspecifics, particularly if it is sudden, erratic, or advancing towards the horse [3]. This may overcome even familiar stimuli and provoke unexpected responses. Habituation can reduce the intensity of reactions to aversive stimuli and be facilitated by desensitisation. For example, horses may be gradually exposed to an aversive stimulus with increasing intensity while ensuring that they remain in a calm state [55]. The horse learns that a calm response is more relevant than an aroused and fearful response.

Sensitisation *is when an individual's response intensity is increased.* In contrast to desensitisation, repeated exposure to arousing stimuli results in an increased likelihood of a response of increased speed or intensity to other stimuli as well. Being able to recognise this process when it is occurring is important when managing a horse during exposure to potentially threatening stimuli. It is worth noting that sensitisation to pressure cues from the rider is often desirable in equitation [56].

Operant conditioning *describes training using rewarding or aversive consequences.* Such consequences are shown to be either reinforcing or punishing by their effect on the preceding behaviour. Therefore, if behaviour reduces in frequency, duration or intensity, the behaviour has been punished, whereas if it increases in frequency, duration or intensity, it has been reinforced. A behaviour can be punished by *applying* a noxious stimulus, which is known as positive punishment, or by *removing* a desired stimulus, which is known as negative punishment. Similarly, positive reinforcement describes the addition of a desired stimulus, and negative reinforcement the taking away of a noxious stimulus. It is important to understand which of the four operant processes is taking place. Punishments suppress behaviour, and may have suppressive effects not only on the behaviour that was punished, but also on contextual aspects and the horse's future willingness or unwillingness to offer new behaviours [57]. In contrast, reinforcement encourages behaviours, particularly approach behaviours, when the reinforcement used is positive reinforcement [58]. Negative reinforcement is used extensively in horse-riding, with physical pressure being applied to parts of the horse's body and being released when the desired behaviour is performed [57] (*i.e.*, the release of pressure is reinforcing). It is therefore imperative that the release of pressure be prompt, consistent, and easily achieved by the horse [3]. Negatively reinforced responses that rely on aversive stimuli (such as most rein or leg signals) should be continuously checked and maintained in order to avoid problem behaviours that may lead to reduced welfare [31] (*i.e.*, so that the horse does not habituate to the aversive stimulus).

Shaping *is the gradual step-by-step building of behaviours, by reinforcing each step.* The targeted behavioural goal is achieved by rewarding successive approximations so that each step should differ only slightly from the previous step [57]. The advantage of this approach is that it enables horses to have many successes on the way to learning the final behaviour, which is likely to encourage positive emotional states associated with training [31].

Classical conditioning *uses cues and signals to trigger and elicit behaviours*. They must be precisely timed to coincide with the start of the desired behaviour, and in that way, over time, they become cues that predict and trigger the desired behaviour [59].

It is critical to use learning theory appropriately and with skill. The implications of using it incorrectly are broad. The use of aversive stimuli in training to punish a behaviour are likely to negatively impact the horse's mood [60]. Where an animal's experiences are frequently aversive, it will expect more aversive experiences. This may make it more flighty and defensive. Furthermore, incorrect use of operant conditioning and errors in shaping can lead to frustration, which can, in turn, lead to aggression or conflict behaviours [31]. Understanding how habituation, sensitisation, and classical conditioning work and having the ability to correctly identify when these processes are taking place enables riders and trainers to avoid horses developing negative associations with training-related stimuli and encourages horses to build positive or neutral associations. Negative associations lead to negative moods [50], triggering potentially dangerous problems that have already been discussed, such as vigilance, hyper-reactivity, and conflict.

3.3. *Train Easy-to-Discriminate Signals*

Operant and classically conditioned signals should be unique and easily discriminated, particularly for signals that modulate behaviours in opposite directions [25]. Some examples are up/down gait transitions and faster/slower gait variations. Signals that are blurred or ambivalent may lead to horses becoming confused and distressed, particularly if the consequences of error are aversive, such as the application of punishment, increased pressure, or apparently inescapable pressure [61]. This could produce frustration and conflict with its associated dangers, but may also, paradoxically, lead to an incorrect response, particularly when training behaviours that are relatively novel. When both the correct and incorrect response have been reinforced, confusion may manifest as either a default to commonly practised behaviours or a reversion to stress-related behaviours, such as escape, aggression or apathy. Furthermore, similar signals for different responses may lead to the horse offering the opposite response to what was requested. For example, if the conditioned stimulus for moving the legs both "faster" and "slower" are similar (because the rider's legs are involved in both responses), then an incorrect discrimination is likely to lead to the horse increasing or decreasing speed when asked to do the exact opposite. Such confusion can lead to dangers for humans and can have welfare implications for horses.

3.4. *Shape Responses and Movements*

Training should begin by reinforcing any rudimentary attempts at the target behaviour [62,63] that of course is completely unknown to the horse. Expecting the horse to extrapolate the correct response can lead to confusion and frustration, and subsequent compromised welfare [64]. As discussed above, this has implications for the safety of riders and trainers as it increases the likelihood of aggressive or escape behaviours, but also may lead to decreased reliability in responding to signals, which could put both horse and rider in the path of immediate danger, particularly if the horse does not stop or slow when signalled [5].

3.5. *Elicit Responses One-at-a-Time*

Cues or signals should be given individually, with a clear separation between them. This ensures that contradictory or conflicting signals are not given simultaneously, which can lead to inhibition of both signalled behaviours [65], as well as behaviours declining in strength and, potentially, signals being confused. The timing of signals should also be considered so that signals closely align with the behaviours they are cueing. For example, horses have four fundamental gaits. In walk, there are four beats, in trot, two beats, and in canter, three beats. The optimal time to elicit a response is when the leg is in swing phase because on the contrary stance phase, the limbs are preoccupied by mechanical constraints [66]. The implications for this are that there are four moments in walk swing phase in which to elicit an accelerating, decelerating or turning limb response, two moments in trot, and three in canter. This principle has the same implications for safety as shaping responses and using signals that are easy to discriminate between.

3.6. *Train Only One Response per Signal*

Each signal should elicit only a single response, so that it is clear to the horse which response is being signalled. In the dressage domain, rein stimuli are often used both to make the horse arch its neck, and for deceleration signals. In addition, rein cues signal the horse to turn, so tightening reins to bend the horse's neck could also mean deceleration, or an ambiguous turn signal. Finally, the rider's legs are frequently used for all the various locomotory effects as well as for turning and so called "bending" the horse's body. There is also a strong potential here for confusion, and this use of one signal for several possible responses also violates the next principle of consistency.

Training only one response per signal is especially critical when separating acceleration and deceleration signals, and signals for speed and direction. Ambiguous signals lead to confusion in the horse, and a variety of broader effects on behaviour. Evidence in humans shows that ambiguity is avoided as aversive [67], and it can produce context-specific responses in animals [68] that may result in horses performing behaviours that are unexpected by the rider and may be punished, or produce behaviours from the human that are unexpected by the horse. This may in turn lead to increased conflict behaviours, incorrect behaviour at critical moments, hesitation, frustration, aggression, and ongoing anxiety surrounding being ridden [20], all of which may produce horse behaviours that endanger riders. One response may have multiple signals, but those signals should be exclusive to the one response and not be used to elicit other responses [31].

3.7. *Form Consistent Habits*

Consistency is a powerful tool. From a training perspective, consistency in signals and what they mean across different contexts also leads to consistency in the horse's responses [25]. Clearly, this outcome is desirable, and enhances rider and trainer safety by promoting reliability in the horse's readiness and ability to perform behaviours smoothly and without hesitation when signalled [69]. Consistency in the trainer's approach to shaping so that successive steps always follow a similar pattern no matter what behaviour is being trained helps a horse to predict how training sessions will progress and the likely next steps that will be reinforced. This will reduce frustration, encourage positive affective states and promote persistence in shaping sessions so that the horse will not become frustrated or show a reduction in response rate when criteria change during shaping.

The same approaches that make consistency in and between training sessions beneficial may also promote safety outside training sessions. Consistency in the sequence of activities handlers and carers conduct around horses, the way they move, the way they vocalise, and when and how they interact with horses can allow these animals to predict the chain of events that are relevant to them [53]. This may be a double-edged sword, as it can produce sensitisation to stimuli in some circumstances, but where there are no stimuli producing powerful avoidance responses, it may have the opposite effect and habituate horses to everyday activities and routines [54]. Horses that can predict the immediately following events and know them to be safe are animals that will be more relaxed, less vigilant to potential negative stimuli, less prepared to take evasive or defensive actions or to attempt escape, and may be more exploratory and less fearful [23].

Inconsistent training and behaviour around horses can lead to the development of ambiguous signals that are difficult for horses to reliably interpret and respond to. This decreases their control of outcomes, leading to insecurity, which leads to diminished feelings of safety. This may compromise rider and trainer safety by creating horses that are skittish and unpredictable as they try to anticipate and adapt to surprising behaviour from humans.

3.8. Train Persistence of Responses (Self-Carriage)

Self-carriage refers to the maintenance of learned behaviours that should not need constant signalling or correction, but which the horse will continue performing until signalled otherwise [31,70]. If ongoing signals are required throughout the performance of a behaviour, it can have detrimental effects on signalling. It may lead to dull responses where signalling becomes meaningless background "noise" [59]. It may also lead to sensitivity and hyper-reactiveness to other stimuli. If a signal involves tactile pressure, and is not released when the horse begins to perform the correct behaviour, the horse does not know how to escape the pressure, and may experiment with undesired and dangerous behaviours, such as bucking [71]. Rider safety can be compromised in both scenarios. Equitation science encourages the use of sparse signalling as well as self-carriage of behaviours so that signalling does not become constant [3]. This is in the interests of the horse, as it does not put the animal in states of conflict, but it is also in the interests of the rider by avoiding hyper-reactivity or a drop in responsiveness.

3.9. Avoid and Dissociate Flight Responses (Because They Resist Extinction and Trigger Fear Problems)

Research has suggested that horses displaying a fear response (either flight or fight) feature prominently in horse-related injuries to humans [72]. Flight responses are related to seeking safety, and as such, have unique characteristics that have become adapted to keep horses safe and to find safety in the future. Flight responses are resistant to extinction, because threats to safety are serious and the potential cost of judging a stimulus safe when it is not could be injury or death [27]. In contrast, the cost of judging a stimulus unsafe when it is safe is likely to be less serious, such as missing opportunities for access to or learning about important resources (e.g., nutritious food, shelter, or mates). Flight responses are also prone to spontaneous recovery, even after alternative behaviours have been trained [27]. So, avoidance behaviours may continue even when the emotional need for avoidance is no longer driving the behaviour.

Flight responses are associated with a host of physiological and cognitive changes that are extremely effective at helping horses remove themselves from perceived danger and avoiding that danger in the future, but those same changes can be problematic in training and riding, as well as making horses unsafe [20]. High arousal and increased muscle tone make a horse very physically responsive to stimuli so they are primed to flight at the first hint of potential danger [73]. This is likely to give them the best chance of finding safety in the face of threats, but it is also likely to result in skittishness and unpredictable responses as they process stimuli through the filter of their state of high alertness. Furthermore, high arousal is damaging to the execution of problem-solving skills and concentration [50]. When an animal is in acute distress, such as that brought on by an urge to take flight, problem behaviours such as bolting, aggressive displays, distance-increasing behaviours, or apathy are likely to emerge [64]. These behaviours can be dangerous for horses and for humans working with or around them, and can also create instant negative associations with stimuli that the horses perceive when their need to escape becomes powerful [64]. The horse is also likely to become firmly entrenched in any behaviour that seemed to assist escape, which can mean that a horse learns to engage in dangerous behaviour to avoid perceived threats [71]. It may be that just the appearance of those perceived threats (e.g., someone approaching with a lead rope) can provoke avoidance behaviours that have been successful in the past even before there is any further indication of potential threat [43].

It is believed that when horses are often exposed to threatening stimuli, they can become chronically stressed [74] and that chronically elevated cortisol concentrations may result in health problems that reflect compromised immunity [75] but it can also come with a host of other potential problems that are likely to negatively impact performance and rider safety. These include learning and memory deficits, conflict behaviours becoming ritualised, redirected aggression, and long-term insecurity leading to problems such as separation-related distress, stereotypies, fear of conspecifics and heightened neophobia [76]. Chronically stressed horses may also develop a negative expectation bias, which can result in reduced interest in accessing reinforcers, and a subsequent cycle of negative emotional states perpetuated by negative expectation bias and avoidance and escape behaviours. This cascade may prevent adjustment to expectations and lead to risk aversion that discourages the explorations that might give them access to reinforcers that could improve mood [74].

Thus, chronic stress can result in a horse that may be volatile and unpredictable, may behave aggressively, and present extreme escape behaviours that can unseat and injure riders. Such horses may also have difficulties learning alternative behaviours due to their negative expectations and their preoccupation with performing behaviours that have successfully delivered escape and safety in the past. Furthermore, high glucocorticoid concentrations associated with chronic stress can affect cognition and memory, enhancing memory consolidation for avoidance learning, but hampering recall of other memories [77]. Equitation science demands avoidance of provoking flight responses, which, arguably, present the most dangerous aspects of horse behaviour. Horses attempting to take flight may fail to notice humans in their way, may be compelled to barge past them, or may view them as an obstacle to escape that heightens their distress or provokes human-directed aggression [20]. It is important to realise that horses in such states are often incapable of responding to signals in the way they have been trained, and are probably prioritising their own safety. This usually means increasing their distance from fear-inducing stimuli by whatever means are available. Where they are unable to do this with escape behaviours, they may attempt to do so through aggression [37]. Furthermore, their focus on imminent danger is likely to lead them to misinterpret, signals, stimuli, and human behaviours that they may be both familiar and comfortable with as threatening.

Attempts at training alternative behaviours without addressing the cause of the flight response may introduce further conflict or may cause it to be expressed in new ways. In addition, the horse may identify additional threats, such as humans stalking or chasing it, and this risks the horse making a single-trial association between its high arousal state and stimuli it has formerly been comfortable with, including humans, mounting blocks, or tack [58].

Flight responses in horses are clearly dangerous on several levels. Equitation science emphasises avoiding provoking them in the first place, but where this has occurred, it is important to realise that such horses have probably also associated their threatened safety with stimuli or events that occurred or were present at the time of the expression of these flight responses [3]. Dissociating those signals from flight responses may be difficult and time consuming, with many trials needed, but it must be achieved, and with minimal further threats to the horse's perceived safety. This is why equitation science also recommends that the original source of a flight response and conflict be ascertained and addressed. For example, dysfunctions in negative reinforcement of deceleration responses require that these responses be retrained through the correct use of negative reinforcement or combined negative and positive reinforcement.

3.10. Demonstrate Minimum Levels of Arousal Sufficient for Training (To Ensure Absence of Conflict)

As discussed, arousal levels are associated with performance. The nature of the association depends on the task to be performed. If the task is complex, requiring assessment of options, problem-solving, precision, or self-control, the relationship can be described as curvilinear or an "inverted-U", (*i.e.*, the shape of a graph with performance on the y-axis and arousal on the x-axis), with poor performance when arousal is low and the quality of the performance increasing until arousal is at a moderate level before declining as arousal peaks [50]. At low arousal, performance is poor because arousal is linked to interest and motivation, which are needed for the investment of energy into a related task. Performance increases with arousal to reach its maximum at moderate arousal, where the optimal level of arousal promotes interest in performing the task and speed in performance without the detrimental effects of high arousal. Increases in arousal beyond its optimal level lead to ever-poorer performance, while at the same time facilitating bursts of speed or strength. This can be explained by attributes of animals in high arousal that may impair performance, such as increased muscular tonus reducing the capacity for precise movements, and the sharpening of focus onto a single task or stimulus that may make animals less attentive and unable to respond to other stimuli [78]. This means that highly aroused horses may struggle to process signals not directly related to the salient stimuli, hampering problem-solving and the detection of and appropriate response to other external stimuli [3]. It also amplifies the veracity of the earlier mentioned Principle 5, that responses should be elicited singly.

There are several theories of arousal [79,80], and many attempt to account for deviations in this inverted-U pattern. It is posited that where the required task is simple and highly relevant to the consequences of performing it, higher arousal will generally improve performance in a linear fashion and there will be no dramatic decline at higher levels [79]. For example, where a horse's goal is to escape a noxious stimulus, increased arousal will increase heart rate and blood pressure, serving to increase speed. The horse's focus will narrow to fixate on escape routes. Such horses are now in a good state to get to safety as quickly as possible, but if they are required to perform a complex task in order to do so, such as negotiating a maze, they will need a lower level of arousal to accomplish their goal quickly and efficiently.

High arousal states may improve performance for simple, energetically costly behaviours, but it is possible that they come with additional risks other than lack of precision and impaired problem-solving and decision-making skills. Arousal is a state of readiness, so it follows that at higher levels it may come with ever-increasing likelihoods of very active behaviours, and an ever-increasing dependency on them to solve any problems or threats that may be encountered. In a large prey animal such as a horse, this could make them more prone to dangerous behaviours related to flight or aggression, regardless of why they initially became aroused. It is possible that being in the presence of an attachment figure may help them to moderate arousal [81]. As such, equitation science encourages trainers to aim for promoting the minimum arousal level required to perform the target behaviour [82]. Not only will this support the horse's ability to adapt to changing conditions appropriately and respond correctly even to unanticipated signals, but it will also reduce the likelihood of dangerous flight responses and aggressive behaviour in the case of conflict or when startled, and thus keep trainers and riders safer [20].

4. Conclusions

Horses can weigh more than 500 kg and are prone to strong flight responses. The danger they present when highly aroused and fixated on creating distance between themselves and perceived threats cannot be overstated. Those working with horses may inadvertently trigger flight responses, or be seen as a threat to be escaped from or driven away by horses, depending on environmental conditions such as available space or the presence of other threats. Humans can behave in ways that confuse, frustrate and frighten horses, or a combination of all three. Minimising the likelihood and strength of flight responses by managing a horse's arousal and emotional state is one element of the ISES 10 Principles of Equitation Science that may serve to guide human behaviour around horses to promote human safety. These principles also address training goals that serve to minimise frustration and confusion in horses during horse-riding and husbandry with best practice use of learning theory and an understanding of equine ethology. These principles are likely to enhance human safety by promoting consistency and responsiveness in horses and avoiding conflict during the training of new behaviours and signalling of known behaviours.

Acknowledgments: The authors acknowledge the invaluable help of the Council of the International Society for Equitation Science in drafting the principles of ethical equitation.

Author Contributions: Melissa Starling prepared the manuscript and Andrew McLean and Paul McGreevy provided additional material, comments, and references.

Conflicts of Interest: Two authors—Paul McGreevy and Andrew McLean—are founding members of the International Society for Equitation Science (ISES).

References

1. Oxford Dictionaries Community. *Oxford Dictionary of English*, 3rd ed.; Stevenson, A., Ed.; Oxford University Press: Oxford, UK, 2010.

2. McGreevy, P.D. The advent of equitation science. *Vet. J.* **2007**, *174*, 492–500. [CrossRef] [PubMed]
3. McGreevy, P.D.; McLean, A. *Equitation Science*; Wiley Blackwell: Oxon, UK, 2010.
4. Jones, B.; McGreevy, P.D. Ethical equitation: Applying a cost-benefit approach. *J. Vet. Behav. Clin. Appl. Res.* **2010**, *5*, 196–202. [CrossRef]
5. McGreevy, P.D.; Harman, A.; McLean, A.; Hawson, L. Over-flexing the horse's neck: A modern equestrian obsession? *J. Vet. Behav. Clin. Appl. Res.* **2010**, *5*, 180–186. [CrossRef]
6. Thompson, K.; McGreevy, P.; McManus, P. A critical review of horse-related risk: A research agenda for safer mounts, riders and equestrian cultures. *Animals* **2015**, *5*, 561–575. [CrossRef] [PubMed]
7. De La Guérinière, F.R. *École De Cavalerie*; Xenephon Press: Cleveland, OH, USA, 1992; Volume 1. (In French)
8. Decarpentry, G.; Barlte, N. *Academic Equitation*; J.A. Allen & Co.: London, UK, 1971.
9. Wright, M. *The Jeffery Method of Horse Handling: An Introduction to a New Approach to the Handling of Horses*; R.M. Williams: Prospect, Australia, 1973.
10. Boot, M.; McGreevy, P.D. The X files: Xenophon re-examined through the lens of equitation science. *J. Vet. Behav. Clin. Appl. Res.* **2013**, *8*, 367–375. [CrossRef]
11. McGreevy, P.D.; McLean, A.N. Roles of learning theory and ethology in equitation. *J. Vet. Behav. Clin. Appl. Res.* **2007**, *2*, 108–118. [CrossRef]
12. Mills, D.S. Applying learning theory to the management of the horse: The difference between getting it right and getting it wrong. *Equine Vet. J. Suppl.* **1998**, *30*, 44–48. [CrossRef]
13. McLean, A.N. *The Mental Processes of the Horse and Their Consequences for Training*; University of Melbourne: Melbourne, Australia, 2005.
14. Hume, D. *The Natural History of Religion*; Stanford University Press: Stanford, CA, USA, 1956.
15. Darwin, C. *The Expression of Emotions in Man and Animals*; Oxford University Press: New York, NY, USA, 2002.
16. Freud, S. *Civilization and Its Discontents*; Norton: New York, NY, USA, 1989.
17. Creighton, E. Equine learning behaviour: Limits of ability and ability limits of trainers. *Behav. Process.* **2007**, *76*, 43–44. [CrossRef] [PubMed]
18. Goodwin, D.; McGreevy, P.; Waran, N.; McLean, A. How equitation science can elucidate and refine horsemanship techniques. *Vet. J.* **2009**, *181*, 5–11. [CrossRef] [PubMed]
19. McGreevy, P.D.; McLean, A.N. Punishment in horse-training and the concept of ethical equitation. *J. Vet. Behav. Clin. Appl. Res.* **2009**, *4*, 193–197. [CrossRef]
20. Hawson, L.A.; McLean, A.N.; McGreevy, P.D. The roles of equine ethology and applied learning theory in horse-related human injuries. *J. Vet. Behav. Clin. Appl. Res.* **2010**, *5*, 324–338. [CrossRef]
21. McGreevy, P.; Hawke, C.; Celi, P.; Downing, J. Learning and teaching animal handling at the University of Sydney's Faculty of Veterinary Science. *J. Vet. Med. Educ.* **2007**, *34*, 586–597. [CrossRef] [PubMed]
22. McGreevy, P.; Oddie, C.; Burton, F.; McLean, A. The horse-human dyad: Can we align horse training and handling activities with the equid social ethogram. *Vet. J.* **2009**, *181*, 12–18. [CrossRef] [PubMed]
23. McGreevy, P.D.; Henshall, C.; Starling, M.J.; McLean, A.N.; Boakes, R.A. The importance of safety signals in animal handling and training. *J. Vet. Behav. Clin. Appl. Res.* **2014**, *9*, 382–387. [CrossRef]
24. Waring, G.H. *Horse Behavior: The Behavioral Traits and Adaptations of Domestic and Wild Horses, Including Ponies*; Noyes: Park Ridge, NJ, USA, 1983.
25. McGreevy, P.; Boakes, R.A. *Carrots and Sticks: Principles of Animal Training*; Cambridge University Press: Cambridge, UK, 2007.
26. D'Amato, M.R.; Fazzaro, J.; Etkin, M. Anticipatory responding and avoidance discrimination as factors in avoidance conditioning. *J. Exp. Psychol.* **1968**, *77*, 41–47. [CrossRef] [PubMed]
27. Le Doux, J.E. *Scientific American*; Nature Publishing Group: London, UK, 1994; pp. 32–39.
28. Bolles, R.C. Species-specific defense reactions and avoidance learning. *Psychol. Rev.* **1970**, *77*, 32. [CrossRef]
29. Rees, L. *The Horse's Mind*; Ebury Press: London, UK, 1977.
30. Maier, S.F.; Seligman, M. Pavlovian fear conditioning and learned helplessness: Effects on escape behaviour (a) the CS-U.S. contingency and (b) the independence of the U.S. and voluntary responding. In *Punishment and Aversive Behaviour*; Campbell, B.A., Church, R.M., Eds.; Appleton-Century-Crofts: New York, NY, USA, 1969.
31. McLean, A.N.; McLean, M. *Academic Horse Training: Equitation Science in Practice*; Australian Equine Behaviour Centre: Victoria, Australia, 2008.

32. McGreevy, P.D.; McLean, A. Behavioural problems with the ridden horse. In *The Domestic Horse: The Origins, Development and Management of Its Behaviour*; Mills, D.S., McDonnell, S.M., Eds.; Cambridge University Press: Cambridge, UK, 2005; pp. 196–211.
33. Dollard, J.; Doob, L.W.; Miller, N.E.; Mowrer, O.H.; Sears, R.R. *Frustration and Aggression*; Yale University Press: New Haven, CT, USA, 1939.
34. Zimmerman, P.; Koene, P.; Van Hooff, J. The vocal expression of feeding motivation and frustration in the domestic laying hen, Gallus gallus domesticus. *Appl. Anim. Behav. Sci.* **2000**, *69*, 265–273. [CrossRef]
35. Clubb, R.; Mason, G. Animal welfare: Captivity effects on wide-ranging carnivores. *Nature* **2003**, *425*, 473–474. [CrossRef] [PubMed]
36. Veissier, I.; Andanson, S.; Dubroeucq, H.; Pomies, D. The motivation of cows to walk as thwarted by tethering. *J. Anim. Sci.* **2008**, *86*, 2723–2729. [CrossRef] [PubMed]
37. McGreevy, P. *Equine Behavior—A Guide for Veterinarions and Equine Scientists*; Saunders: London, UK, 2004.
38. Cooper, J.J.; McGreevy, P.D. Stereotypic behaviour in the stabled horse: Causes, effects and prevention without compromising horse welfare. In *The Welfare of Horses*; Waran, N., Ed.; Springer Science & Business Media: Dordrecht, The Netherlands, 2002; Volume 1.
39. Cooper, J.; McDonald, L.; Mills, D. The effect of increasing visual horizons on stereotypic weaving: Implications for the social housing of stabled horses. *Appl. Anim. Behav. Sci.* **2000**, *69*, 67–83. [CrossRef]
40. McDonnell, S.M. Practical review of self-mutilation in horses. *Anim. Reprod. Sci.* **2008**, *107*, 219–228. [CrossRef] [PubMed]
41. McLean, A.N.; McGreevy, P.D. Revisting the ISES training principles. In Proceedings of the Conference of the International Society for Equitation Science, Vancouver, BC, Canada, 6–8 August 2015.
42. Waters, A.J.; Nicol, C.J.; French, N.P. Factors influencing the development of stereotypic and redirected behaviours in young horses: Findings of a four year prospective epidemiological study. *Equine Vet. J.* **2002**, *34*, 572–579. [CrossRef] [PubMed]
43. McGreevy, P.D. *Why Does My Horse...?*; Souvenir Press: London, UK, 1996.
44. Freire, R.; Buckley, P.; Cooper, J.J. Effects of different forms of exercise on post inhibitory rebound and unwanted behaviour in stabled horses. *Equine Vet. J.* **2010**, *41*, 487–492. [CrossRef]
45. McGreevy, P.D.; French, N.P.; Nicol, C.J. The prevalence of abnormal behaviours in dressage, eventing and endurance horses in relation to stabling. *Vet. Rec.* **1995**, *137*, 36–37. [CrossRef] [PubMed]
46. De Lahunta, A.; Glass, E.N.; Kent, M. *Veterinary Neuroanatomy and Clinical Neurology*; Saunders: Philadelphia, PA, USA, 1983.
47. Kandel, E.R. *In Search of Memory: The Emergence of a New Science of Mind*; W.W. Norton & Company: New York, NY, USA, 2006.
48. Panksepp, J.; Fuchs, T. The basic neuroscience of emotional experiences in mammals: The case of subcortical FEAR circuitry and implications for clinical anxiety. *Appl. Anim. Behav. Sci.* **2011**, *129*, 1–17. [CrossRef]
49. McBride, S.D.; Mills, D.S. Psychological factors affecting equine performance. *BMC Vet. Res.* **2012**, *8*, 180. [CrossRef] [PubMed]
50. Starling, M.; Branson, N.; Cody, D.; McGreevy, P. Conceptualising the impact of arousal and affective state on training outcomes of operant conditioning. *Animals* **2013**, *3*, 300–317. [CrossRef] [PubMed]
51. Mendl, M.; Paul, E.S. Do animals live in the present? *Appl. Anim. Behav. Sci.* **2008**, *113*, 357–382. [CrossRef]
52. Haselton, M.G.; Nettle, D. The paranoid optimist: An integrative evolutionary model of cognitive biases. *Personal. Soc. Psychol. Rev.* **2006**, *10*, 47–66. [CrossRef] [PubMed]
53. Payne, E.; Boot, M.; Starling, M.; Henshall, C.; McLean, A.; Bennett, P.; McGreevy, P. Evidence of horsemanship and dogmanship and their application in veterinary contexts. *Vet. J.* **2015**, *204*, 247–254. [CrossRef] [PubMed]
54. Christensen, J.W.; Rundgren, M.; Olsson, K. Training methods for horses: Habituation to a frightening stimulus. *Equine Vet. J.* **2006**, *38*, 439–443. [CrossRef] [PubMed]
55. Wolpe, J. *The Practice of Behavior Therapy*; Pergamon Press: Oxford, UK, 1969.
56. Waran, N.; McGreevy, P.; Casey, R.A. Training methods and horse welfare. In *The Welfare of Horses*; Waran, N., Ed.; Kluwer Academic Publishers: Dordrecht, The Netherlands, 2002; pp. 151–180.
57. Lieberman, D.A. *Learning: Behaviour and Cognition*; Brooks/Cole Publishing: Pacific Grove, CA, USA, 1993.
58. Henshall, C.; McGreevy, P.D. The role of ethology in round pen horse training—A review. *Appl. Anim. Behav. Sci.* **2014**, *155*, 1–11. [CrossRef]

59. Warren-Smith, A.K.; McLean, A.N.; Nicol, H.I.; McGreevy, P.D. Variations in the timing of reinforcement as a training technique for foals (*Equus caballus*). *Anthrozoos Multidiscip. J. Interact. People Anim.* **2005**, *18*, 255–272. [CrossRef]
60. McLean, A.N. The positive aspects of correct negative reinforcement. In Proceedings of the International Society of Anthrozoology Conference, Glasgow, UK, 6–9 October 2004.
61. McLean, A.N.; McGreevy, P.D. Horse-training techniques that may defy the principles of learning theory and compromise welfare. *J. Vet. Behav. Clin. Appl. Res.* **2010**, *5*, 187–195. [CrossRef]
62. Morgan, W.G. The shaping game: A teaching technique. *Behav. Ther.* **1974**, *5*, 271–272. [CrossRef]
63. Domjan, M. *The Principles of Learning and Behavior*, 6th ed.; Wadsworth Cengage Learning: Belmont, CA, USA, 2010.
64. Kiley-Worthington, M. *The Behaviour of Horses: In Relation to Management and Training*; J.A. Allen: London, UK, 1987.
65. McLean, A.N. Overshadowing: A silver lining to a dark cloud in horse training. *J. Appl. Anim. Welf. Sci.* **2008**, *11*, 236–248. [CrossRef] [PubMed]
66. Maes, L.; Abourachid, A. Gait transitions and modular organization of mammal locomotion. *J. Exp. Biol.* **2013**, *216*, 2257–2265. [CrossRef] [PubMed]
67. Rangel, A.; Camerer, C.; Montague, P.R. A framework for studying the neurobiology of value-based decision making. *Nat. Rev. Neurosci.* **2008**, *9*, 545–556. [CrossRef] [PubMed]
68. Bouton, M.E. Context, ambiguity, and unlearning: Sources of relapse after behavioral extinction. *Soc. Biol. Psychiatry* **2002**, *52*, 976–986. [CrossRef]
69. Loch, S. *The Classical Rider: Being at One with Your Horse*; Trafalgar Square: North Pomfret, VT, USA, 1977.
70. Clayton, H.M. The mysteries of self-carriage. *USDF Connect.* **2004**, 14–17.
71. McLean, A. *The Truth about Horses*; Quarto Publishing: London, UK, 2003.
72. Williams, F.; Ashby, K. *Horse Related Injuries*; Sands and McDougall Printing Pty. Ltd.: North Melbourne, Australia, 1997; Volume 23, pp. 1–16.
73. Morris, D. *Horsewatching*; Jonathan Cape: London, UK, 1988.
74. Hall, C.; Goodwin, D.; Heleski, C.; Randle, H.; Waran, N. Is there evidence of learned helplessness in horses? In Proceedings of the 3rd International Equitation Science Symposium, East Lansing, MI, USA, 13–14 August 2007.
75. Pell, S.M.; McGreevy, P.D. A study of cortisol and beta-endorphin levels in stereotypic and normal Thoroughbreds. *Appl. Anim. Behav. Sci.* **1999**, *64*, 81–90. [CrossRef]
76. McGreevy, P.D.; McLean, A.N. Roles of learning theory and ethology in equitation. *J. Vet. Behav. Clin. Appl. Res.* **2007**, *2*, 108–118. [CrossRef]
77. Parker, M.; Redhead, E.S.; Goodwin, N.D.; McBride, S.D. Impaired instrumental choice in crib-biting horses (*Equus caballus*). *Behav. Brain Res.* **2008**, *191*, 137–140. [CrossRef] [PubMed]
78. Mendl, M.; Burman, O.; Parker, R.; Paul, E.S. Cognitive bias as an indicator of animal emotion and welfare: Emerging evidence and underlying mechanisms. *Appl. Anim. Behav. Sci.* **2009**, *118*, 161–181. [CrossRef]
79. Hanoch, Y. When less is more: Information, emotional arousal and the ecological reframing of the Yerkes-Dodson law. *Theor. Psychol.* **2004**, *14*, 427–452. [CrossRef]
80. Mair, R.G.; Onos, K.D.; Hembrook, J.R. Cognitive activation by central thalamic stimulation: The Yerkes-Dodson law revisited. *Dose-Response* **2011**, *9*, 313–331. [CrossRef] [PubMed]
81. Payne, E.; DeAraugo, J.; Bennett, P.; McGreevy, P.D. Exploring the existence and potential underpinnings of dog-human and horse-human attachment bonds. *Behav. Process.* **2015**. [CrossRef] [PubMed]
82. DeAraugo, J.; McLean, A.; McLaren, S.; Caspar, G.; McLean, M.; McGreevy, P. Training methodologies differ with the attachment of humans to horses. *J. Vet. Behav. Clin. Appl. Res.* **2014**, *9*, 235–241. [CrossRef]

animals

MDPI

Article

Factors Influencing the Safety Behavior of German Equestrians: Attitudes towards Protective Equipment and Peer Behaviors

Christina-Maria Ikinger *, Jana Baldamus and Achim Spiller

Department of Agricultural Economics and Rural Development (DARE), Georg-August University of Göttingen, Platz der Göttinger Sieben 5, 37073 Göttingen, Germany; Jana.Baldamus@t-online.de (J.B.); a.spiller@agr.uni-goettingen.de (A.S.)

* Correspondence: christina.ikinger@agr.uni-goettingen.de; Tel.: +49-0551-39-4827

Academic Editor: Kirrilly Thompson
Received: 30 October 2015; Accepted: 3 February 2016; Published: 18 February 2016

Simple Summary: The handling and riding of horses can be quite dangerous. Although the use of protective gear among equestrians is increasing, a high number of incidents occur and the voluntary use of safety equipment is described as inconsistent to low. Therefore, this study looks at the safety behavior of German equestrians and at factors influencing this behavior to decrease the high number of horse-related injuries. The results reveal that attitudes towards safety products as well as the protective behavior of other horse owners and riding pupils from the stable are key factors that might alter the safety behavior of equestrians.

Abstract: Human interactions with horses entail certain risks. Although the acceptance and use of protective gear is increasing, a high number of incidents and very low or inconsistent voluntary use of safety equipment are reported. While past studies have examined factors influencing the use of safety gear, they have explored neither their influence on the overall safety behavior, nor their relative influence in relation to each other. The aim of the present study is to fill this gap. We conducted an online survey with 2572 participants. By means of a subsequent multiple regression analysis, we explored 23 different variables in view of their influence on the protective behavior of equestrians. In total, we found 17 variables that exerted a significant influence. The results show that both having positive or negative attitudes towards safety products as well as the protective behavior of other horse owners or riding pupils from the stable have the strongest influence on the safety behavior of German equestrians. We consider such knowledge to be important for both scientists and practitioners, such as producers of protective gear or horse sport associations who might alter safety behavior in such a way that the number of horse-related injuries decreases in the long term.

Keywords: protective behavior; horse; equestrians; horse-related accidents; safety equipment; risk; injury prevention; multiple regression analysis

---

1. Introduction

According to estimates of the Food and Agricultural Organization of the United Nations FAOSTAT for the year 2013, there are approximately 60 million horses worldwide [1]. Equestrianism today encompasses both recreational as well as professional activities that are becoming increasingly popular in many parts of the world [2–4]. Human interactions with horses, whether while handling a horse on the ground or mounted, entail certain risks [5–8]. The literature shows a high rate of horse-related injuries and fatal accidents [6,9–11]. Part of the risk is related to the rider's ability to predict the behavior of the horse. Equestrianism is the only sport that involves a non-human partner that is not only much larger and stronger than its human teammate, but also able to achieve high speeds of up to 65 km/h. Additionally, as a herd and prey animal, the horse has a natural, ethologically predictable flight reflex when facing potentially frightening situations and environments [3–7,12]. To prevent responses from the horse to external frightening stimuli that might endanger the rider, the rider needs to be able to correctly assess and control such flight behavior [13]. Numerous studies stress that horse riding is as dangerous as—or even more hazardous than—other medium- and high-impact sports like motorcycle riding, skiing, automobile racing, football, or rugby [5,7,9,11,14]. Although descriptions of the incidence rate of horse-related injury vary and tend to be underreported, most studies agree that the severity and fatality rate of equestrian injuries can be high [2,10,15]. Differences in risk assessment stem from the unknown number of participants in equestrian sports as well as the unknown number of hours of horse-related activity per equestrian [10]. These numbers are difficult to estimate but reported rates show that one in five equestrians will suffer a severe injury during their riding career [10] and that, compared to other sports, equestrianism has the highest mortality rate with an annual death rate of 1 in 1 million population [11,14,16].

The most common horse-related injuries involve the head, the spine, or the extremities [5,14,17–19]. Head injuries are the most serious and most deadly [9,19,20]. In addition to education about horse behavior, safe riding practices, and the proper handling of horses, as well as an increase in personal safety awareness, the use of protective equipment such as helmets or protective vests are commonly recommended strategies to avoid serious horse-related injuries [2,4,7,19,21]. Several studies have already examined and confirmed the effectiveness of helmets within different sports regarding their potential to prevent both frequency and severity of injury, indicating a head injury risk reduction of 60–88%, depending on the type of sport [2,7,22–24]. Regarding the equestrian context, previous studies noted a considerable decrease in both the severity and frequency of head injuries due to an increased use and improved design of protective helmets [4,10,18,25]. Although an increasing acceptance and awareness regarding the utility of helmets has already been shown, numerous studies report a high number of incidents and very low or inconsistent voluntary use of safety equipment [4,6,7,14,18,19,26,27]. Most studies regarding helmet use report low rates, with fewer than 40% of riders wearing helmets at the time of injury [3,7,9,12,14]. In contrast to the studies on helmets, there is still a need for research to confirm the effectiveness of protective vests, which are also less commonly used than helmets [2,10,19]. However, the wearing of protective vests is often generally recommended and they are also becoming increasingly popular [2,7].

Some studies have also looked at the reasons behind the refusal to wear protective equipment, in particular regarding equestrian helmets. Multiple psychological, social, and cultural barriers seem to be present [12]. Studies have reported a negative attitude towards helmets among some riders. These equestrians think of helmets as uncomfortable, unnecessary, silly-looking, too expensive, or restricting rider movement [19,26,28]. Some equestrians seem to perceive that horse-riding is not a risky pastime or that the risk is somehow controllable due to experience with and knowledge about horses or familiarity with certain situations and environments [26,27]. Social influences exerted by trainers, family members, peers, and the media may contribute to poor helmet use among riders, too. However, these same influences can also promote the use of protective equipment and act as role models, in particular for young equestrians, simply by adopting the use of protective equipment [26,27]. Furthermore, a certain risk perspective that accepts the unavoidable risk inherent associated with equestrianism regardless of riding experience seems to enhance the use of safety equipment [27]. These results underline the complexity regarding the relationship between risk perception, safety knowledge, attitudes, and protective behavior [27].

A large number of retrospective scientific studies that deal with horse-related risk are now available. These studies have focused on the epidemiology of equestrian-related injuries, identification of risk factors, and the higher-risk groups. They also explore the use and efficacy of technical interventions and safety equipment. In comparison to these studies, research has rarely investigated factors associated with preventing injuries [6,7,19,27,29]. Therefore, examining the preventative measures adopted by equestrians and identifying the influences that promote the adoption of safe practices would be useful [3,6]. Such influencing factors have received comparatively little attention to date. Previous research has tended to focus on the adoption of specific protective equipment (e.g., helmets, vests, boots) or the risk of injury rather than the overall adoption of safety-oriented behavioral practices by equestrians [2,17,26,27]. The present study aims to fill this gap by identifying potential influencing factors and quantifying their impact on the safety behavior of equestrians. An understanding of the drivers for adoption of safety behavior not only informs the scientific discourse but also provides information that can prove useful for producers of equestrian safety equipment or for policy makers regarding, for example, the decision on mandatory helmet use. Administrators of equestrian associations can also use this information to more effectively influence the adoption of safe behaviors by their members. In other fields, such as car driving [30,31], motorcycle riding [32], or bicycle helmet use [33], several studies have already demonstrated that the identification of influencing factors on and determinants of risk behavior provides important information and indications for the design of safety campaigns and for the development of other countermeasures to reduce risk-taking behavior. Identifying influencing factors to alter horse-related safety behavior might therefore also be helpful to reduce the number of horse-related injuries and might further provide useful insights for other high-risk sports and activities where the use of protective gear is recommended, such as cycling, climbing, or skiing.

2. Methods

2.1. Hypothesis Development

The aim of the present study is to look at potential influencing factors and their impact on the overall safety behavior of equestrians. For this purpose, we measured overall safety as an index including the use of specific protective equipment and the adoption of additional safety measures related to equestrianism (see Section 2.2). In the following, we will briefly present the factors considered here and will derive hypotheses about their potential influence on the protective behavior from the literature.

Several patient surveys repeatedly mention gender as a demographic aspect in connection with horse-related risk. Their results show that the majority of the injured equestrians were young females and conclude that female gender represents a considerable risk factor [2,5,7,17,18,20]. This might be due to the fact that mainly women practice equestrianism [2,7]. However, a multiple regression analysis searching for factors predicting equestrian injury found that the fact that women are more often involved in equestrian activities does not seem to be a reason why women are injured more frequently [17]. Yet, despite the fact that females are getting injured more often, it has also been noted that male equestrians tend to suffer from more serious injuries [4] and less frequently use helmets and other protective gear [26]. In conformity with the more general finding that men are generally less risk-averse than women [34], we assume that:

H1: Female equestrians demonstrate more pronounced safety behavior than their male counterparts.

In view of the link between age and horse-related risk, young riders aged between 10 and 35 constitute the most vulnerable group [4,5,7,18,19]. The share of riding accidents in the total number of fatal sport accidents among children and teenagers is estimated at up to 25% [4]. Some studies found that adolescents take more risks than older individuals, which leads to less pronounced safety behavior [35,36]. In contrast, a study on helmet use within the equestrian community revealed that, due to their lack of experience, young riders perceive themselves as being at a greater risk of injury and tend to wear a helmet more often, while older equestrians are less likely to wear protective gear and, if at all, only do so in certain circumstances that are considered potentially hazardous, such as riding a strange or young and unexperienced horse [26]. Hence, we expect that:

H2a: Older equestrians will show less pronounced safety behavior.

H2b: The age group of children and teenagers protects itself more intensively than older equestrians.

However, perceived social responsibility and the aspiration to act as a behavioral role model positively influence adult behavior regarding safety equipment such as the use of safety helmets. Research has shown this relationship within different kinds of sports and leisure activities, for instance cycling [37], winter sports [38], and also within equestrianism [27]. Scientific research on risk preferences further shows that men and women experience a considerable increase in risk aversion when they have a child, whereas this increase is largest shortly after giving birth and disappears when the child becomes older [39]. Therefore, we assume a positive relationship:

H3: Equestrians with children show more pronounced safety behavior.

Another factor related to horse-related risk is riding experience. Several studies noticed that novice and unexperienced riders constitute a particularly sensitive risk group [4,7,10,15,17,29,40]. In volume terms, a study revealed that the probability for a horse-related injury is three times higher for novice riders compared to moderately experienced riders, five times higher compared to experienced riders, and eight times higher compared to professional riders. In line with this, previous research also found that more years of experience have a decreasing effect on the incidence rate of horse-related injuries [10]. However, several studies agree that experience alone does not necessarily moderate the severity of injuries [4,10] and professional riders' accidents were found to result in more serious injuries, which may be due to the fact that they train and compete at an increased level of difficulty [7]. Regarding the willingness to wear protective equipment, less experienced equestrians tend to—according to the higher risk they are exposed to—protect themselves more often by means of a safety helmet than experienced riders [10]. Among more skilled equestrians, experience seems to act as a popular argument for foregoing a riding helmet, whereas some even seem to hold the prejudice that wearing protective headgear is automatically associated with being an inexperienced rider and has to be avoided [26,27]. In view of these findings, we expect the following relationships:

H4a: Novice riders protect themselves more than experienced equestrians.

H4b: The more experience a rider has, the less effort s/he puts into protecting himself/herself.

In relation to the opinion that experience can replace other protective measures to prevent serious injury, it can make a difference whether one has already been personally involved in a horse-related accident [27]. Looking at other sports, in the context of alpine ski racers and sky divers some studies have already shown that witnessing the injury of a teammate can lead to fear, which in turn might lead to increased safety behavior [41,42]. We therefore also assume for the equestrian context that:

H5a: The more severe a directly experienced riding accident, the more pronounced the protective behavior.

H5b: The more severe a directly observed riding accident, the more pronounced the protective behavior.

Another risk-related factor is the preferred riding style, as different forms of horse riding can be more dangerous than others due to speed, the need to jump obstacles, or unfamiliar surroundings [15,26]. A study on the most hazardous riding activities revealed that riding outdoors was particularly prone to accidents, followed by dressage, show-jumping, and eventing [19]. Previous research further notes that equestrians practicing one of the English-style riding disciplines (dressage, show-jumping, and eventing (formerly military)) more frequently wear protective helmets [14,26]. Conforming to this finding, helmets are being increasingly employed within a competitive setting [7], with the English-style disciplines traditionally being those with the strongest competitive orientation [43]. For example, in the case of show-jumping, which is also perceived as a higher-risk activity that requires extra protection, riders consider a hard helmet as common [7,17].

Similar thoughts apply to eventing, which represents another equestrian discipline associated with high injury rates [11]. Due to these well-known elevated risks, we expect that equestrians who participate in show-jumping and eventing have a higher acceptance and necessity for safety equipment and we therefore hypothesize that:

H6a: Eventers and show-jumpers show more pronounced protective behavior.

As riding helmets are also known as traditional equipment within dressage riding—except for the rather small group of upper-level dressage riders, who prefer to wear a soft, non-protective hat—and dressage furthermore belongs to the English riding disciplines, which have been observed to have a higher helmet use rate, and, as they are more competition-oriented, are also more likely to accept the use of protective vests [7,14,26]. We expect that:

H6b: Dressage riders show more pronounced protective behavior.

In contrast to the English-style riders, Western riders are less likely to wear a protective helmet because they consider brimmed hats or Stetsons as appropriate headgear within the traditional Western riding culture [7,14,26]. Western riding also belongs to the so-called recreational riding styles, which Havlik [7] found to have much lower helmet use rates and to be less likely to wear protective vests compared to the more competition-oriented English disciplines. We assume that:

H6c: Western riders show less pronounced protective behavior.

As it is still difficult to enforce helmet use in the non-competitive, recreational, or training setting, particularly in riding styles where the classical helmet and the protective vests are not part of the traditionally accepted attire, we further suggest that [7]:

H6d: More recreational riding styles show less pronounced protective behavior.

As already mentioned above, a dangerous but very popular riding activity consists in riding outdoors, which is principally independent of the preferred riding discipline. Horse and rider usually are in more or less unknown surroundings and have less control over their environment, which is one of the reasons why riding outdoors is considered one of the most dangerous pastimes on horseback [6,14,19]. Depending on the surroundings, riding outdoors can also be quite dangerous; in the case of a fall, the ground is likely to be harder than the usually soft riding surfaces, or in the case of the uncontrollable flight of the horse one might hit low branches or even pedestrians or cars. Due to the high risk associated with riding outdoors, we assume that equestrians pay more attention to safety behavior when riding in the countryside compared to riding in familiar environments such as a domestic riding hall or a riding ground and it is postulated:

H6e: Equestrians who spend much of their time with horses outdoors show more pronounced safety behavior.

Several studies also discuss whether there are certain characteristics of the horse, such as breed, type, temperament, or gender, which might represent a risk factor regarding horse-related injuries [6,17,26]. For example, different breeds are associated with certain personalities or specific behaviors such as thoroughbred horses, which rate as rather anxious and excited horses [44] and are less suitable for unexperienced equestrians [4,6]. Research found that typical sport horses such as warmblood horses and thoroughbreds are more reactive [44,45], and hence riding them tends to imply a higher risk. Warmblood horses and thoroughbreds are breeds traditionally used by competition-oriented English-style riders, who show a generally higher helmet and protective vest use rate [7,26]. Therefore, we assume for the overall safety behavior that:

H7: Riders of sport horses such as thoroughbred and warmblood horses show more pronounced safety behavior.

Previous research has already shown the influence of other equestrians on safety behavior like helmet use or other protective measures, e.g., by trainers and other horse owners or riding pupils at the stable [26,27]. This effect is likely to also hold true for the overall protective behavior and therefore we expect that:

> *H8: The influence of other horse owners or riding pupils from the stable who attach great importance to safety has a positive impact on the protective behavior of equestrians.*

Another important aspect regarding equestrian protective behavior relates to the attitude towards specific safety equipment and to horse-related risk perception. For example, previous studies found that negative attitudes towards equestrian helmets—regardless of whether they were perceived as useless, unattractive, or uncomfortable—will exert a negative influence on the willingness to use a protective helmet, whereas a positive attitude can exert a positive influence [26,27]. This finding probably also applies to other items of safety equipment. Accordingly, we assume that:

> *H9a: A positive attitude towards protective equipment is linked to more pronounced safety behavior.*

> *H9b: A negative attitude towards protective equipment is linked to less pronounced safety behavior.*

Finally, in line with findings in the general literature on risk perception that indicate that perceptions of risk can increase preventive behavior [46,47], we address whether the risk perception of the equestrians is influencing protective horse-related behavior [27]. In this context, we expect that the more risky horse-riding is perceived to be by equestrians, the more likely they are to wear protective gear:

> *H10a: A sensitive risk perception in general leads to more pronounced safety behavior.*

> *H10b: A positive horse-related risk perception leads to more pronounced safety behavior.*

2.2. Materials and Methods

The data for the present study were collected through an online survey that was open for about six weeks from April to June 2015. The aim of the study was not only to gain information about the protective behavior of equestrians but also to test the effectiveness of a potential campaign for equestrian safety.

The survey comprised four major sections. The first part concerned questions regarding the general horse-related behavior such as riding discipline, experience, skill level, or horse ownership. The second part aimed to collect information about safety-related equestrian behavior such as risk perception and the attitude towards protective gear, the possession and use of safety equipment, former accidents, *etc.* We designed the third part to test the effectiveness of five different safety campaigns and the fourth part comprised questions on general socio-demographic characteristics. We measured most items on a five-point Likert scale. We further utilized single choice, multiple choice, semantic differential, and open questions. The survey used EFS Survey software and was promoted on various German websites and social networks related to equestrian sports such as equestrian journals, horse sport, and breeding associations. In total, 2572 equestrians participated in the survey.

To measure the overall safety behavior, we computed an index based on two factors. The first factor involved the wearing of concrete safety gear such as helmets, safety vests, airbag vests, and combinations of safety and airbag vests in different riding situations (dressage work, jumping/eventing, riding outdoors, riding in the riding hall, riding on the riding arena, and riding unknown horses). We measured the intensity of this behavior on a five-point Likert scale of 1 = Never to 5 = Always. As airbag vests and combinations of safety and airbag vests are worn less frequently (see Section 3.1), we double weighted their score. The second factor concerned additional safety measures such as ensuring that the horse has enough access to a free-range area, that the horse has a balanced and reliable character, the use of safety stirrups, riding outdoors only in a group instead of alone, and lunging the horse before it is ridden. We also weighted these activities according to their prevalence within the sample, with the less frequent activities being more heavily weighted so that high index levels indicated very pronounced safety behavior. The calculated safety behavior index had a theoretical range of 0 to 107 (see Section 3.1).

2.3. *Analysis*

We analyzed the data with the statistics software IBM SPSS Statistics 23. For the purpose of the present study both uni-, bi-, and multivariate procedures were applied. As risk perception and the attitude towards protective equipment were measured by various statements, initially a factor analysis was used for dimensional reduction. In a second step we conducted a multiple regression analysis with the safety behavior index as a dependent variable to identify and determine the impact of significant influence factors.

3. Results and Discussion

3.1. *Sample Description*

Of the 2572 respondents, 5.2% were male and 94.8% were female and the average age was 32.5 years (SD: 11.9 years; min.: 12 years; max.: 75 years); 21.4% had one or more children. A large proportion of the respondents had completed their high school education (73.0% had passed the German high school examination). With regard to the basic population of German equestrians, reliable and comparative data is scarce. The only data available originate from the Allensbach Institute for Demoscopy (AWA). Compared to their data from 2014, our sample has a higher-than-average number of females (AWA: 22% male; 78% female) and a higher proportion of respondents who had completed high school education (AWA: 41% with a German high school exam) [48]. However, as comparative data are missing it is difficult to assess whether the AWA figures are also representative. Due to the lack of available data, it is therefore not possible to determine whether our sample is representative for the basic population of equestrians in Germany. Yet, it is a large sample that is suitable for a first exploratory study within the equestrian context.

Related to equestrianism, the participants reported that they had been practicing horse-riding for 19.9 years on average (min.: 1 year, max.: 62 years). Table 1 shows the skill level and preferred riding styles of the participants.

Table 1. Skill level and riding styles.

Skill Level	%
Beginner	30.2
Intermediate	49.0
Advanced	20.8
Riding Styles *(Multiple Answers Were Possible)*	**%**
Outdoors	82.7
Dressage	75.0
Show jumping	38.9
Eventing (former military)	13.9
Western riding	12.6
Riding of gaited horses	11.5

Approximately 98.1% of the participants reported that they had already witnessed a riding accident, with 15.1% ending up in a severe or very severe injury and 68.0% in slight or very slight injury. Another 98.1% reported having been injured themselves, with 15.8% serious and very serious injuries and 64.1% light and very light injuries. The most often affected body parts are shown in Table 2.

The possession and use of safety equipment is shown in Tables 3 and 4. The calculated safety behavior index was 22.9 on average and varied from 0 to 71.5 (SD: 8.5). The Kolmogorov–Smirnov test on normal distribution was highly significant, rejecting the hypothesis of normal distribution of the index. However, the graphic representation showed a good approximation of normal distribution.

Table 2. Most often affected body parts.

Affected Body Parts	%
Upper extremities	28.8
Lower extremities	15.0
Pelvis	13.0
Spine	10.1
Head	8.1

Table 3. Possession of safety gear and other safety measures.

Safety Equipment (n = 2572)	%
Protective helmet	97.7
Safety stirrups	46.7
Protective vest	44.1
Airbag vest	3.3
Combination of protective and airbag vest	0.9
None	2.0
Other Safety Measures (n = 2570)	**%**
Horse has enough access to a free-range area	84.9
Select horses that show a more predictable behavior	57.7
Riding outdoors only in a group	32.1
Lunging the horse before it is being ridden	10.8

Table 4. Percentage of riders always or often using safety equipment in different riding situations (top-2-box).

Riding Situation	Safety Equipment			
	Protective Helmet	*Safety Vest*	*Airbag Vest*	*Combination of Safety/Airbag Vest*
Jumping/eventing	96.0 % (n = 1911)	78.1 % (n = 917)	88.7 % (n = 62)	81.0 % (n = 21)
Riding of unknown horses	95.6 % (n = 2312)	43.8% (n = 934)	49.2 % (n = 65)	45.0 % (n = 20)
Riding outdoors	92.4 % (n = 2500)	40.4 % (n = 1087)	55.1 % (n = 78)	33.3 % (n = 21)
On the riding arena	83.0 % (n = 2487)	17.2 % (n = 1054)	32.4 % (n = 74)	9.1 % (n = 22)
Dressage work	82.7 % (n = 2406)	14.7 % (n = 1023)	31.1 % (n = 74)	4.8 % (n = 21)
In the indoor riding hall	82.1 % (n = 2385)	15.1 % (n = 1022)	29.2 % (n = 72)	9.1 % (n = 22)

Note: items were measured on a scale of 1 (always) to 5 (never); the use of the respective safety product is illustrated here as top-2-box including 1 (always) and 2 (often).

3.2. Preliminary Factor Analysis

We conducted an exploratory factor analysis (principal component analysis with orthogonal rotation (varimax)) to reduce the statements regarding the attitude towards safety equipment and risk perception to a lower number of factors. It revealed seven factors that are shown in Table 5.

Both the Kaiser–Meyer–Olkin value (0.887) and the Bartlett's test of sphericity prove that the present data set is suitable for the application of a factor analysis [49]. We measured the reliability of each factor with the Cronbach's Alpha value, where values higher than 0.6 indicate a reliable factor, and values higher than 0.5 are accepted in the early stages of research [50,51]. We attained a

Cronbach's Alpha higher than 0.6 for all factors except for factor 7, which only reaches a value of 0.5. However, as its content is quite interesting and the value is not too low, we decided to keep it for further investigation. During reliability analysis regarding factor 6, the statement that aimed at investigating the influence of peer groups "I always wear a helmet/vest, because everybody in our stable does it" did also load on this factor but did not fit well regarding the content of the factor. Elimination of this statement did not decrease the reliability of the factor so we removed it from factor 6.

Table 5. Results of the exploratory factor analysis.

Factor	Variable/Statement	Factor Loading
Factor 1: Perception of general riding risks (Cronbach's Alpha = 0.869)	Grooming the horse is particularly dangerous.	0.845
	Doing ground work with the horse is particularly dangerous.	0.790
	Riding in the riding hall is particularly dangerous.	0.738
	Grooming the hind legs of a horse is particularly dangerous.	0.732
	Riding in the riding arena is particularly dangerous.	0.721
	The loading of horses is particularly dangerous.	0.531
Factor 2: Negative attitude towards safety equipment (Cronbach's Alpha = 0.740)	Safety equipment such as helmets or vests is simply uncomfortable.	0.731
	Safety equipment such as helmets or vests look unflattering.	0.725
	Safety products are too expensive; I prefer spending the money on my horse.	0.576
	I want to relax when practicing my hobby and don't want to think about risks.	0.574
	I believe that safety products do not really protect in the most serious cases.	0.541
	Nothing has ever happened to me when riding; I think the risk is often exaggerated.	0.538
	Safety equipment is just for kids.	0.516
Factor 3: Perception of special riding risks (Cronbach's Alpha = 0.736)	Jumping in the country is particularly dangerous.	0.774
	Jumping over an obstacle is particularly dangerous.	0.689
	Riding on/beside a country road is particularly dangerous.	0.556
	A prize-giving ceremony during an equestrian event is particularly dangerous.	0.505
	Carriage rides are particularly dangerous.	0.411
Factor 4: Perception of general health risks (Cronbach's Alpha = 0.717)	Insufficient exercise is particularly dangerous.	0.708
	Insufficient sleep is particularly dangerous.	0.701
	Stress is particularly dangerous.	0.679
	Eating lots of fat and sugar is particularly dangerous.	0.644
	Being exposed to the sun unprotected for a long time is particularly dangerous.	0.517
	Alcohol consumption is particularly dangerous.	0.450

Table 5. *Cont.*

Factor	Variable/Statement	Factor Loading
Factor 5: Perception of extreme and motor sport risks (Cronbach's Alpha = 0.602)	Motorbike riding is particularly dangerous.	0.735
	Extreme sports like sky diving and cliff climbing are particularly dangerous.	0.704
	Fast driving is particularly dangerous.	0.569
Factor 6: Positive attitude towards safety equipment (Cronbach's Alpha = 0.635)	Riding without helmet is particularly dangerous.	0.597
	Riding without a safety/airbag vest is particularly dangerous.	0.511
	Bicycle riding without a helmet is particularly dangerous.	0.466
Factor 7: Risk averse perception and behavior (Cronbach's Alpha = 0.501)	I'm willing to spend money on safety equipment.	0.752
	I'm fully aware of the risk of horse riding.	0.708

Note: KMO: 0.887, Bartlett's test of sphericity highly significant ($p < 0.001$), explained variance: 51.91%.

3.3. Multiple Regression Analysis

We computed a multiple regression analysis to test the influence of the discussed factors (see Section 2.1) on the safety behavior of equestrians (for an overview see Table 6). We selected the forced entry method as stepwise regressions are often criticized to be influenced by random variation in the data causing non-replicable results. The model is able to explain 39.2% of the variance of the dependent variable. The ANOVA further confirms that the model, overall, is a significantly good prediction of the safety behavior index as dependent variable [51].

We examined the quality of the multiple regression analysis in terms of multicollinearity, autocorrelation or residuals, and heteroscedasticity. Multicollinearity can represent a serious problem when there is a strong correlation between two or more predictors. We examined both the correlation matrix and the variance inflation index (VIF). The correlation matrix showed only correlations smaller than 0.7 and the VIF was substantially smaller than 10 for all of the predictors. These results show that collinearity does not apply for this model. We looked at the Durbin–Watson statistic to detect the presence of autocorrelation in the residuals. The attained value was very close to 2, showing the assumption of independent errors as fulfilled. We used the plot of standardized residuals against standardized predicted values to test the assumption of homoscedasticity. It revealed that the graph looked slightly like a funnel, indicating moderate heteroscedasticity. To check for cases that might be influencing the regression model, we calculated both Cook's distance and Mahalanobis distance. Both criteria were fulfilled and less than 5% of cases had standardized residuals above 2 (and 1.6% had standardized residuals above 2.5). In conclusion, there does not seem to be a major problem with influential cases [51].

In total, we included 23 variables in the regression model. Thereof, 17 variables showed a significant influence, while we found no significant influence on the horse-related safety behavior of the variables gender (H1), age (H2a), skill level (H4a), severity of witnessed horse-related accident (H5b), gaited horse riding (H6d), and riding outdoors (H6e). The two variables concerning the general risk perception regarding health risks and high risk and motorsports both exerted a significant influence, yet it was in the opposite direction than expected (H10a). The three variables with the highest influence on the safety behavior proved to be a positive attitude towards safety equipment (Beta = 0.32; $p < 0.001$), followed by the influence of the stable mates (Beta = 0.2; $p < 0.001$), and a negative attitude towards safety equipment (Beta = −0.18; $p < 0.001$).

Table 6. Results of the multiple regression analysis.

Variables	Unstand. Coefficients		Stand. Coefficients	Sig.	VIF	Hypothesis
	B	Std. Error	Beta			
(Constant)	18.38	1.37		0.000		
Gender	−0.33	0.71	−0.01	0.650	1.17	H1 not confirmed
Age	0.03	0.02	0.04	0.186	3.07	H2a not confirmed
Children/teenager *vs.* adults	−1.279	0.48	−0.05	0.007	1.41	H2b confirmed
Having children	0.44	0.21	0.04	0.033	1.41	H3 confirmed
Beginners *vs.* intermediate/advanced riders	−0.51	0.34	−0.03	0.133	1.18	H4a not confirmed
Riding experience	−0.05	0.02	−0.06	0.016	2.18	H4b confirmed
Severity of own accident	0.40	0.13	0.05	0.003	1.11	H5a confirmed
Severity of witnessed accident	0.18	0.13	0.03	0.154	1.09	H5b not confirmed
Show-jumping	1.45	0.35	0.08	0.000	1.43	H6a confirmed
Eventing (former military)	2.93	0.44	0.12	0.000	1.19	H6a confirmed
Dressage	0.81	0.40	0.04	0.039	1.41	H6b confirmed
Western riding	−1.51	0.48	−0.06	0.002	1.27	H6c confirmed
Gaited horse riding	0.25	0.50	0.01	0.613	1.28	H6d not confirmed
Riding outdoors	0.40	0.39	0.02	0.307	1.07	H6e not confirmed
Breed (sport horse)	0.96	0.34	0.06	0.005	1.44	H7 confirmed
I always wear a helmet/vest, because everybody in our stable does it.	1.31	0.13	0.20	0.000	1.28	H8 confirmed
Positive attitude towards safety equipment (Factor 6)	2.76	0.19	0.32	0.000	1.84	H9a confirmed
Negative attitude towards safety equipment (Factor 2)	−1.55	0.15	−0.18	0.000	1.15	H9b confirmed
Perception of general health risks (Factor 4)	−0.31	0.15	−0.04	0.040	1.13	H10a not confirmed
Perception of extreme and motor sport risks (Factor 5)	−0.56	0.16	−0.07	0.001	1.28	H10a not confirmed
Perception of general riding risks (Factor 1)	0.38	0.15	0.05	0.013	1.17	H10b confirmed
Perception of special riding risks (Factor 3)	0.53	0.15	0.06	0.000	1.07	H10b confirmed
Risk averse perception and behavior (Factor 7)	0.85	0.15	0.10	0.000	1.08	H10b confirmed

Note: Method = forced entry, R^2 = 39.2%, ANOVA = 0.000; Durbin-Watson: 2.06; Cook's distance: 0.00; Mahalanobis distance: 22.99; dependent variable = safety behavior index.

3.4. Discussion

On the basis of previous findings from the literature, we derived 10 hypotheses regarding the influence of various factors on the protective behavior of equestrians, which we subsequently tested by means of a multiple regression analysis. Concerning the influence of gender, we hypothesized that female equestrians demonstrate more pronounced safety behavior compared to male equestrians because they are generally more risk-averse [34] and also tend to be more at risk [2,5,7,17,18,20]. However, we detected no significant effect of gender. Possible causes for this finding may be that horse riding is not classified as hazardous by female equestrians as it is "just" a sport like any other with which they grew up, so they have gotten used to and thus displace the risk inherent to equestrianism. Also, it may be the case that the argument that protective helmets and other safety equipment do not look good plays a more important role for female equestrians and leads to a higher level of rejection [26]. Or it might also be possible that the more risk-averse females do not practice horse-riding at all. If only less risk-averse women choose to practice horse riding, this might be a possible reason why there was no relationship found between gender and safety behavior. It might be that there is no or only a small relationship between gender-specific risk perception and horse-related safety behavior. A significant debate is underway as to whether the high share of women among equestrians may bias the findings that female riders represent a high-risk group for injury [2,7]. In their literature review on horse-related injuries, Hawson *et al.* [4] also found that the younger age groups were predominately female, while the share of men was higher among the older age groups. However, the smaller share of men among the younger age groups was still more likely to get seriously injured. The authors discussed that for certain age groups, especially younger patients, the age- and gender-related distribution of equestrian accidents seems to correspond to that of the basic riding population. Yet this does not seem to hold true for all age groups. For example, the fact that older males are more likely to suffer more serious injury than females in the same age group cannot be explained by the demographic distribution [4]. However, as the share of male equestrians in the sample is quite low (about 5%), the high share of female participants may have somewhat distorted the results, although this is not very likely regarding the large sample size. Looking at the data in more detail, we could find at least some small differences between male and female equestrians. Male riders were older on average, were more often earning their money through riding, and less often rode the same horse but rather different horses. We further found male riders to practice eventing more frequently, while the share of female equestrians riding outdoors was higher compared to their male counterparts. Regarding the use of protective equipment, we found that male participants owned and used a riding helmet less frequently, but were more likely to own an airbag vest. Regarding the perception of horse-related risk, male riders perceived the general riding risks as more dangerous, whereas the female equestrians perceived the special riding risks as being more problematical. This shows that the relationship between gender and protective behavior seems to be highly complex and quite manifold, so that it might be difficult to detect a clear relationship here. It follows that there is a need for further research to clarify the underlying relationships between gender-related risk, risk perception, and safety behavior.

Regarding the influence of age, we hypothesized that younger equestrians in general (H2a) and children and teenagers in particular show more pronounced safety behavior (H2b), but the present results only confirmed the second hypothesis. The finding that young riders tend to be a high-risk group [4,5,7,18,19] and wear helmets more often seems to be reflected in their safety behavior [26]. However, whether this behavior is self-imposed and possibly influenced by a high risk perception on the part of the young riders or rather mandated by parents or trainers remains to be clarified. Also, the requirements for junior riders to wear helmets are more closely mandated by associations, which might also play a role in this context [52]. The finding that older equestrians are less likely to wear protective gear could not be confirmed here [26]. One reason for this could be that the relationship between age and safety behavior is not completely linear. Looking at the correlation between age and the safety behavior index confirms this assumption as we found no significant relationship between them. Although the safety behavior probably decreases with growing years of experience [10,26,27], it might start to decrease—analogous to growing risk aversion when having a child—due to certain events in life such as the birth of a child, increasing domestic or job responsibilities, or getting older in general [39]. A study on attitudes and behaviors towards helmet use revealed similar findings, showing that those were affected by perceived social responsibility and care not only for other riders, but also for relatives, families, and friends [27]. In line with this, the results of the present study confirm that adult equestrians with children show a more pronounced safety behavior compared to those without children, proving Hypothesis 3.

Hypothesis 4a and 4b refer to the relationship between riding experience and safety behavior. Regarding the skill level, the results did not confirm that novice equestrians show a more pronounced safety behavior compared to more advanced or professional riders, although the results did confirm that, in terms of years of experience, the more experienced riders show a less pronounced safety behavior (see Table 6). The finding that experienced riders make less of an effort to protect themselves may have several causes. Firstly, more experienced riders may already have had some probably minor accidents in the past and underestimate the risk for more serious injuries. Also, it could be that as experience increases the control over the horse will improve, which might in turn lead to a false sense of security. In line with the findings by Haigh and Thompson [27], the perception of being able to control the horse and read equine behavior could favor the rejection of safety equipment. As shown by previous research, another possible reason could lie within the fact that some of the more experienced riders seem to relate the use of protective headgear with being an inexperienced rider [26,27] and therefore try to evade the use of protective equipment or other protective measures that might characterize them as newbies. In contrast to more experienced riders, the finding that novice riders do not see the necessity of wearing protective equipment might be influenced by the fact that they might not have experienced or observed a serious fall yet. However, in line with findings indicating that less experienced equestrians tend to—according to the higher risk they are exposed to—protect themselves more often by means of a safety helmet than experienced riders [10], the supposed negative relationship between increasing experience and decreasing protective behavior could again be confirmed. It has to be noted that the classification of riding skills was based on the participants' self-assessment, which can lead to inaccuracies. Due to the growing heterogeneity within equestrianism, it is increasingly difficult [53], especially across different riding disciplines, to have a reliable skill classification. Hasler *et al.* [17] suggest that a comparable educational level-injury risk index that tracks the true improvement in skills could be a more reliable measure of the relationship between experience and safety behavior. Furthermore, it is possible that not all advanced equestrians reject wearing safety equipment and prejudices might be getting slowly removed. Yet, both novice riders—due to the high number of accidents [4,7,10,15,17,29,40]—and more experienced riders—due to the increased severity of sustained injuries and reduced willingness to exert safety behavior [7,10]—represent rather important target groups for sensitization regarding equestrian-related risks. There is a need for more detailed information regarding the risk awareness and attitudes of these target groups to better assess their behavior. One option to sensitize these target groups would be to use an experienced rider giving a testimonial who serves for both novice as well as advanced riders as an inspiring model giving advice on the safe handling of horses through seminars or courses. Sports celebrities are widely used in classical advertising with the aim of improving awareness and recall, driving sales, and influencing behavior. However, it must be noted that the particular celebrity has to comply with certain conditions and has, for example, to match the product or the topic to be effective [54,55]. More specifically, previous research results have shown that the use of celebrities within public health campaigns can be able to influence health-related attitudes, beliefs and risk behavior and seems to be a promising possibility within the present context [56,57].

Relating to horse accident experience, the high share of each of the 98% of equestrians who stated they had already experienced or witnessed a horse-related accident is consistent with the high number of equestrian injuries found in the literature [9–11,14]. The results of the multiple regression analysis confirmed Hypothesis 5a, which assumed a positive influence of the severity of a personally experienced accident on the individual's protective behavior. However, we found no significant influence for the severity of an observed accident (H5b). It is plausible that accidents that are experienced firsthand arouse a greater awareness of horse-related risks. If the accident has only been experienced indirectly by observation, the personal distance towards the risk seems to be larger and the corresponding risk might be more easily dismissed. Yet this differs from O'Neil's [41] findings that the injury of alpine ski racers did have a psychological impact on the respective teammates. One reason for this could be that individuals might think they are safe and believe that misfortunes only ever happen to others. However, witnessing traumatic injuries can distort such beliefs. To cope with such traumatic events, people engage in different coping strategies [58]. One of these strategies might consist of mental distancing from the injury. In line with this, a study on helmet use showed that it is being frequently argued by equestrians that one could do without safety equipment as the risk is perceived to be controllable [27]. Such a false belief can be more easily maintained in the case of a witnessed accident as the riders themselves have not lost control and can talk themselves into believing that they might have reacted differently. The extent to which a rider can maintain this false belief might further depend on the observed severity of the injury. However, this is only speculation and it would be interesting to explore whether there is a certain degree of injury necessary to realize a change in safety behavior and what additional factors might play a role regarding the influence of observed accidents. As research in this field is said to focus too heavily on the effects on the direct victim and has frequently neglected the potential effects of serious injuries on the witness, it would be interesting to take a deeper look at this phenomenon within equestrianism and other kinds of sports [58].

Previous research has also found that equestrians who practice one of the more competitive-oriented English-style riding disciplines wear helmets more frequently [7,14,26,43]. The present results confirmed that show-jumpers (H6a), eventers (H6a), and dressage riders (H6b) show more pronounced protective behavior. Also, other studies have showed that the less frequent use of helmets, which is characteristic of Western riding [7,14,26] and other more leisure-oriented disciplines, has [7]—as proposed in hypothesis 6c—a negative effect on the overall protective behavior. In contrast to the riding of gaited horses, which also belongs to the more leisure-oriented riding disciplines, we found no significant relationship. The present results could only partially confirm Hypothesis 6d, which assumes that more leisure-oriented disciplines put less effort in safety behavior. Regarding the relationship between protective behavior and riding outdoors, which several studies consider as one of the most dangerous pastimes on horseback and which often includes a high share of leisure-oriented equestrians of different disciplines, again no significant relationship could be found (H6d) [6,14,19]. It is becoming clear that there are huge variations between the several leisure-oriented riding disciplines and their respective safety behavior. Such information would be helpful for horse sport associations to identify and communicate with more vulnerable groups that have a greater need for safety education. Western riders represent such a vulnerable group, as they often refuse to wear protective helmets as these are not considered appropriate Western-style equipment. In recent years, some producers have tried to develop Western-style protective helmets, but they were not successful on the market. Whether this was due to the look of the helmet, its wearing comfort, or other reasons remains unknown. For the Western riding associations, this implies that it is necessary to work on the development of protective gear that is better accepted by the Western riding culture as well as education about horse-related risk and the advantages and effectiveness of protective gear.

Hypothesis 7 assumed a significant relationship with the breed of the horse. As expected, we could observe a positive relationship between the riding of sport horses such as warmblood or thoroughbred horses and more pronounced safety behavior. It is unclear whether the observed positive relationship between riding of sport horses such as warmblood or thoroughbred horses and more pronounced safety behavior is due to a perception that these breeds are associated with a greater degree of unpredictable behavior and higher risk [44,45].

In line with the findings from the literature, the present results indicate that the influence of social groups and peer groups can positively influence helmet use [26,27]. Yet, the present study only looked at the influence of one social group, namely other horse owners and riding students from the stable. To keep the number of influencing factors manageable, we looked first at the more general group of other horse owners and riding pupils from the stable. That specific group is likely to have a close and horse-relevant contact with an equestrian and hence also the possibility to exert a strong influence on the latter. In view of the strength of the respective impact, this variable constitutes a quite important influence factor, as it exerts the second highest influence overall and provides a valuable starting point for the promotion of safety behavior. Here, especially trainers, stable managers, and horse sport associations are asked to inform their pupils, members, and clients about safety aspects concerning equestrianism, to reduce safety-related prejudices and to establish a positive security culture among the riders in a stable. As we did not further differentiate the group of horse owners and riding pupils from the stable, it would be useful to ask which individuals and subgroups, such as trainers or equestrian idols and also non-riding friends and family members, influence safety behavior at all and which of these groups exert a particularly high impact. Such knowledge could provide further starting points to enhance preventive behavior.

From all the factors examined within the multiple regression analysis, we identified the attitude towards protective gear as the most influential factor. As expected, a positive attitude exercised a positive influence (H9a) and a negative attitude exercised a negative influence (H9b) on protective behavior. The positive attitude exerted the highest impact overall and the negative attitude had the third-largest influence. The attitude towards safety equipment seems to represent a key aspect when trying to increase the use of safety products. Producers of safety equipment and horse sport associations should continue to try to find out more about the underlying reasons for these attitudes to identify potential enablers and barriers. Deficient design, untraditional appearance, lack of comfort, and doubts about the effectiveness of safety products seem to be major reasons for rejection. Therefore, the research into this area must be continued, especially on safety vests as their effectiveness to reduce horse-related injuries has still to be confirmed and has already been questioned within other sports [2]. Furthermore, the design and comfort of safety equipment might be other important aspects to look at. In the case of helmets, the design can contribute decisively to the decision to wear safety equipment [26]. Potentially, if the producers of safety equipment would work on the comfort and look of the product, it might turn into a rather desirable fashion item that riders would more often voluntarily use.

Finally, the last hypothesis expected both a positive influence of risk perception in general (H10a) and horse-related risk perception in particular (H10b) on protective behavior. However, the present results only confirmed the second hypothesis. Surprisingly, we detected a small but significant negative relationship. We measured risk perception in general as the personal risk perception of basic and high risk situations. The basic situation comprised moderate health risks such as having too much stress or insufficient sleep or exercise. The high-risk situations included activities such as extreme sports and motor sports. The reasons for the observed negative relationship with protective behavior are not clear so far and a more detailed examination of the connection between risk perception in general and horse-related risk perception in particular is necessary. Perhaps the phenomenon of risk suppression is a possible explanation. In this sense, more risk-averse people in general, especially when they have finally decided to participate in a high-risk sport or hobby such as equestrianism, might willingly suppress the associated risk, which might be considered quite harmless compared to other high-risk sports. Certain findings in the recent literature partially confirm this phenomenon [26,27], which shows that equestrians generally state that they believe they can control horse-related risk. Future research projects need to scrutinize this assumption. Perception of the level of danger associated with the equestrian activity in general and specific riding situations positively influenced the protective behavior patterns reported by equestrians. This relationship engenders a responsibility of horse sport associations to educate their members about the inherent horse-related risks to produce more safety-oriented behavior in equestrians.

3.5. Limitations

As already discussed in the sample description (see Section 3.1), it is unclear whether the sample is representative so the generalizability of the present study may be limited. The present study is subject to self-selection bias in that certain types of respondents participated in the survey. Those with a high interest in horse-related safety may be overrepresented. Therefore, the extent to which the transfer of the results of the present study to the German equestrian population may be limited. However, given the high number of participants and the finding that sociodemographic variables did not exert a strong influence, it is likely that these results provide an important first approximation in this area of study.

The study's methodological limitations include selected method of analysis and the calculation of the index to measure safety behavior. Although multiple regression analysis is a method able to identify the relationship among two or more variables, this does not automatically imply that this relationship is also causal [59]. However, based on logical considerations, this affects only a small number of variables, such as show-jumping. For instance, it could be that some equestrians might only dare to jump at all because they practice very pronounced safety behavior. Such a relationship could further be linked to the phenomenon of risk compensation, in connection with which it is being discussed whether the wearing of protective gear can also exert the opposite effect on safety behavior, such as the use of protective equipment like helmets or safety vests, giving a false sense of security and promoting risk-prone behavior instead of reducing it [2,7,60]. Another methodological limitation concerns the dependent variable. The safety behavior index mainly covered the use of protective equipment in different riding situations, which might imply that factors concerning protective gear exert a disproportionally high impact. Since several studies showed that not only horse riding itself is dangerous, but also simply handling the horse can lead to serious injuries, such as trampling, being kicked, or being bitten [5,8], it would be interesting to look at the specific protective behavior when handling a horse and compare it to the behavior when riding.

Moreover, it has to be noted that it is difficult to judge the quality of protective gear as it might be useful in the case of serious injury but might not avoid dangerous situations overall [4,14]. It may even be possible, as already discussed above, that some kind of risk compensation is at work such that the wearing of protective gear can also result in riskier behavior [2,7,60]. Future research regarding the use and effectiveness of protective gear but also the impact of additional measures that can reduce horse-related risk could provide additional useful information. As already proposed in the literature, an important additional safety measure constitutes the improvement of the predictability of horses through better education and understanding of equine learning and behavior patterns, building on recent findings from research on horse ethology and equitation science, as it is a commonly cited cause of human injury [4]. In this context, improving riders' competence in physical skills such as fitness, balance, the proper application of aids, and falling techniques should make them more resilient to injury and falls; establishing clear rules and legislation requiring the mandatory wearing of approved safety gear and increasing general awareness of horse-related risk for both individuals and the general public are further possible measures to reduce horse-related risks. The establishment of good practices and a comprehensive safety management within stables will ensure a safe environment [6,8,12,19].

4. Conclusions

The present comprehensive study examined the potential influence and impact of 10 different factors on the safety behavior of equestrians by means of a multiple regression analysis. It should be noted that the relationships between the respective variables are quite complex. The results show that the attitudes towards safety products as well as the protective behavior of other horse owners and riding pupils from the stable are key aspects in altering the safety behavior of equestrians. The obtained outcomes could help horse sporting associations, politicians, and producers of horse-related safety gear find additional starting points for the promotion of risk preventive behavior and identify important high-risk groups that should be made more aware of the various advantages of protective gear. The findings herein may also inform other high-risk sports administrators seeking promotion of more pronounced safety behavior in their participants.

Author Contributions: The study was conceived and managed by Jana Baldamus, Christina-Maria Ikinger and Achim Spiller. The questionnaire development and data collection was conducted mainly by Jana Baldamus. Analysis, presentation, and interpretation of the findings were undertaken by Christina-Maria Ikinger, who also wrote the paper. Achim Spiller supervised and advised the statistical analysis and editing of the paper.

Conflicts of Interest: The authors declare no conflict of interest.

References

1. FAOSTAT. Download Data. Available online: http://faostat3.fao.org/download/Q/QA/E (accessed on 15 October 2015).

2. Hessler, C.; Eckert, V.; Vettorazzi, E.; Meenen, N.; Jürgens, C.; Schult, M.; Flamme, C.; Herberhold, H.-J.; Madert, J.; Ekkernkamp, A.; *et al.* Effectiveness of safety vests in pediatric horseback riding. *Klin. Padiatr.* **2012**, *224*, 443–447. [CrossRef] [PubMed]
3. Guyton, K.; Houchen-Wise, E.; Mayberry, J. Equestrian injury is costly, disabling, and frequently preventable: The imperative for improved safety awareness. *Am. Surg.* **2013**, *79*, 76–83. [PubMed]
4. Hawson, L.A.; McLean, A.N.; McGreevy, P.D. The roles of equine ethology and applied learning theory in horse-related human injuries. *J. Vet. Behav. Clin. Appl. Res.* **2010**, *5*, 324–338. [CrossRef]
5. Sorli, J.M. Equestrian injuries: A five year review of hospital admissions in British Columbia, Canada. *Inj. Prev.* **2000**, *6*, 59–61. [CrossRef] [PubMed]
6. Thompson, K.; McGreevy, P.; McManus, P. A critical review of horse-related risk: A research agenda for safer mounts, riders and equestrian cultures. *Animals* **2015**, *5*, 561–575. [CrossRef] [PubMed]
7. Havlik, H.S. Equestrian sport-related injuries: A review of current literature. *Curr. Sports Med. Rep.* **2010**, *9*, 299–302. [CrossRef] [PubMed]
8. Northey, G. Equestrian injuries in New Zealand, 1993–2001: Knowledge and experience. *N. Z. Med. J.* **2003**, *116*, 1–8.
9. Hessler, C.; Schilling, B.; Meenen, N.M.; Lockemann, U.; Püschel, K. Risks in sport riding—A critical survey of safety standards in sport riding. *Sportverl. Sportschad.* **2010**, *6*, 100–110.
10. Mayberry, J.C.; Pearson, T.E.; Wiger, K.J.; Diggs, B.S.; Mullins, R.J. Equestrian injury prevention efforts need more attention to novice riders. *J. Trauma Acute Care Surg.* **2007**, *62*, 735–739. [CrossRef] [PubMed]
11. Paix, B.R. Rider injury rates and emergency medical services at equestrian events. *J. Sports Med.* **1999**, *33*, 46–48. [CrossRef]
12. Thompson, K.; Matthews, C. Inroads into equestrian safety: Rider-reported factors contributing to horse-related accidents and near misses on australian roads. *Animals* **2015**, *5*, 592–609. [CrossRef] [PubMed]
13. McGreevy, P.; McLean, A.; Buckley, P.; McConaghy, F.; McLean, C. How riding may affect welfare: What the equine veterinarian needs to know. *Equine Vet. Educ.* **2011**, *23*, 531–539. [CrossRef]
14. Ball, C.G.; Ball, J.E.; Kirkpatrick, A.W.; Mulloy, R.H. Equestrian injuries: Incidence, injury patterns, and risk factors for 10 years of major traumatic injuries. *Am. J. Surg.* **2007**, *193*, 636–640. [CrossRef] [PubMed]
15. Silver, J.R.; Parry, J.M. Hazards of horse-riding as a popular sport. *Br. J. Sp. Med.* **1991**, *25*, 105–110. [CrossRef]
16. Pounder, D.J. "The grave yawns for the horseman." Equestrian deaths in South Australia 1973–1983. *Med. J. Aust.* **1984**, *141*, 632–635. [PubMed]
17. Hasler, R.M.; Gyssler, L.; Benneker, L.; Martinolli, L.; Schötzau, A.; Zimmermann, H.; Exadaktylos, A.K. Protective and risk factors in amateur equestrians and description of injury patterns: A retrospective data analysis and a case-control survey. *J. Trauma Manag. Outcomes* **2011**. [CrossRef] [PubMed]
18. Moss, P.S.; Wan, A.; Whitlock, M.R. A changing pattern of injuries to horse riders. *Emerg. Med. J.* **2002**, *19*, 412–414. [CrossRef] [PubMed]
19. Hessler, C.; Eckert, V.; Meiners, J.; Jürgens, C.; Reicke, B.; Matthes, G.; Ekkernkamp, A.; Püschel, K. Ursachen, Verletzungen, Therapie und Präventionsmöglichkeiten von Unfällen im Reitsport. *Unfallchirurg* **2014**, *117*, 123–127. [CrossRef] [PubMed]
20. McCrory, P.; Turner, M. Equestrian injuries. *Med. Sport. Sci.* **2005**, *48*, 8–17. [PubMed]
21. Thomas, K.E.; Annest, J.L.; Gilchrist, J.; Bixby-Hammett, D.M. Non-fatal horse related injuries treated in emergency departments in the United States, 2001–2003. *Br. J. Sports Med.* **2006**, *40*, 619–626. [CrossRef] [PubMed]
22. Thompson, D.C.; Rivara, F.; Thompson, R. Helmets for preventing head and facial injuries in bicyclists (Review). *Cochrane Database Syst. Rev. Rev.* **1999**, *4*, 1–31.
23. Thompson, D.C.; Rivara, F.P.; Thompson, R.S. Effectiveness of bicycle safety helmets in preventing head injuries—A case-control study. *J. Am. Med. Assoc.* **1996**, *276*, 1968–1973. [CrossRef]
24. Sulheim, S.; Holme, I.; Ekeland, A.; Bahr, R. Helmet use and risk of head injuries in alpine skiers and snowboarders. *J. Am. Med. Assoc.* **2006**, *295*, 919–924. [CrossRef] [PubMed]
25. Chitnavis, J.; Gibbons, C.; Hirigoyen, M.; Parry, J.L.; Simpson, A. Accidents with horses: What has changed in 20 years? *Injury* **1996**, *27*, 103–105. [CrossRef]
26. Condie, C.; Rivara, F.P.; Bergman, A.B. Strategies of a successful campaign to promote the use of equestrian helmets. *Public Health Rep.* **1993**, *108*, 121–126. [PubMed]

27. Haigh, L.; Thompson, K. Helmet use amongst equestrians: Harnessing social and attitudinal factors revealed in online forums. *Animals* **2015**, *5*, 576–591. [CrossRef] [PubMed]
28. Nelson, D.E.; Rivara, F.P.; Condie, C. Helmets and horseback riders. *Am. J. Prev. Med.* **1994**, *10*, 15–19. [PubMed]
29. Newton, A.M.; Nielsen, A.N. A review of horse-related injuries in a rural Colorado hospital: Implications for outreach education. *J. Emerg. Nurs.* **2005**, *31*, 442–446. [CrossRef] [PubMed]
30. Shope, J.T. Influences on youthful driving behavior and their potential for guiding interventions to reduce crashes. *Injury Prev.* **2006**. [CrossRef] [PubMed]
31. Horswill, M.S.; Helman, S. A behavioral comparison between motorcyclists and a matched group of non-motorcycling car drivers: Factors influencing accident risk. *Accident Anal. Prev.* **2003**, *35*, 589–597. [CrossRef]
32. Wong, J.T.; Chung, Y.S.; Huang, S.H. Determinants behind young motorcyclists' risky riding behavior. *Accid. Anal. Prev.* **2010**, *43*, 275–281. [CrossRef] [PubMed]
33. Berg, P.; Westerling, R. Bicycle helmet use among schoolchildren—The influence of parental involvement and children's attitudes. *Inj. Prev.* **2001**, *7*, 218–222. [CrossRef] [PubMed]
34. Weber, C.S. Cultural Differences in Risk Tolerance. *IWE Work. Paper*, 2013. Available online: http://www.economics.phil.uni-erlangen.de/forschung/workingpapers/riskaversion.pdf (accessed on 23 September 2015).
35. Steinberg, L. Risk taking in adolescence—New perspectives from brain and behavioral science. *Curr. Dir. Psychol. Sci.* **2007**, *16*, 55–59. [CrossRef]
36. Furby, L.; Beyth-Marom, R. Risk taking in adolescence: A decision-making perspective. *Dev. Rev.* **1992**, *12*, 1–44. [CrossRef]
37. Dannenberg, A.L.; Gielen, A.C.; Beilenson, P.L.; Wilson, M.H.; Joffe, A. Bicycle helmet laws and educational campaigns: An evaluation of strategies to increase children's helmet use. *Am. J. Public Health* **1993**, *83*, 667–674. [CrossRef] [PubMed]
38. Ruedl, G.; Pococco, E.; Raas, C.; Blauth, M.; Brucker, P.U.; Burtscher, M.; Kopp, M. Influence of adult role modeling on child/adolescent helmet use in recreational sledging: An observational study. *Wien. Klin. Wochenschr.* **2015**, *10*, 1–5. [CrossRef] [PubMed]
39. Görlitz, K.; Tamm, M. Parenthood and risk preferences. *Ruhr Econ. Papers* **2015**, *552*, 1–10. [CrossRef]
40. Fikri, A.-Z.; Rao, S. Factors affecting the severity of horse-related injuries. *Injury* **2003**, *34*, 897–900.
41. O'Neill, D.F. Injury contagion in alpine ski racing: The effect of injury on teammates' performance. *J. Clin. Sport Psychol.* **2008**, *2*, 278–292.
42. Kerr, J.H. Sudden withdrawal from skydiving: A case study informed by reversal theory's concept of protective frames. *J. Appl. Sport Psychol.* **2007**, *19*, 337–351. [CrossRef]
43. Ikinger, C.; Münch, C.; Wiegand, K.; Spiller, A. Reiterleben Reiterwelten—Zielgruppen zwischen Reitweisen, Motiven und der Liebe zum Pferd. 2013. Available online: http:www.uni-goettingen.de/de/document/download/1988e74b5e6a7bf92bf38381a71a47f0.pdf/2013-04%20reitsportstudie_screen.pdf (accessed on 23 September 2015).
44. Lloyd, A.S.; Martin, J.E.; Bornett-Gauci, H.L.I.; Wilkinson, R.G. Horse personality: Variation between breeds. *Appl. Anim. Behav. Sci.* **2008**, *112*, 369–338. [CrossRef]
45. Bachmann, I.; Stauffacher, M. Prävalenz von Verhaltensstörungen in der Schweizer Pferdepopulation. *Schweiz. Arch. Tierheilkd.* **2002**, *144*, 356–368. [CrossRef] [PubMed]
46. Van der Pligt, J. Risk perception and self-protective behaviour. *Eur. Psychol.* **1996**, *1*, 34–43. [CrossRef]
47. Brewer, N.T.; Chapman, G.B.; Gibbons, F.X.; Gerrard, M.; McCaul, K.D.; Weinstein, N.D. Risk Meta-analysis of the relationship between risk perception and health behavior: The example of vaccination. *Health Psychol.* **2007**, *26*, 136–145. [CrossRef] [PubMed]
48. Ikinger, C.; Wiegand, K.; Spiller, A. Facts and Figures on German Equestrians. *Discuss. Pap.* **2014**, *1408*, 1–57.
49. Backhaus, K.; Erichson, B.; Plinke, W.; Weiber, R. *Multivariate Analysemethoden—Eine Anwendungsorientierte Einführung*, 12th ed.; Springer-Verlag: Berlin, Germany, 2010.
50. Nunally, J.C. *Psychometric Theory*, 2nd ed.; McGraw-Hill: New York, NY, USA, 1978.
51. Field, A. *Discovering Statistics Using IBM SPSS Statistics*, 4th ed.; Sage Publications Ltd: London, UK, 2013.

52. FEI. Protective Headgear Requirements at FEI Events—Summary of the FEI Requirements on Protective Headgear. Available online: https://www.fei.org/sites/default/files/FEI%20Protective%20Headgear%20requirements%20-%20April2015_0.pdf (accessed on 7 December 2015).
53. Münch, C. Pferdemarkt der Zukunft. *DGfZ Schr.* **2014**, *64*, 185–188.
54. Ruihley, B.J.; Runyan, R.C.; Lear, K.E. The use of sport celebrities in advertising: A replication and extension. *Sport Mark. Q.* **2010**, *19*, 132–142.
55. Pringle, H.; Binet, L. How marketers can use celebrities to sell more effectively. *J. Consum. Behav.* **2005**, *4*, 201–214. [CrossRef]
56. Brown, W.J.; Basil, M.D.; Boçarnea, M.C. The influence of famous athletes on health beliefs and practices: Mark McGwire, child abuse prevention, and androstenedione. *J. Health Commun.* **2003**, *8*, 41–57. [CrossRef] [PubMed]
57. Brown, W.J.; Basil, M.D. Media celebrities and public health: Responses to "Magic" Johnson's HIV disclosure and its impact on AIDS risk and high-risk behaviors. *Health Commun.* **1995**. [CrossRef]
58. Day, M.C.; Bond, K.; Smith, B. Holding it together: Coping with vicarious trauma in sport. *Psychol. Sport Exerc.* **2013**, *14*, 1–11. [CrossRef]
59. Tabachnick, B.G.; Fidell, L.S. *Using Multivariate Statistics*, 5th ed.; Allyn & Bacon: Boston, MA, USA, 2007; pp. 117–194.
60. Hagel, B.; Meeuwisse, W. Risk compensation: A "side effect" of sport injury prevention? *Clin. J. Sport. Med.* **2004**, *14*, 193–196. [CrossRef] [PubMed]

animals

MDPI

Article

Look Before You Leap: What Are the Obstacles to Risk Calculation in the Equestrian Sport of Eventing?

Denzil O'Brien

South Australian Spinal Cord Injury Research Centre, Hampstead Rehabilitation Centre, 207 Hampstead Rd., Northfield, SA 5085, Australia; denzil.obrien@optusnet.com.au; Tel.: +61-883-980-367

Academic Editor: Kirrilly Thompson
Received: 27 October 2015; Accepted: 2 February 2016; Published: 16 February 2016

Simple Summary: This paper examines a number of methods for calculating injury risk for riders in the equestrian sport of eventing, and suggests that the primary locus of risk is the action of the horse jumping, and the jump itself. The paper argues that risk calculation should therefore focus first on this locus.

Abstract: All horse-riding is risky. In competitive horse sports, eventing is considered the riskiest, and is often characterised as very dangerous. But based on what data? There has been considerable research on the risks and unwanted outcomes of horse-riding in general, and on particular subsets of horse-riding such as eventing. However, there can be problems in accessing accurate, comprehensive and comparable data on such outcomes, and in using different calculation methods which cannot compare like with like. This paper critically examines a number of risk calculation methods used in estimating risk for riders in eventing, including one method which calculates risk based on hours spent in the activity and in one case concludes that eventing is more dangerous than motorcycle racing. This paper argues that the primary locus of risk for both riders and horses is the jump itself, and the action of the horse jumping. The paper proposes that risk calculation in eventing should therefore concentrate primarily on this locus, and suggests that eventing is unlikely to be more dangerous than motorcycle racing. The paper proposes avenues for further research to reduce the likelihood and consequences of rider and horse falls at jumps.

Keywords: horses; eventing; risk; falls; injury; riders; human–animal relationships; human–horse relationships

---

1. Background

Equestrian sport is unique. It involves a relationship between two beings, one of which is not human [1]. The horse is a large four legged prey animal whose successful evolution has resulted from its strong flight instincts [2–4]. Humans have sought to use and control horses for thousands of years [5], and over time have extended these uses and controls from simply riding and harnessing horses to involving them in a range of challenging activities: chariot-racing, bull-fighting, thoroughbred racing, jumps racing and buck-jumping, to name just a few.

Eventing is the horse sport usually characterised as the most dangerous of all the modern mainstream equestrian sports [6–8]. It is often described as the triathlon for horses [1,8,9], the ultimate test of horse and rider, based on a military tradition. In fact, the French name of the sport is *concours complet*: the complete contest. Eventing demands equine attributes similar to those of a warhorse: obedience, agility and grace on parade (represented by the dressage phase), courage, strength, fitness and speed in battle (the cross-country phase), and a level of fitness and effective recovery which will allow resumption of normal duties immediately after battle (the show-jumping phase) [1,9–11].

Eventing is one of the more difficult sports to explain to the uninitiated [10]. First, it has several different names: eventing, horse trials, three-day eventing, and one-day eventing. In this paper, the sport is referred to as eventing. A full explanation of the sport, its scoring and its rules can be found on the website of the Fédération Equestre International (FEI), the peak body for equestrian sports [9].

In the cross-country phase of the sport, horses gallop across open country on a predetermined course, jumping big obstacles within set time limits, incurring penalties for completing too quickly or too slowly. The obstacles may include walls, steps, jumps into and out of lakes and creeks, almost vertical slides, palisades, tables, ditches with fences set within them, banks, keyholes, log piles, sunken roads, and jumps made out of everyday objects such as the flatbeds of trucks, or large carved animals, and often a combination of several of these individual elements [9,10].

The cross-country phase of eventing is generally viewed as the riskiest for both horse and rider [12,13]. These risks include the full range of unwanted outcomes one might expect from an activity which involves riding a horse at speed over fixed obstacles: death, head and brain injury, spinal injury, crush injuries and fractures, as well as minor injuries such as sprains, bruising and abrasions [12–16]. The complex inter-species relationship which exists in all horse-related activity [8,17] is taken to extremes in eventing.

However, until the late 1990s there had not been a great deal of evidence-based research on exactly how risky eventing was [12,13]. The horse world's attention became focused abruptly, however, when in 1999 five eventing riders died in competition in the UK within a few months of each other [18–20], and another died in the USA [20]. Five of these six deaths involved the horse somersaulting over the jump, the rider falling forwards with continued momentum after the horse's forward motion was abruptly stopped, and the horse subsequently landing on top of the rider (known as a rotational horse fall or a somersault fall [21]).

The FEI responded urgently and promptly to these deaths, and instituted wide-ranging reviews into the sport, with rule changes focusing immediately on reducing the possibility of a rotational horse fall. The 2000 report of the International Eventing Safety Committee [22] stated the following:

> *A fundamental conclusion which pervades every detailed recommendation is that everything should be done to prevent horses falling: this single objective should greatly reduce the chances of riders being seriously injured, as well as significantly improving the safety of competing horses. (p. 2)*

Despite this prompt response, an almost continuous stream of rule and format changes, new requirements governing the construction of fences, and changes to the roles of officials, a further 38 eventing riders have died in or after competition between 2000 and the time of writing (October 2015), including two in Pony Club competitions [20,23,24]. At least 27 of these rider deaths resulted from a rotational horse fall [24].

2. Defining Risk in Human—Horse Interactions

All interactions with horses are potentially hazardous. Horses are much larger and heavier than humans, while being sentient, sensitive and prone to a well-developed flight instinct as a direct result of their evolution from a prey animal [2–4]. They can travel very quickly, and stop and change direction in less than a second; they can bite, kick, crush and squash—and that is before one has even mounted them [25]. Once one is astride, the ambit of hazard widens to include falling from height, often with speed as an additional hazard, as well as rapid changes in momentum and direction [25].

The notion of risk itself involves much discussion about definitions and meaning, and about methods for calculating, evaluating and mitigating risk in any given situation [26,27]. While one way to calculate risk in a particular activity is to count the number of unwanted outcomes such as injuries or deaths associated with the activity, simply enumerating them is not particularly helpful in determining the actual risks involved for participants, or in managing and reducing those risks, since injuries in sport result from a complex interaction of multiple factors and events [28]. Simple analyses such as counting the number of injuries over a period of time ignore many other contributing factors, and may not be useful in developing appropriate risk-management policies [28]. If one accepts that risk is always present when humans interact with horses [1–4], and if one accepts that these risks can and should be reduced or eliminated, then an appropriate process to achieve this is to first identify the context within which risk occurs [29], then identify the risks, assess them, control them, and review those controls [30]. Through this process, risk can be quantified, and this quantification can then be used to mitigate risk and help to shape sports injury prevention policies [30–34]. However, before this process can be undertaken, one must first be able to accurately identify cases appropriate to the question.

2.1. Data on Human—Horse Interactions

By and large, data about human injury and mortality resulting from human–horse interactions are sourced predominantly from hospital separations records, coronial reports, trauma registries, emergency department records, surveillance programs, surveys and literature reviews.

Research into horse-related injuries tends to focus on the following:

- horse-related injuries in general without specifying a particular equestrian sport, using data from sources such as those outlined above: hospital separations data, injury surveillance programs, coronial data, trauma registries and emergency department data collections [7,35–43];
- reviews of the broader literature on overall equestrian-related injury [8,15,44];
- measurement of the incidence of specific horse-related injuries such as spinal injury or maxillofacial injuries, again using sources such as coronial data, hospital separations data, trauma registries, and emergency department data collections, but not focusing on a specific equestrian sport [16,45]; and
- measurement of the incidence of specific injuries such as traumatic brain injury which are associated with keynote sports but which may or may not be an outcome of horse sport [46–48].

2.2. Challenges for the Researcher

There are a number of challenges facing the researcher who seeks accurate and comparable data on overall horse-related injury, let alone injury resulting from specific horse sports such as eventing.

2.2.1. Data Keeping

In Australia, national information on the epidemiology of overall horse-related injury and mortality has until recently been minimal and fragmented. Cripps [37] was able to provide quite detailed information on horse-related deaths and hospitalisations in Australia using national mortality and hospital separation datasets, but was unable to identify the specific type of horse-related activity, the place of occurrence or the mechanism of injury because of limitations in hospitals' data coding at the time. He reported that in the year 1996–1997, there were 3124 hospitalisations for horse-related injury, and using the Australian Bureau of Statistics mortality unit data collection, identified 410 horse-related deaths between 1979 and 1998. On this basis he estimated an average of just over 20 horse-related deaths a year at that time, and this figure has been consistently cited since.

The 2014 report by Safe Work Australia (SWA) [30] identifies a total of 11,635 hospital admissions for horse-related incidents between 1 July 2008 and 30 June 2011, an average of 3878 a year, with 40% of these occurring during "sports." However, "trail or general horseback riding" account for 80% of these so-called sports-related injuries, and neither trail riding nor general horseback riding technically qualifies as a sport, based on the definition of sport as "an organised group activity centred on a contest between at least two parties" [49]. The SWA report also identifies 98 horse-related deaths between July 2000 and June 2012, an average of just over eight per year. Seventy-four per cent of these deaths involved a fall from a horse. However, these data also do not reveal whether the fatality occurred during sport (other than horse-racing), and there is no information on mechanism of injury.

The 2014 report from the Australian Skills Quality Authority (ASQA) [50], initiated after the horse-related death of a young trainee jillaroo in 2009, reports 1,568 hospitalisations resulting from equestrian activities for the 12 months 2011–2012, markedly lower than the average annual number reported by Cripps [37] or the SWA report [30]. ASQA also accessed recent data from the National Coronial Information System, identifying 132 horse-related fatalities between 1 July 2000 and 31 December 2013, an average of just under eight per year. This average is comparable to the findings of the SWA report, but both are less than half the average reported by Cripps [37]. Further, the SWA report points out that 34 horse-related deaths occurred in the 18-month period between 1 July 2012 and 31 December 2013, highlighting the problems which may arise from comparing averages rather than rates. Watt and Finch [45] comment on the difficulties in interpreting and comparing published data on injury in general, because of non-standardised data collection and/or analysis methods. They emphasise that such difficulties are exacerbated in analysis of sport-related injury because of lack of consensus on appropriate definitions. Such lack of consensus is also evident in the 2014 Safe Work Australia report [30].

2.2.2. Data Capture

This leads to a further challenge facing researchers in the area of general horse-related injury: the very wide scope of activities revolving around horses which militates against accurate coding of horse-related injury. "Horse-related activity" includes leading, grooming, feeding or just being with a horse; riding for fun, on a road or in a paddock; going to a competition on the weekend; mustering cattle or sheep; loading and unloading a horse from a float or truck; shoeing or trimming its hooves; training a horse for a specific purpose or competition, including dressage, eventing, show-jumping, endurance riding, polocrosse, driving, vaulting or camp-drafting; competing in one of these sports; horse racing; Western riding; breeding and raising horses; and so on. It is highly unlikely that any hospital injury coding system will capture all of these activities. The World Health Organization's International Classification of Diseases (ICD) is the most widely used hospital coding system, and in many Western countries, the only coding system in use. The version used in Australia, the ICD-10-AM, includes reasonably comprehensive activity codes, but they do not differentiate between sport and recreation activities, or between professional sport and backyard games [43]. Activity codes relating to sport and recreation are often missing in hospital separation records and as a result there is significant underestimation of the number of cases involved [44]. Even with recent revisions, which allow more codes to describe a specific mechanism and circumstance of injury, there is still consistent allocation of external causes and circumstances of injury to "other" and "unspecified," which may lead to considerable under-reporting [45]. The ICD-10-AM is not structured to permit the fine-grained coding protocols which will generate accurate information on which specific horse-related activity has resulted in the injury, precisely what injury mechanism has caused it, in which precise location the injury occurred, or which circumstances caused it. This means that it is difficult and indeed unusual for research to compare like with like, since often the specific type of horse-related activity cannot be accurately identified. Such difficulties are not unique to horse-related activities: similar challenges were identified by researchers seeking accurate case records for snow sport injury in New Zealand [51].

2.2.3. Non-Nuanced Research

Yet another challenge in accessing accurate and useful data is that researchers may themselves assess horse-related risk based on possibly inadvertent characterisation of horses, their size and their speed. Researchers may conflate data from the full range of horse-related activity, from recreational horse-riding to competitive equestrian sport, probably because of the difficulties outlined above in accessing specific activity-related data, and also because the researchers themselves may not be personally familiar with the range of horse characteristics and behaviour, nor of individual horse sports. Thus, for example, in a much-cited literature review of equestrian sport-related injury [15], horses are described as weighing "an average of 1500 lbs" (680 kg) and travelling "at up to 40 mph"

(64 kph). Given that horses come in all sizes, from the tiny (about 18 inches or 44 cm) to the enormous (about 82 inches or over 2 m) [52,53], average weight is not a particularly useful characterisation when measuring risk for riders.

Similarly, citing the top speed of horses when assessing risk is also not useful [15]. While horses *can* travel at up to 40 mph (64 kph) and more, this is unusual and cannot be sustained for any length of time. Such speed is restricted to specialist conditions such as thoroughbred sprint racing, quarter-horse racing, polo, and perhaps barrel racing, in which explosive bursts of speed are inherent, rather than recreational riding or competitive equestrian sport. For example, the fastest recorded race speed for a winner over 402 m (approximately a quarter of a mile) is 70.76 kph, over a period of 20.57 s [54], and the fastest recorded race speed for a winner of the Melbourne Cup, one of the world's premier 3200-m races, is 58.32 kph over 3 min and 16.3 s [55]. Citing the average speed from these two examples—64.54 kph—is not useful in measuring risk for riders, since neither recreational nor eventing riders would gallop at such speeds. Indeed, in either a recreational or a competitive sport setting, it is not so much the speed of the horse which contributes to the likelihood of a rider fall, but a sudden acceleration or deceleration, combined with surprise, which will result in a rider falling off (such as the horse breasting a fence and catapulting the rider forwards) [56]. In fact, slower horse speed in eventing may contribute to the risk of a rotational fall resulting in a horse landing on a rider [57].

3. Defining Risk in the Sport of Eventing

Given the challenges outlined above in accessing data on general horse-related human injury and fatality, it is not surprising that these challenges are greatly increased when one seeks data specifically related to eventing, and eventing-related injury and death.

3.1. Data on Eventing and Injury

One of the major issues for data collection in the sport of eventing is that unless a rider suffers a serious injury requiring hospital admission, medical attention at a trauma centre, or at a hospital emergency department, the injury will be unreported, and will not be captured in the current statistical collection processes.

Specifically in relation to eventing, while the possibility of catastrophic injury or death certainly exists, the great majority of injuries are minor [58]. In an Australian national surveillance project collecting data on rider and horse injuries in eventing from 2002 to 2006 [21], almost all riders who responded to questionnaires about injuries incurred in their cross-country falls characterised their injuries as minor, even while a few of these also reported that they sought medical attention later for serious injuries such as concussion and fractures. One rider responded in the negative to the first question which asked whether they had been injured in their recent fall, and then later in the questionnaire reported that he had later sought medical attention for concussion and broken ribs. Some reports of concussion and fractures were clearly self-diagnoses, since the respondents did not report that they had then gone on to seek medical attention. It is possible that this mischaracterisation of the serious nature of the riders' injuries is a reflection of the complex inter-species relationship which exists between horse and rider, in which eventing riders express more concern for their horses' safety than for their own [59]. In any case, the project did reveal a tendency among eventing riders to underestimate the risks involved in the sport, and to under-report any injuries. The surveillance project collected information on 1732 rider falls, in which 374 riders reported at least one injury, ranging from abrasions and bruises through dislocations to concussion and fractures. Because multiple responses were possible, it is not feasible to quantify the exact number of individual injuries. However, of the 1732 falls reported, only 129 were categorised as resulting in serious injury, being fracture, concussion or loss of consciousness. Given that only 900 of the possible 1732 riders who fell completed their questionnaires, it is likely that a similar proportion of riders who did not return their questionnaires were also injured.

3.2. Challenges for the Researcher

Data Capture

As mentioned above, hospital admission coding systems are not capable of capturing accurate data about general horse-related injury [43,44]. Data collection in the sport of eventing is further complicated by the fact that until relatively recently (2008), a rider who fell during the cross-country phase could remount and continue, even if they were injured [60,61]. Many records of eventing-related injuries were consequently lost to the event organisers and ultimately to data collectors, because they were not reported. In 2008 the rule was changed so that even a simple, non-injurious rider fall at a jump on the cross-country course would result in elimination, and in 2009 this was extended to any fall on course, jump related or not [62]. During the scope of the Australian national surveillance project [21], riders were able to remount and continue, and the project recorded several instances of riders who had incurred quite serious injuries such as concussion or fractures, which later required medical attention or even hospital admission, but who remounted and continued the competition.

Famously, two Australian eventing riders won Olympic gold medals while competing with serious injuries: Bill Roycroft fell during the cross-country phase at the Rome Olympic Games in 1960, suffering multiple injuries including a broken collarbone and severe concussion. He remounted and finished the course, and was then airlifted to hospital. The following day he checked himself out of hospital in order to compete in the show-jumping phase, as the Australian team was down to only two combinations, and needed three to qualify for the medal. He competed with his arm in a sling. Gillian Rolton fell when her horse slipped on grass at the Atlanta Olympic Games and she fractured her collarbone and some ribs. She remounted and continued, and fell again at a water complex as she was unable to use her left arm at all to control her horse. She again remounted and completed the course, successfully jumping another 15 obstacles, riding one-handed. She refused painkillers in case she was required to ride the following day in the show-jumping phase, but fortunately was not required, and was able to get the medical attention she needed. Her heroic gesture won her a second gold medal. Under the current rules, both these competitors would have been eliminated [9].

Even at the present time, it is difficult at the sport-based level to capture all injury records from the sport of eventing. While the FEI requires comprehensive injury reporting from their international event officials and uses these data for their annual reporting on injury rates and severity [58], the organisation concedes that in the past, injury reports were completed by fence judges, based on their own opinion of the severity of the injury incurred [58]. At a national level in Australia at least, there does not yet appear to be any consolidated data which researchers can access. The Australian peak national body, Equestrian Australia (EA), appointed a national safety officer after the FEI Eventing Safety Forum in 2008 [62], but a search of EA's website using the keywords "eventing safety," "risk management," and "eventing falls data" reveals no report from the national safety officer or any other report on the topic, other than a call for expressions of interest in the voluntary position [63]. In addition, in the days when a rider was able to remount and continue, there was strong motivation for riders to conceal injuries so that they could continue in the competition, particularly at the higher levels such as World Championships and Olympic Games. Bill Roycroft and Gillian Rolton both demonstrate that the prospect of an Olympic Gold Medal is a first-class painkiller.

4. Quantifying Risk in the Sport of Eventing

Given the background difficulties outlined above in accurate case identification and access to comparable data in the topic of horse-related injury in general, it is not surprising that there is a lack of consensus on how best to quantify and calculate risk in the sport of eventing.

There have been many different approaches to the identification and quantification of risk in eventing outlined in the literature, using many disparate denominators, but usually based on the assumption that rider injury is the unwanted outcome usually employed for calculating risk [12–15,21]. These approaches have most commonly included:

- Measuring the number of falls as a proportion of the number of participants, assuming that rider and horse/rider falls are the most common cause of rider injury. This process generates a fall incidence rate [21,58].
- Measuring the number of rider injuries as a proportion of the number of participants, again assuming that rider and horse/rider falls are the most common cause of injury. This process generates an injury incidence rate [12,21,58].
- Establishing surveillance and monitoring projects, assuming that more universal capture of data on participants engaged in the actual sport, with details of their actual experience, will provide an opportunity for more fine-grained analysis. Such projects can generate information on a number of factors, including demographic data, situational data, information about the horse/rider combination, and factors contributing to the fall, and usually involve on-ground data collection followed up by questionnaires or surveys [14,21]. This process can generate both fall and injury incidence rates.
- Calculating the number of injuries per hour spent in the activity. This process generates an injury rate per hour [12,14].

4.1. Fall Incidence Rate

This particular denominator, which measures the number of rider falls as a proportion of the number of competitors, has been used by the FEI since 2000, when the international body began collecting falls and injury data from all international-level events held under its aegis. This denominator is useful, in that the assessment of risk in eventing is based on the assumption that rider injuries occur mainly as a result of a rider fall, and so calculating the rate of rider falls can provide an insight into the likelihood of rider injury.

The most recent data from the FEI [58], reporting on all competition years from 2005 to 2014, reveal a total of 8556 rider falls from 152,821 starters, a rate of one fall for every 18 starters. These figures include riders whose fall did not involve a jump—*i.e.*, falls "on the flat" (n = 538).

The Australian national surveillance project [21] reported a rate of one fall for every 34 starters (1732 falls from 58,557 starters) and similarly included falls on the flat, which were not differentiated from falls at fences, because at that time falls on the flat were not penalised and as such were often not even recorded. The authors acknowledged that the figure of 1732 rider falls was very likely an underestimate because of problems with data capture.

These two data sources, widely different in scope and number, reveal a marked difference in rate, with the FEI's calculation being approximately twice that of the Australian national surveillance project.

4.2. Injury Incidence Rate

This denominator, which measures the number of rider injuries as a proportion of the number of competitors, was used in a yardstick analysis by Dr Bruce Paix, an Australian anaesthetist and trauma and recovery specialist who was personally involved in eventing and was the medical officer in attendance at 35 events held in South Australia between 1990 and 1998, about 10% of all such events (including Pony Club events) held in that state during that time. He published his findings [12] to considerable publicity. Using data from the events at which he officiated, involving 4220 competitors, with 37 injured riders, he first estimated the injury incidence rate per competitor at 0.88%. He then compared this incidence rate with other published injury incidence data from other sports, specifically a report which estimated the injury incidence rate for motor racing (both car and motorcycle) participants at England's Brands Hatch circuit as 0.24% per motorcycle racing competitor [64]. Paix consequently arrived at the conclusion that eventing was more dangerous than motorcycle racing, a conclusion which attracted a great deal of attention and one which has since been consistently cited in the injury literature (55 citations according to Google Scholar at the time of writing).

In relation to injury severity, Paix found that just over 70% of injured eventing riders (26 out of 37) were referred to hospital, with nearly half (12 out of 26) being admitted. If one were using hospital admission as a denominator for serious injury, this would demonstrate a rate of serious injury: 32% of all injured competitors, and 0.28% per competitor.

The FEI also reports annually [58] on the injury incidence rate per competitor, arriving at a rate of 0.65% per competitor, then further broken down by level of injury rather than level of competition. The FEI combines "serious" and "fatal" injuries when reporting on the numbers of riders injured, and does not define "serious" injury. The 2015 report reveals 311 seriously or fatally injured riders in the period 2005 to 2014, from a total of 8556 falls from 152,821 competitors, a considerably larger dataset than that available to Paix (4220 competitors) [12]. The total number of riders injured (slight and serious/fatal) was 978 for the period, revealing that 32% of injured riders were seriously or fatally injured, identical to Paix's results [12].

The Australian national surveillance project [21] calculated an injury rate of 0.63% per competitor (58,557 competitors, 374 riders reporting at least one injury), almost the same as the overall rate reported by the FEI. This project also reported 31% with serious injuries such as fractures or concussion (n = 119), with only 23 riders reporting that they were admitted to hospital.

These three data sources show remarkably similar results for the number of seriously injured riders as a proportion of all injured riders, with two [21,58] also showing a similar injury incidence rate per competitor (0.65% and 0.63%, respectively). Paix's finding of an injury incidence rate of 0.88% is noticeably higher. This injury incidence rate denominator is particularly useful, as it can further illuminate the risk of a rider injury in eventing, and is also capable of identifying the risk of serious injury.

4.3. Surveillance and Monitoring Projects

This method collects all relevant records within a given time period, and might include the number and level of competitions, the total number of competitors, the number of competitors at each level, the number of reported injuries and/or fatalities, and in the case of eventing, the number of rider and horse falls and the number of jumping efforts in the competition. From such data, an incidence rate can be calculated, again using a number of different denominators such as falls per number of starters and injuries per number of falls. Frequently, surveillance projects combine on-site competition data collection with follow-up surveys and questionnaires.

The Australian national surveillance project [21] was one such surveillance project. As mentioned previously, there were some difficulties in data capture and although there was a high return rate for questionnaires sent to riders known to have had a fall cross-country, overall it was understood that the known number of rider and horse falls was considerably lower than those which actually occurred. However, on the basis of the recorded 1,732 rider and horse/rider falls it was possible to estimate a rate of rider falls at 3 per 100 starters over the 5-year period, 1.2 rider falls per 1000 jumping efforts, and an injury rate of 0.2 per 1000 jumping efforts.

The study by Ekberg and colleagues [14] is a similar national surveillance and monitoring study, capturing injury information from all members of the Swedish Equestrian Federation with eventing as their primary discipline (n = 513), in a one-year retrospective study. The survey attracted an eventual return rate of 70% (n = 357), collecting information on all traumatic injury events, whether in competition or training. The authors reported that 62.8% of injury events occurred during training, and 37.2% during competition. The study sought to capture the incidence of traumatic injury events measured against hours of activity (see 4.4. below), and did not report on injuries as a proportion of the number of starters.

4.4. Calculating Injury Rate against Time Spent "in the Saddle"

Paix [12] uses measurement of time spent "in the saddle" to calculate various levels of risk for horse riders. Specifically for eventing, he uses the average time taken to complete a cross-country course (7.5 min) to calculate an overall injury rate of one per 14 h of cross-country riding in competition, and extrapolates this to calculate that the cross-country phase of eventing is over 70 times more dangerous than horse-riding in general. He further estimated that for riders at "the highest level" (inferred as equivalent in difficulty to Olympic or World Championship competition), their rate of one injury per 5.5 h (and at an incidence of 2.2% per competitor per event) is over 180 times that for all forms of horse-riding combined.

Ekberg and colleagues [14] also calculated the incidence of traumatic injury events using the number of injury events reported (n = 143), divided by the total person-time at risk, calculating such incidence as 0.54 injury events/1000 eventing hours for novice riders, and 0.35 injury events/1000 eventing hours for qualified riders.

Extensive searches revealed only these two studies which use the specific denominator of "time in the saddle" and so it is difficult to comment on the usefulness of such a denominator. It is certainly worth further exploration, as it is probable that time in the saddle contributes to overall risk in the sport, simply through exposure over time.

All these various approaches assist in illuminating the scope of risk for riders in a sport already well recognised for being "dangerous" [8,12,17]. From these different approaches, it is possible to identify a range of different ways in which to calculate risk for riders in eventing, and these are difficult to assess against each other, employing as they do completely different denominators. There are implications for equestrian sports' governing bodies whose responsibilities include making the sport as safe as possible, since such studies individually may not prove overly helpful in formulating an overall risk assessment, risk management and risk reduction strategy. Furthermore, labelling a sport as "more dangerous" or "the most dangerous" may have considerable and unintended impacts on participation, and so it is important that any such labels are applied only when the data support them. As outlined above, much of the published research on risk in eventing uses different denominators for calculation, and often compares eventing with other sports which are quite different. For example, if injury risk in eventing is associated primarily with the horse or the rider falling [12,21,22], and that risk is concentrated around the jump itself, and the action of the horse jumping [12,21,22,24,57,58,65], then risk cannot be said to be equally distributed around the cross-country course. In motorcycle racing, for example, risk of injury from falling off the motorcycle would seem to be more equally distributed around the whole racecourse since motorcycle racing involves one continuous action (going as fast as possible) around a relatively uniform track. On the face of it, motorcycle racers are at risk of falling and injuring themselves at almost every minute of a race. Another different aspect of eventing is that the course itself is often over undulating ground, with twists and turns, and frequent changes of pace as riders approach a jump, negotiate the jump, and then cover the ground to the next jump. Thus, a comparison between eventing and motorcycle racing does not seem appropriate in calculating risk for participants.

Table 1 summarises the findings from a very small sample of published work on the risks in eventing, showing the different denominators used in the calculations. It demonstrates the difficulties facing researchers and policy makers who may wish to identify a useful and standard baseline (*i.e.*, the same unit of measurement) for calculation of risk, when no such baseline exists.

Table 1. Reports of Fall and injury rates in eventing, 1990–2014.

Denominator	**FEI [58] (2005–2014) n = 152,821**	**O'Brien and Cripps [21] (2002–2006) n = 58,557**	**Paix [12] (1990–1998) n = 4220**	**Ekberg *et al.* [14] (2007) n = 357**
Fall rate per number of starters	1/18 starters	1/34 starters	Not reported	Not reported
Injury rate per number of starters	Slight injury: 1/229 Serious injury: 1/506 Fatal: 1/16,980	Overall rate: 1/156 Serious injury: not reported Fatal injury: not reported	Not reported	Not reported
Injury incidence per starter	Slight and serious injury: 0.65% Fatal: 0.00061%	Overall rate: 0.63% Serious and fatal: Not reported	Overall rate: 0.88% Serious injury: 0.28% Fatal: not reported	Not reported
Injury rate per eventing hour	Not reported	Not reported	1/14 eventing hours overall, 1/5.5 eventing hours "at the highest level" of competition	0.54 injury events/1,000 eventing hours (novice riders) 0.35 injury events/1,000 eventing hours (qualified riders)
Fall rate per number of jumping efforts	Not reported	1.2 falls/1,000 jumping efforts	Not reported	Not reported
Injury rate per number of jumping efforts	Serious injury: 1/15,687 jumping efforts	Overall: 0.2/1,000 jumping efforts	Not reported	Not reported

5. Where Is the Locus of Risk?

While acknowledging that all these approaches can provide valuable information, it may be more useful to explore where the *locus* of risk is in eventing.

Most observers agree that the primary locus of risk in eventing is during the cross-country phase of the sport, where speed and jumping combine, rather than in the dressage or show jumping phases [1,10,57,65]. But this is too general a field from which to identify a precise locus of risk. Multiple factors may combine to produce a fall event which results in a rider's injury or death: the individual experience of the rider and the horse, and their experience as a combination; the fitness of the rider and the horse; the weather; the light; the ground conditions; the approach to the jump; the capacity of the rider and the horse to judge distance accurately; the skill of the course designer and the jumps builder; and sometimes just plain luck. In a single accident event, it will always be very difficult to exclude any of these contributing factors, or indeed to accurately measure the contribution of each variable and their interactions (clustered data). Sophisticated multilevel statistical analysis such as hierarchical linear modelling is needed in clustered data, and is increasingly being used in analysis of other "risky" sports such as jumps racing [66–68].

However, what almost all rider and horse falls in eventing have in common are: the jump itself, the action of the horse jumping, the consequences of the horse jumping, or its failure to jump. Paix's much-cited article [12] states that "most of" the 37 rider injuries which occurred during the period of study occurred as a result of the rider falling off the horse, or from horse and rider both falling, mostly while jumping an obstacle. No precise numbers are reported, however. The FEI reports that of the 8556 rider falls recorded between 2005 and 2014, 94% of them occurred at a jump [58]. A study by Stachurska and colleagues [69], in identifying risk factors associated with falls cross-country, also focused on the jumps themselves, pointing to factors such as successive elements of combinations, narrow jumps, brush-type jumps, and jumps with alternative routes, across all levels of competition, from novice (one star) to Olympic and World Championship (four star) level. Other studies [57,65] have identified additional jump-related risks associated with jumps with a drop landing and jumps with approaches out of water, as well as riders knowing they were in the lead after dressage, and riders who received tuition. The most recent research commissioned by the FEI has determined that other jump-related factors increase the likelihood of a horse fall: corner fences, square spreads, upright post and rails, jumps into or out of water, downhill terrain, and some combinations of these factors [70].

Mechanically, falls of riders and horses in eventing are usually the result of a sudden loss of the forward momentum of the horse and the continuation of the forward momentum of the rider [71,72]. As mentioned previously, horses can fall "on the flat" during eventing, but compared to the proportion of falls which occur at jumps, falls on the flat do not present the same degree of "danger" for riders [58]. There are at least six scenarios for a fall of a rider during a cross-country course, including:

- away from an obstacle (that is, "on the flat"): the horse slips and falls and the rider continues onwards or sideways, or—rarely—lands under the falling horse;
- the horse refuses to jump, stopping in front of the obstacle, and the rider continues onwards or sideways;
- the horse attempts to clear the jump but hits it, the horse's forward momentum ceases abruptly and the rider again continues onwards;
- the horse fails to negotiate the jump successfully and the rider is caught up in the ensuing chaos;
- the horse stumbles or falls on landing after jumping, and the rider again continues onwards or is trapped underneath the fallen horse; or
- the horse hits the jump with its chest or front legs, having failed to clear the jump, and its forward momentum carries its body onwards and forwards as it somersaults over the jump. The rider, propelled ahead of the horse's momentum, lands in in the vicinity of the place where the horse is itself going to land shortly thereafter.

The author, with help from the aforementioned Dr Bruce Paix, has identified and collected information on 59 confirmed rider deaths in eventing since 1993, across all levels of the sport, from Pony Club to Regional Championship [24] (see Appendix Table A1). The collection began in conjunction with the Australian national surveillance project [21], and used search phrases and terms including "rider death eventing," "cross-country death," "eventing deaths," and "rider deaths." The data collection is based on online articles from recognised newspapers and magazines (see, for example, [18,19]), often followed up by on-line newspaper reports of coronial findings. The fatalities involved 25 males and 34 females ranging in age from 12 to 64, with a median age of 32. There were fatalities at 15 international (FEI)-level competitions, 34 national-level competitions and three Pony Club competitions. In seven cases, there was insufficient information to accurately determine whether the competition was at the international or national level. In 32 cases, the number of the jump is identified, with a range of jump number 2 to jump number 26, with a median of jump number 10. However, there are 25 cases in which the jump number is either unknown or is not adequately identified (for example, "last," "second last," "halfway through"). Since the number of jumps and jumping efforts in any one course may vary depending on the level of competition, and within the maximum and minimum number of jumping efforts required at that level of competition [9], it is not possible to extrapolate any meaningful information from these data.

Figure 1 shows the total number of known rider deaths between 1993 and 2015 (n = 59), and those known to have been the result of a rotational horse fall (n = 41). In eight cases, there is no verifiable information on whether the horse fall was rotational or not.

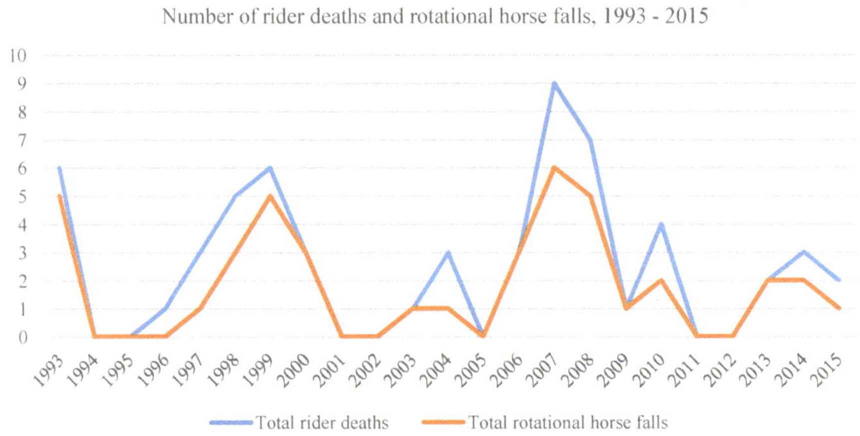

Figure 1. Rotational horse falls and rider fatalities 1993 to 2015.

While risk in eventing is unevenly distributed, there is a very strong demonstrated bias towards jumping efforts (that is, the number of times a horse is required to jump an obstacle in any one eventing course) [21,58]. It would suggest, therefore, that the primary locus of risk is the jump itself, and the action of the horse jumping, and that, as the Hartington Report stated, the best chance of reducing risk for riders and horses in eventing consists of reducing the chance of the horse falling [22]. Since the horse is most likely to fall at a jump, this suggests that the greatest risk occurs at the jump. In this case, the focus of future research should be on this locus of risk, and its relationship with the other factors which contribute to risk in the sport. Calculations of risk in eventing should concentrate first on measuring the number of rider and horse/rider falls and the number of rider injuries in relation to the number of jumping efforts performed by the horses.

The FEI has led the way in developing changes to the infrastructure of the sport in order to reduce risk for riders and horses, and these changes revolve primarily around the jump, its construction, and the rules governing how it can be jumped [73]. First, after the sudden cluster of six rider deaths in 1999, five of them resulting from rotational horse falls, the FEI immediately changed the rules governing how a rider might re-present a horse to a jump if it had refused, or not approached the jump correctly. Until this time, as long as the horse was not deemed to have stepped backwards, the stop was not considered a refusal, and therefore incurred no penalties. The rider could simply ask the horse to jump from a virtual standstill after an initial failure to jump, greatly increasing the likelihood of the horse breasting the fence because of lack of momentum, dislodging the rider forwards, and then somersaulting over the jump onto the rider on the other side. The rule changes now require a rider to circle away and attempt the jump again, incurring penalties, or opt to take a longer, slower option with alternative obstacles, which will also incur penalties. Secondly, the FEI has overseen and financed the development and introduction of significant technical interventions in jumps construction, with the invention of so-called "frangible pins," a system which deconstructs the jump when it is subjected to the force exerted by a horse colliding with it (*i.e.*, at the start of what might otherwise be a rotational horse fall). Interestingly, the most recent research from the FEI has found an increased risk for horses falling at fences with frangible pins [70], definitely an unforeseen consequence of a research-based safety innovation, and clearly a focus for further research. The FEI has also supported research and development in the areas of helmet manufacture and body protectors, ensuring that comparable safety standards across all jurisdictions could allow the FEI to mandate the use of particular helmets and body protectors [74]. Thirdly, there have also been changes to the qualification requirements for riders and horses, with the goal of ensuring higher levels of skill before higher levels of competition [73]. In addition, the FEI's comprehensive data collection from all international and national-level events [58] has provided a base upon which to assess risk with a greater degree of accuracy than simply counting numbers of rider falls and injuries.

6. Conclusions

As outlined previously, a multitude of factors interact to result in a fall of rider and/or horse in the sport of eventing, and the risk of such falls is neither evenly distributed nor constant. However, given that more than 90% of these falls happen at the jump itself [58], future research should then focus appropriately. The FEI is already undertaking a major research project on the factors which contribute to falls and injuries, first by examining the role played by specific fence types in horse falls [70], and then concentrating on rider and horse qualifications. However, this paper's contention that the primary locus of risk is the jump itself should be further tested by analysis of larger datasets than those held in the current FEI database, which includes only data on international-level events. Systems should be established to ensure accurate recording of the circumstances surrounding all falls of riders and/or horses at all competition levels, in every eventing country in the world, including information on as many variables as possible. Such data will provide the basis for future multilevel analyses which may help unravel the many variables at play in such a complex set of clustered data.

Eventing is an expensive sport for organisers. If further research can support the contention that the primary locus of risk is the jump itself, then this will encourage organisers to concentrate their limited resources on jump-related interventions in the first place, while not ignoring other interventions.

Other avenues for future research include determining what is an acceptable level of risk in the sport of eventing, and for which group such a level of risk is acceptable. What is an acceptable risk level for the riders? For the horses? For the public? For the organisers? Each group will almost certainly give a different assessment, and these assessments must play a role in arriving at an overall assessment of risk, and an acknowledgement of a degree of acceptance of risk. The acceptability of risk within a specific sport is dependent on the perceptions of the participants [31] and on those of the observers as well. As far as the riders are concerned, one research study [59] suggests that eventers are generally more concerned about injury to their horses than to themselves, and that riders identify the horse itself as the source of risk, at the same time expressing the view that an effective partnership between horse and rider is the best form of risk mitigation. Thus the horse is represented as "both source and saviour of risk" [59].

This paper has not dealt at all with the risks for horses in eventing, but at least 74 horses are known to have died in competition or immediately afterwards, since June 2005, an average of more than seven a year [23]. The fate of eventing horses is not well documented, as their deaths have been rarely reported until relatively recently, and they may very well sustain an injury which requires euthanasia away from the competition. No central record system exists for eventing horses, other than their competition record. In the current climate of strong public antipathy to the perceived cruelty to animals [75–78], combined with ambivalence about the use of animals for entertainment [61,79,80], the issue of horse fatalities in the sport of eventing will undoubtedly soon attract public attention, as it has in the sport of jumps racing [75,76,78]. The unpublished reports of Harrison and colleagues [67,68] found a total of 113 horse deaths in jumps racing in the Australian State of Victoria between 1995 and 2005, an average of 10 a year. The recent report [78] on Australian jumps racing, conducted now only in Victoria and South Australia, reveals 10 horse deaths in jumps racing between 2012 and 2014, an average of five a year. The fate of thoroughbred racehorses, whether flat racers or jumps racers, has until very recently also been unknown once the horse leaves the racetrack for the last time. In 2015, Racing Victoria Ltd changed its rules to require trainers to notify them of the death of a horse in training [78]. No such requirement currently exists for eventing horses. Accurate data on horse injuries and deaths in the sport of eventing might go some way to addressing concerns about the lives of eventing horses after competition.

Eventing can never be totally risk-free for riders or horses. Risk is inherent to the activity, and those who engage in the sport and those who manage it know and understand this. If risk assessment serves to reduce and minimise risk, then any calculation of risk in eventing in comparison with other activities should first be based on comparing like with like. If risk in eventing is unevenly distributed and not constant, then using comparisons with other activities in which risk is both evenly distributed and relatively constant—such as motorcycle racing—is not likely to throw light on the topic. Although more research needs to be done on the topic, risk calculation based on the number of rider injuries and deaths as a proportion of the number of jumping efforts would indicate that eventing is unlikely to be as dangerous as motorcycle racing. The focus of risk calculation in eventing should be first on the jump, and the action of the horse jumping, and on preventing horses from falling at jumps. It is by focusing on the locus of risk—the jump—that eventing will have a sporting chance.

Acknowledgments: The author acknowledges the support of Associate Professor Kirrilly Thompson, CQU, in the development of this paper, and thanks Bruce Paix for providing additional data on rider deaths. She also thanks Holly Bowen for formatting the paper for submission.

Conflicts of Interest: The author declares no conflict of interest.

Appendix 1.

Table A1. Fatalities in eventing, 1993—2015.

Date	Name	Age	M/F	Location	Injury Mechanism	Level	Rotational Horse Fall	Circumstances	Fence #and Type
29/05/1993	Richard Adams	23	M	England (Windsor)	Fall	NAT Advanced 3DE	Yes	Horse hit fence then landed on rider (source: internet)	#14 table
5/06/1993	Malcolm Munro-Kerr	42	M	England (Lowesby)	Hit head on ground	NAT Preliminary	Not reported	Fell. No other information (source: internet)	#19 third element, post and rails
6/06/1993	Verna Hiltebeitel	24	F	USA (Douglassville)	Crush	Not reported	Yes	Horse somersaulted, landed on rider (source: Bruce Paix (BP), internet)	Not reported
16/06/1993	Mark Holliday	23	M	England (Hexham)	Crush	Not reported	Yes	Horse somersaulted, landed on rider (source: internet)	Not reported
18/08/1993	Vanessa Weaver	40	F	England (Tythrop)	Crush	Not reported	Yes	Horse's legs became entangled in rails, rider fell, horse somersaulted. landed on rider (source: internet)	Sloping rails
3/10/1993	Karen Smart	28	F	England (Asgarsby)	Crush	*Pony Club* Hunter Trials	Yes	Horse refused at jump, somersaulted over on top of rider, landed on top of her (source: internet).	#15 Upright of telegraph poles
22/04/1996	Natalie Dunlop	14	F	Australia (Penola)	Head *vs.* tree	NAT Grade III	No	Horse bolted, rider thrown at high speed into tree (Source: BP; internet)	Not reported. Possibly not fence related
18/05/1997	Anna Savage	36	F	Australia (Naracoorte)	Blunt trauma	NAT Open I/M	No	Horse stumbled in ditch, not fence-related ("near end of course"). (Source: BP; personal information)	*Not fence related*, near end of course
6/09/1997	Sam Moore	35	M	England (Blenheim)	Crush	FEI CCI ***	Yes	Horse lacking energy. Horse hit fence (keyhole), rider catapulted, horse somersaulted, landed on rider's chest (source: internet)	#19 keyhole
20/09/1997	Amanda Warrington	29	F	USA (Fair Hill)	Blunt trauma	FEI CCI ***	No	Fell off, major head injuries. Unconscious for several months, died *28/12/1997* (source: internet)	Table
14/03/1998	Linda Riddle	37	F	USA (Trojan)	Not reported	Not reported	Yes	Referred to in Internet forum: "Linda Riddle died when her horse misjudged the bounce . . . competing advanced." Horse hit the second element and the rider flew over the fence, the horse then jumped on the rider. (source: BP; internet)	Bounce, rails and ditch
22/03/1998	Tasha Khouzam	18	F	Australia (Jupiter Creek)	Crush	NAT Novice	Yes	Horse hit fence, somersaulted over on top of rider (source: BP; internet)	Log fence, 4′ wide

Table A1. *Cont.*

Date	Name	Age	M/F	Location	Injury Mechanism	Level	Rotational Horse Fall	Circumstances	Fence #and Type
13/04/1998	David Foster	43	M	Ireland (Rathmolyon)	Crush	NAT Advanced	Yes	Horse somersaulted over on top of rider at water jump (source: internet)	Not reported
20/06/1998	Keith Taylor	32	M	USA (Radnor)	Crush	International' (unconfirmed) Level not reported	Not reported	Horse made an error in jumping and fell, landing on rider. (source: BP, internet)	Coffin/ditch
11/10/1998	Roberta Scioscia	43	F	USA (Tryon)	Unknown	NAT Preliminary	Not reported	Death notice in *NYTimes* (online), 11/10/1998. (source: internet)	# 3, Log,
1999 †	Ken Machette	58	M	USA (Trojan)	Crush	NAT Preliminary	Yes	The horse jumped from a standstill, flipped and landed on rider. "Broke Ken in half." (source: BP, internet)	Log drop into water
15/05/1999	Peta Beckett	33	F	England (Savernake)	Blunt, crush	NAT Novice	Yes	Horse hit fence (bounce, offset rails), somersaulted over on top of rider (source: internet)	#7 & 8 double (bounce), rails
27/06/1999	Robert Slade	30	M	England (Wilton Park)	Crush	NAT Intermediate	Yes	Attempted bounce fence with insufficient impulsion, horse fell on rider (source: BP, internet)	#17 double (bounce)
22/08/1999	Polly Phillips	30	F	Scotland (Thirlestane Castle)	Crush	International" (unconfirmed) Scottish Champion-ships	Yes	Horse hit fence hard, somersaulted over on top of rider (source: BP, internet)	#10 Oxer
4/09/1999	Simon Long	38	M	England (Burghley)	Crush	FEI CCI ****	Yes	Horse refused at direct element (water), was immediately turned to attempt the RH option, hit fence with both front legs and somersaulted onto rider. (Source: BP, internet)	#20 jump out of water
26/09/1999	Peter McLean	20	M	England (Somerleyton)	Crush	NAT Intermediate	No	Parallel rail fence, horse stumbled on landing, fell on rider (source: internet)	Oxer
9/04/2000	Mark Myers	32	M	Australia (Fig Tree Pocket)	Crush	NAT Intermediate	Yes	Horse hit second element of bounce, rail into water. Horse somersaulted over onto rider, rider already dead when medical team arrived. (source: internet)	#10 (approx)
16/04/2000	Rhonda Mason	35	F	Australia (Camperdown)	Crush	NAT Pre-novice	Yes	"Horse was on the forehand, and galloping downhill, and just didn't see the fence because its head was too low" (eyewitness, Internet forum). Horse hit fence, stumbled on landing and rolled onto rider (source: internet)	#15 4-rail roost
30/04/2000	Jemima Johnson	38	F	England (Wilmslow)	Crush	NAT Advanced	Yes	Horse failed to make the height, hit log pile on the rise, somersaulting over onto rider (source: internet)	Not reported

Table A1. *Cont.*

Date	Name	Age	M/F	Location	Injury Mechanism	Level	Rotational Horse Fall	Circumstances	Fence #and Type
25/08/2003	Samantha Hudson	33	F	England (Moreton-in-Marsh)	Crush	NAT Novice	Yes	Horse hesitated, hit fence (log over ditch) with chest, somersaulting over onto rider (source: internet)	#14
24/07/2004	Cindy Burge	41	F	USA (Rebecca Farm)	Crush	NAT Level unknown	No	Not fence related. Horse slipped on turn, rolled onto rider. *Died 28/7/04* (source: internet)	*Not fence related*
29/08/2004	William Booth	51	M	USA (Senator Bell Farm)	Blunt trauma	NAT Level unknown	No	Horse refused at stone wall, throwing rider into wall (source: internet)	Not reported
5/09/2004	Caroline Pratt	41	F	England (Burghley)	Crush (in water)	FEI CCI ****	Yes	Horse clipped jetty, somersaulted over onto rider in shallow water (source: internet)	#26 (third last)
21/08/2006	Sherelle Duke	28	F	England (Brockenhurst Park)	Crush	NAT Advanced	Yes	Horse hit second element of double, somersaulting over onto rider (source: internet)	Not reported
23/11/2006	Mia Erikkson	17	F	USA (Galway Downs)	Crush (in water)	FEI CCI **	Yes	Slow rotational fall. Horse fell on her after rotational fall. *Sister Shana died in riding accident in 2003* (source: internet)	#19
8/12/2006	Kim Hyung-Chil	47	M	Qatar (Doha)	Crush	FEI CCI *	Yes	Horse caught front legs on upper part of jump, somersaulted over onto rider (source: internet, YouTube)	#8
17/02/2007	Amanda Bader	32	F	USA (Ocala)	Blunt trauma	NAT Prelim	No	Fell from her horse, sustained head injury. *Died 28/2/07* (source: internet)	Not reported
14/03/2007	Amelie Cohen	30	F	France (Fontainbleu)	Crush	NAT Novice	Yes	Horse somersaulted at fence and landed on top of rider (source: internet)	#7
18/04/2007	Jo-Anne Williams	34	F	England (Sapey)	Blunt trauma	NAT Novice	No	Rider sustained serious head injury when her horse hit the top of a bench jump (source: internet)	#8
8/05/2007	Julie Silly	17	F	France (Jardy)	Crush	NAT Novice	Yes	Horse fell heavily on rider after rotational fall at "house fence" (source: internet)	#3b
21/07/2007	Elin Stalberg	19	F	Sweden (Bollnas)	Unknown	NAT Intermediate	Yes	Horse somersaulted over fence (corner fence), landed on rider (source: internet)	Third last
31/07/2007	Anke Wolf	40	F	Germany (Neu Wulmstorf)	Crush	NAT Novice	Yes	Horse somersaulted over last fence, landed on top of rider (source: internet)	Last
4/08/2007	Tina Richter-Vietor	32	F	Germany (Schenefeld)	Blunt trauma	FEI CIC **	No	Horse fell after catching a leg in fence, rider thrown right, horse fell left (source: internet)	#2
2/09/2007	Maia Boutanos	29	F	France (Moulin)	Crush	NAT Novice	Yes	Portable fence moved on impact, fell over, causing horse to fall on the rider (source: internet)	#5b

Table A1. *Cont.*

Date	Name	Age	M/F	Location	Injury Mechanism	Level	Rotational Horse Fall	Circumstances	Fence #and Type
19/11/2007	Eleanor Brennan	21	F	USA (Ocala)	Crush	FEI CCI **	Yes	Horse struck fence (table), rolled over onto its neck, trapping rider. Horse died instantly (source: internet)	#18
29/01/2008	Shannon Bloomfield	12	F	England (Malt Hill Farm)	Crush	*Pony Club*	Yes	Horse clipped the jump, and landed on the child, kicking her in the head as it got to its feet (source: internet).	Not reported
6/04/2008	Franz Graf	62	M	Austria (Aspang)	Crush	NAT Prelim	Yes	Horse hung a foreleg at fence (table) and somersaulted over onto rider (source: internet).	Last
20/04/2008	Karen Rodgers	41	F	Ireland (Ballindenisk)	Crush	NAT Advanced	Yes	Horse fell at fence (water) and landed on rider (source: internet).	Second last
9/08/2008	Emma Jonathan	23	F	England (Hartpury)	Crush	FEI CCI **	Yes	Horse struck fence and flipped vertically, landing on rider (source: internet)	#19
30/09/2008	Debbie Atkinson	unknown	F	USA (Kentucky)	Not reported	NAT Intermediate	Not reported	'Fell at an intermediate cross-country fence on *2/9/2006*. Rider died 2 years later, *30/9/2008* (source: internet)	Not reported
15/08/2008	Stephen Moore	45	M	Ireland (Antrim)	Not reported	Unknown	Not reported	Fell at event *2 August 2008*, suffered injuries, discharged from hospital, died of unknown causes. *Twin brother Sam died at Blenheim in 1997* (source: internet)	Not reported
30/10/2008	Jade South	15	F	England (Redmarley)	Crush	*Pony Club*	Yes	Horse hesitated at fence, rider pushed on. Horse hit portable fence with front legs, fence lifted 10" off the ground, horse somersaulted over fence with rider still in the saddle. Landed on top of rider (source: internet).	Not reported
26/04/2009	Ian Olding	47	M	England (Belton)	Crush	NAT Advanced	Yes	Rotational fall at table element of combination, horse landed on rider (source: internet)	#13, halfway
18/03/2010	Dirk Grouwels	48	M	Belgium (Zutendaal)	Probably crush, as rider was killed instantly after horse somersaulted.	NAT Possibly unaffiliated event	Yes	Horse refused at first part of combination, rider insisted and the horse somersaulted over the fence and landed on rider (source: internet).	Not reported
24/05/2010	Elena Timonona	15	F	Russia (Krasny Voskhod)	Crush	NAT CNC *	Yes	Rotational fall at an upright fence. YouTube video shows horse failed to lift front legs, went straight over the fence and landed fully on rider. Rider does not move after fall. Horse trotted away (source: YouTube, internet)	#16

Table A1. *Cont.*

Date	Name	Age	M/F	Location	Injury Mechanism	Level	Rotational Horse Fall	Circumstances	Fence #and Type
14/09/2010	Robin Donaldson	64	M	England (Beckwithshaw)	Landed on his head	NAT BE100 (pre-novice)	No	Horse straddled a fence and rider fell over her neck and landed on his head. Fell on *21 August 2010*, died *14 September 2010* (source: internet).	Not reported
18/09/2010	Sebastian Steiner	22	M	Italy (Montelibretti)	Not reported	FEI CCI ***	Not reported	No information available	10C (or 11)
2/03/2013	Bruno Bouvier	60	M	Portugal (Barocca d'Alva)	Not reported	FEI CIC **	Yes	Rotational fall at fence. Horse trod on its own shoe and tripped into brush arrowhead fence, landed on rider (source: internet)	#5 arrowhead
19/08/2013	Tom Gadsby	26	M	UK (Somerford Park)	Crush	FEI CIC *	Yes	Rotational horse fall at fence, table type, landed on rider (source: internet)	#4B table
14/06/2014	Jordan McDonald	30	M	England (Nunney)	Crush	NAT Novice	Yes	Rotational horse fall, landed on rider "The horse struck the jump with its front legs and slowly rotated across the jump and on to the other side, throwing its rider, Jordan McDonald, to the ground and landing full square on top of him." (Coroner's report) (source: horseandhound.co.uk).Coroner found that body protector did not meet British Eventing's standard (source: internet).	#7a
14/06/2014	Benjamin Winter	25	M	Germany (Luhmühlen)	Crush	FEI CCI ****	Yes	Rotational horse fall at fence 20 (second ride in class), landed on rider (source: internet)	#20 Table
12/11/2014	Guillaume Pucci	28	M	France (Le Pouget)	Fell from horse	FEI CIC **	Not reported	Fell *12/11/2014, Died 14/6/2015, after being in a medically induced coma since the fall.*(source: horsetalk.co.nz)	#8
14/02/2015	Francisco Seabra	30	M	Spain (Seville)	Not reported	FEI CIC **	Not reported	Horse and rider fall (source: internet)	#10C 3rd element of water complex
12/04/2015	Sabrina Manganaro	25	F	Italy (Cuceglio)	Fall when horse fell. Possible kick from horse.	NAT Level 3 (equivalent to Preliminary)	Yes	Rotational horse fall. Horse apparently underestimated the size of the fence and fell (source: internet)	#13 "Skinny" sloping log fence

Notes: † Exact date unknown; *, **, ***, **** The number of stars at FEI level events signifies degree of difficulty, with * being the lowest and **** being the highest.

References

1. Wipper, A. The partnership: The horse-rider relationship in eventing. *Symb. Interact.* **2000**, *23*, 47–72. [CrossRef]
2. Mills, D.S.; McDonnell, S.M. *The Domestic Horse: The Origins, Development and Management of Its Behaviour*; Cambridge University Press: Cambridge, UK, 2005; pp. 5–7.
3. Kalof, A. *Looking at Animals in Human History*; Reaktion: London, UK, 2007; pp. 17–22.
4. Christensen, J.W.; Keeling, L.J.; Nielsen, B.L. Responses of horses to novel visual, olfactory and auditory stimuli. *Appl. Anim. Behav. Sci.* **2005**, *93*, 53–65. [CrossRef]
5. Robinson, I.H. The human-horse relationship: how much do we know? *Equine Vet. J.* **1999**, *31*, 42–45. [CrossRef]
6. Brandt, K. A language of their own: An interactionist approach to human-horse communication. *Equine Vet. J.* **2007**, *39*, 422–428. [CrossRef]
7. Bixby-Hammett, D.M.; Brooks, W.H. Neurologic Injuries in Equestrian sports. In *Sports Neurology*; Jordan, B.D., Tsairas, P., Warren, R.F., Eds.; Rockville MD: Aspen, CO, USA, 1989; pp. 229–234.
8. Wolframm, I. *The Science of Equestrian Sports. Theory, Practice and Performance*; Routledge: Oxon, UK, 2014; pp. 119–1219.
9. Fédération Equestre International. About Eventing. 2015. Available online: http://www.fei.org/fei/disc/eventing/about (accessed on 20 May 2015).
10. United States Eventing Association. What Is Eventing? 2010. Available online: http://www.discovereventing.com/?q=node/122 (accessed on 27 November 2105).
11. United States Eventing Association. Discover Eventing. 2010. Available online: http://www.discovereventing.com/?q=node/67 (accessed on 27 November 2015).
12. Paix, B.R. Rider injury rates and emergency medical services at equestrian events. *Br. J. Sport. Med.* **1999**, *33*, 46–48. [CrossRef]
13. Whitlock, M.R. Injuries to riders in the cross-country phase of eventing: The importance of protective equipment. *Br. J. Sport. Med.* **1999**, *33*, 212–214. [CrossRef]
14. Ekberg, T.; Timpka, T.; Ramel, H.; Valter, L. Injury rates and risk-factors associated with eventing: A total cohort study of injury events among adult Swedish eventing athletes. *Int. J. Inj. Contr. Saf. Promot.* **2011**, *18*, 261–267. [CrossRef] [PubMed]
15. Havlik, H.S. Equestrian sports-related injuries: A review of current literature. *Curr. Sports Med. Rep.* **2010**, *9*, 299–302. [CrossRef] [PubMed]
16. Silver, J.R. Spinal injuries resulting from horse-riding accidents. *Spinal Cord* **2002**, *40*, 264–271. [CrossRef] [PubMed]
17. Thompson, K.; McGreevy, P.; McManus, P. A critical review of horse-related risk: A research agenda for safer mounts, riders and equestrian cultures. *Animals* **2015**, *5*, 561–575. [CrossRef] [PubMed]
18. BBC. BBC News Online (UK) Fifth Rider Killed in Eventing Fall September 20 1999. Available online: http://news.bbc.co.uk/2/hi/uk_news/452301.stm (accessed on 27 November 2015).
19. Davison, J. Death of Fifth Rider Shocks Eventing into Crisis Reforms. 1999. Available online: http://www.independent.co.uk/news/death-of-fifth-rider-shocks-eventing-into-crisis-reforms-1120793.html (accessed on 27 November 2015).
20. Paix, B.R.; (Flinders University, Adelaide, South Australia). Personal communication, 2013.
21. O'Brien, D.; Cripps, R.A. *Safety for Horses and Riders in Eventing. The SHARE Database*; Rural Industries Research and Development Corporation: Canberra, ACT, AUS, 2008.
22. Fédération Equestre International. *The International Eventing Safety Committee Report (Hartington Report)*; Fédération Equestre International: Lausanne, Switzerland, 2000.
23. Horsetalk. Eventing Safety. Available online: http://horsetalk.co.nz/2012/04/02/eventing-safety/#axzz2ZqZRRst8 (accessed on 8 October 2015).
24. O'Brien, D.; (South Australian Spinal Cord Injury Research Centre, Adelaide, South Australia). Eventing fatalities, 1993–2015.
25. Strickland, C. Equine-Related Human Injuries. 2000. Available online: http://www.thehorse.com/articles/10102/equine-related-human-injuries (accessed on 20 September 2015).

26. Rohrmann, B. The risk notion: Epistemological and empirical considerations. In *Integrated Risk Assessment: Applications and Regulations*; Stewart, M.G., Melchers, R.E., Eds.; Balkema: Rotterdam, The Netherlands, 1998; pp. 39–46.
27. Peace, C. Risk Processes and Techniques: What Practitioners are Using. *Proceedings of the Society for Risk Analysis (Australia and New Zealand)*, 2014. Available online: http://www.sraanz.org.nz/proceedings.html (accessed on 4 December 2015).
28. Bahr, R.; Holme, I. Risk factors for sports injuries—A methodological approach. *Br. J. Sport. Med.* **2003**, *37*, 384–392. [CrossRef]
29. Australian Football League. *Risk Management for Football Clubs*, 2004. Available online: http://www.aflcommunityclub.com.au/index.php?id=215 (accessed on 4 December 2015).
30. Safe Work Australia. *Guide to Managing Risks When New and Inexperienced Persons Interact with Horses*; Safe Work Australia: Canberra, Australia, 2014.
31. Fuller, C.; Drawer, S. The application of risk management in sport. *Sports Med.* **2004**, *34*, 349–356. [CrossRef] [PubMed]
32. Fuller, C.W. Managing the risk of injury in sport. *Clin. J. Sport Med.* **2007**, *17*, 182–187. [CrossRef] [PubMed]
33. Spinks, A.B.; McLure, R.J. Quantifying the risk of sports injury: A systematic review of activity rates for children under 16 years of age. *Brit. J. Sport. Med.*. [CrossRef] [PubMed]
34. Bowls Queensland. Risk Management Policy. Available online: http://www.bowlsqld.org/BQ-In-Focus/Important-Documents/Policies (accessed on 4 December 2015).
35. Williams, F.; Ashby, K. Horse Related Injuries. In *Hazard*, 23rd ed.; Monash Accident Research Centre: Melbourne, Australia, 1995.
36. Vanarthos, W.J.; Johnson, A.W. Injuries associated with equestrian sports: Epidemiology and mechanism. *Emerg. Radiol.* **1995**, *2*, 29–36. [CrossRef]
37. Cripps, R.A. Horse-related Injury in Australia, 2000. *Australian Injury Prevention Bulletin 24*, Available online: http://www.nisu.flinders.edu.au/pubs/bulletin24/bulletin24.html (accessed on 5 May 2015).
38. Sorli, J.M. Equestrian injuries: A five-year review of hospital admissions in British Columbia. *Injury Prev.* **2000**, *6*, 59–61. [CrossRef]
39. Flood, L; Harrison, J.E. Hospitalised Sports Injury, Australia 2002–2003. 2006. Available online: http://www.nisu.flinders.edu.au/pubs/reports/2006/injcat79.php (accessed on 5 May 2015).
40. Gabbe, B.J.; Finch, C.F.; Cameron, P.A.; Williamson, O.D. Incidence of serious injury and death during sport and recreation activities in Victoria, Australia. *Br. J. Sport. Med.* **2005**, *39*, 573–577. [CrossRef] [PubMed]
41. Smartt, P.; Chalmers, D. A new look at horse related sport and recreational injury in New Zealand. *J. Sci Med. Sport.* **2009**, *12*, 376–382. [CrossRef] [PubMed]
42. Guyton, K.; Houchen-Wise, E.; Peck, E.; Mayberry, J. Equestrian injury is costly, disabling, and frequently preventable: The imperative for improved safety awareness. *Amer. Surg.* **2013**, *79*, 76–83.
43. Kreisfeld, R.; Harrison, J.E.; Pointer, S. Australian sports injury hospitalisations 2011–2012. 2014. Available online: http://www.aihw.gov.au/publication-detail/?id=60129549100 (accessed on 3 January 2016).
44. Finch, C.; Boufous, S. Do inadequacies in ICD-10-AM activity coded data lead to underestimates of the population frequency of sports/leisure injuries? *Injury Prev.* **2008**, *14*, 202–204. [CrossRef] [PubMed]
45. Watt, G.M.; Finch, C.F. Preventing equestrian injuries: Locking the stable door. *Sports Med.* **1996**, *22*, 187–197. [CrossRef]
46. Ueeck, B.A.; Dierks, E.J.; Homer, L.D.; Potter, B. Patterns of maxillofacial injuries related to interactions with horses. *J. Oral Maxill. Surg.* **2004**, *62*, 693–696. [CrossRef]
47. Thurman, D.J.; Branche, C.M.; Sniezeka, J.E. The epidemiology of sport-related traumatic brain injuries in the United States: recent developments. *J. Head Trauma Rehab.* **1998**, *13*, 1–8.
48. Selassie, A.W.; Wilson, D.A.; Pickelsimer, E.E.; Voronca, D.C.; Williams, N.R.; Edwards, J.C. Incidence of sport-related traumatic brain injury and risk factors of severity: A population-based epidemiologic study. *Ann. Epidemiol.* **2013**, *23*, 750–756. [CrossRef] [PubMed]
49. Dunning, E. *Quest for Excitement: Sport and Leisure in the Civilizing Process*; Blackwell: Oxford, UK, 1986; Volume 288, p. 156.
50. Australian Skills Quality Authority. Training in equine programs in Australia. 2015. Available online: http://www.asqa.gov.au/news-and-publications/eleven-recommendations-to-enhance-equine-training.html (accessed on 3 January 2016).

51. Smartt, P.; Chalmers, D.J. Searching hospital discharge records for snow sport injury: No easy run? *Int. J. Inj. Contr. Saf. Promot.* **2011**, *19*, 93–100. [CrossRef] [PubMed]
52. Guinness World Records. Smallest Living Horse—Female. Available online: http://www.guinnessworldrecords.com/world-records/smallest-living-horse (accessed on 4 December 2015).
53. Guinness World Records. BigJake—The World's Tallest Horse-Video. Available online: http://www.guinnessworldrecords.com/news/2012/9/record-holder-profile-big-jake-worlds-tallest-horse-video-44749/ (accessed on 4 December 2015).
54. Guinness World Records. Fastest Speed for a Race Horse. Available online: http://www.guinnessworldrecords.com/world-records/fastest-speed-for-a-race-horse/ (accessed on 4 December 2015).
55. Horse Racing. Melbourne Cup Facts and Statistics. Available online: http://www.races.com.au/melbourne-cup/melbourne-cup-history/facts-and-statistics/ (accessed on 4 December 2015).
56. Dawson, J. Why do Riders Fall? American Association for Horsemanship Safety. Available online: https://asci.uvm.edu/equine/law/articles/instruct.htm/ (accessed on 4 December 2015).
57. Murray, J.K.; Singer, E.R.; Morgan, K.L.; Proudman, C.J.; French, N.P. The risk of a horse-and-rider partnership falling on the crosscountry phase of eventing competitions. *Equine Vet. J.* **2006**, *38*, 158–163. [CrossRef] [PubMed]
58. Fédération Equestre International. FEI Eventing Final Statistics report 2014. Available online: www.fei.org/system/files/FINAL-UPATED%20StatisticsReport%202004%20-%202014.pdf (accessed on 1 July 2015).
59. Thompson, K.; Nesci, C. Over-riding concerns: Developing safe relations in the high-risk interspecies sport of eventing. *Int. Rev. Sociol. Sport* **2013**. [CrossRef]
60. United States Equestrian Federation Eventing Technical Committee. Immediate Eventing Rule Change Makes One Fall on Cross-Country Elimination. Available online: http://www.chronofhorse.com/article/immediate-eventing-rule-change-makes-one-fall-cross-country-elimination (accessed on 4 December 2015).
61. O'Brien, D. Eventing in Crisis? Available online: http://horsetalk.co.nz/saferide/131-eventingincrisis.shtml#axzz2ZqZRRst8 (accessed on 4 December 2015).
62. Equestrian Australia. EOI for National Safety Officer in Eventing. Available online: http://www.equestrian.org.au/news/eoi-national-safety-officer-eventing (accessed on 5 December 2015).
63. Equestrian Australia. Available online: http://www.equestrian.org.au/home (accessed on 4 December 2015).
64. Chapman, M.A.S.; Oni, J. Motor racing accidents at Brands Hatch, 1988/9. *Brit. J. Sport. Med.* **1991**, *25*, 121–123. [CrossRef]
65. Murray, J.K.; Singer, E.R.; Morgan, K.L.; Proudman, C.J.; French, N.P. Risk factors for cross-country horse falls at one-day events and at two-/three-day events. *Vet. J.* **2005**, *170*, 318–324. [CrossRef] [PubMed]
66. Boden, L.A.; Anderson, G.A.; Charles, J.A.; Morgan, K.L.; Parkin, T.D.H.; Clarke, A.; Slocombe, R.F. Risk factors for thoroughbred racehorse fatality in jump starts in Victoria, Australia (1989–2004). *Equine Vet. J.* **2007**, *39*, 422–428. [CrossRef]
67. Harrison, J.E.; Roeger, L.; O'Brien, D. *Safety of Jumping Races*; Flinders University: Adelaide, Australia, 2008.
68. Harrison, J.E.; Roeger, L.; O'Brien, D. *Safety of Jumping Races, Addendum 2007 and 2008 Data*; Flinders University: Adelaide, Australia, 2008.
69. Stachurska, A.; Pięta, M.; Phaff Ussing, A.; Kaproń, A.; Kwiencińska, N. Difficulty of cross-country obstacles for horses competing in Three-Day Events. *App. Anim. Behav. Sci.* **2010**, *123*, 101–107. [CrossRef]
70. Murray, J.K.; Huws, N.; Singer, E.R. Analysis of Risk Factors for Horse Falls in the Cross-Country Test of FEI Eventing: Fence Related Factors (Phase 1). Available online: www.fei.org/system/files/Session4_Eventing%20Barnett.pdf (accessed on 8 October 2015).
71. Young Rider. Falling off a Horse. Available online: http://www.youngrider.com/young-rider-magazine/january-2007/falling-off.aspx (accessed on 4 December 2015).
72. Steinkraus, W. *Reflections on Riding and Jumping: Winning Techniques for Serious Riders*; Trafalgar Square Publishing: North Pomfret, VT, USA, 1997; pp. 83–98.
73. International Federation for Equestrian Sports. FEI Eventing Rules. Available online: http://www.fei.org/fei/regulations/eventing (accessed on 5 December 2015).
74. International Federation for Equestrian Sports. Safety & Personal Protective Equipment. Available online: http://www.fei.org/fei/your-role/medical-safety/safety (accessed on 5 December 2015).

75. The Coalition for the Protection of Racehorses. Available online: http://www.horseracingkills.com/ (accessed on 10 December 2015).
76. The Coalition for the Protection of Racehorses. Making a Killing. Available online: http://www.abc.net.au/4corners/stories/2015/02/16/4178920.htm (accessed on 10 December 2015).
77. Bagshaw, E. New Puppy Farms: "Horrific" Facility Seeks Council Permission to Breed 300 Dogs. Available online: http://www.smh.com.au/nsw/nsw-puppy-farms-horrific-facility-seeks-council-permission-to-breed-300-dogs-20151118-gl2l95.html (accessed on 10 December 2015).
78. Ruse, R.; Davison, A.; Bridle, K. Jump racing safety: Reconciling public debate and Australian thoroughbred jump racing data, 2012–2014. *Animals* **2015**, *5*, 1072–1091. [CrossRef] [PubMed]
79. Smith, S.E. 7 Horrific Tragedies on the Horse Track. Available online: http://www.care2.com/causes/7-horrific-tragedies-on-the-horse-track.html (accessed on 7 January 2015).
80. Smith, J. One Horse Dies Every Three Days on Australian Racetracks Animal Activists. Available online: http://www.theguardian.com/sport/2015/jul/31/one-horse-dies-every-three-days-on-australian-racetracks-animal-activists (accessed on 7 January 2015).

animals

MDPI

Review

Improving the Understanding of Psychological Factors Contributing to Horse-Related Accident and Injury: Context, Loss of Focus, Cognitive Errors and Rigidity

Jodi DeAraugo [1,*,†], Suzanne McLaren [1,†], Phil McManus [2,†] and Paul D. McGreevy [3,†]

[1] Faculty of Health, Department of Psychology, Federation University, University Drive, Mount Helen, Victoria 3350, Australia; s.mclaren@federation.edu.au

[2] School of Geosciences, University of Sydney, Room 435, F09, Madsen Building, New South Wales 2006, Australia; phil.mcmanus@sydney.edu.au

[3] Department of Veterinary Science, University of Sydney, Room 206, R.M.C. Gunn Building, New South Wales 2006, Australia; paul.mcgreevy@sydney.edu.au

* Correspondence: jodi.dearaugo@gmail.com; Tel.: +61-0418-537-193

† These authors contributed equally to this work.

Academic Editor: Kirrilly Thompson

Received: 29 October 2015; Accepted: 2 February 2016; Published: 15 February 2016

Simple Summary: There is a high risk of injury for people involved with horses in their work or recreational pursuits. High risks are particularly evident for racing employees and veterinarians. Elevated risks of injury may be associated with misjudging how to handle situations, reduced attention caused by distractions, taking a general view, and failing to consider other strategies that may reduce risks. To improve safety for humans and horses, it is important to identify safety strategies that are flexible, focused and specific.

Abstract: While the role of the horse in riding hazards is well recognised, little attention has been paid to the role of specific theoretical psychological processes of humans in contributing to and mitigating risk. The injury, mortality or compensation claim rates for participants in the horse-racing industry, veterinary medicine and equestrian disciplines provide compelling evidence for improving risk mitigation models. There is a paucity of theoretical principles regarding the risk of injury and mortality associated with human–horse interactions. In this paper we introduce and apply the four psychological principles of context, loss of focus, global cognitive style and the application of self as the frame of reference as a potential approach for assessing and managing human–horse risks. When these principles produce errors that are combined with a rigid self-referenced point, it becomes clear how rapidly risk emerges and how other people and animals may repeatedly become at risk over time. Here, with a focus on the thoroughbred racing industry, veterinary practice and equestrian disciplines, we review the merits of contextually applied strategies, an evolving reappraisal of risk, flexibility, and focused specifics of situations that may serve to modify human behaviour and mitigate risk.

Keywords: human–horse risk; context-specificity; attention; cognitive error; self-reference

1. Introduction

Safety risks for people working with, riding, or otherwise being in proximity to horses are well documented. These include injury or death and, across the different equestrian disciplines, horse riding is widely considered a high risk activity [1–15], as is working with larger animals [16–18]. The literature identifies racing industry employees, veterinarians and equestrian disciplines at particular risk of injury or death [1–17].

At especially high risk are thoroughbred racing industry employees [1,2,6,7]. For example, in Australia, work-related injury claims from 2002 to 2010 due to injuries sustained from working with horses averaged $A9 million per year for workers in the racing industry [7]. There are high risk injuries [1,2,6,7] in the racing industry with high insurance costs [6,7]. Curry *et al.* [7] reported that 39% of race-day incidents accounted for 52% of the insurance costs. Race-day insurance costs were on average higher ($A33,756) than non-race-day incidents (which averaged $A20,338). Curry *et al.* study reported that 49% of the sample had injuries to lower and upper limbs, with fractures also prevalent [7]. However, head injuries were less common (5.3% of fall injuries and 2.7% of no-fall injuries) but related to greater monetary costs and more days absent from work [7]. Forero Rueda *et al.* [1] reported rates of concussion and head injuries in flat and jumps jockeys in Ireland (23 per 1000 falls), France (32 per 1000 falls) and Britain (19 per 1000 falls) that seemed lower than those reported by Curry *et al.* [7]. Injuries per fall rates from flat racing were between 33% and 44% [1]. Jumps racing per ride had the highest injury rates for all jockeys (*i.e.*, both amateurs and professionals) [1,2] and, in Australia, 17.4%–21.9% of compensation claims were jumps racing jockeys [7]. The average monetary claim for jumps racing jockeys was $45,831 compared to $24,672 for flat racing jockeys [7]. An Australian study by Cowley *et al.* [6] reported that track riders and stable attendants accounted for most of the worker compensation claims (71%) and, among them, 72% cited horse-related injuries due to falls from horses. Other injuries reported were being kicked, struck, hit, crushed or pushed by horses. Fractures and contusions were listed as the primary injuries [6].

Elevated risks of injury are related to the type of race, distance of the race, experience of the horse and jockey (*i.e.*, apprentice and amateur jockeys) [1,2,7]. Hitchens *et al.* [2] considered jockey, horse and environmental variables as influences on the risks in racing. Thus, the racing data show variation in the risk profile of injury, but there are suggestions that inexperienced jockeys, track riders and less accomplished horses elevate the risks of injuries.

Other horse-related professionals at risk appear to be veterinarians. Data indicate that most large-animal vets were at an increased risk of significant injuries compared to companion animal vets [17]: 51% receiving injuries in the workplace that affected the span of their professional working life, and 26% having sustained injuries in the preceding year. Lucas *et al.* [11] cited evidence that the most common injuries (79%) reported by vets were sustained either by being kicked or struck by horses. This was similar to injuries cited for veterinary and animal science students (n = 260): foot/ankle (39.1%), upper leg/knee (34.8%), and hand (13%) being the most common areas injured [12]. The most prevalent ways students were injured were by being trampled or kicked by a hind limb (30.4%), bitten (13%), or falling when riding (8.7%). The most common nature of the 31 injuries reported were bruising/soft tissue injury (91.3%), open wound (17.4%), muscle or tendon injury (8.7%).

Similarly, equestrian disciplines have high rates of injury [3–5]. The equestrian risk (11.2%) was almost equivalent to injuries from all-terrain vehicle injuries (12.2%) and 1.4 times greater than sports-related injuries [5]. Mayberry *et al.* [4] reported that injury is common and serious, most notably in the first one hundred hours of gaining experience. The risk of serious injury for equestrians was reported as 1 in 5 [4]. Despite professionals reporting lower rates than novice riders due to their higher skill level, they were at great risk (94%) of sustaining an injury during their career [4]. Lim reported that less experienced or younger riders were more likely to be hospitalised than experienced riders, possibly because older riders rolled to break the fall [3]. A larger scale review by Hawson *et al.* [10] of the human–horse injury literature stated that the most common risks to non-veterinarians were from falling or being thrown from a horse. Head injuries including concussion and brain injuries are most likely when helmet use is limited [3,6]. Therefore, despite the overall high risk of injury to equestrians, from the limited evidence for injuries requiring hospitalisation it appears that less experienced novice and non-helmeted riders are at greater risk. However, even career professionals are exposed to significant risk with a rate of at least one injury in their career, so exposure to horses seems to be implicated also. These data provide compelling evidence of the need for clear determinants of risk-mitigation models that can better address the risky nature of working with horses and other large animals.

Thompson *et al.* [13] summarised many of the most prevalent horse-related risks for people as including the inherent risks themselves, characteristics of the horse, characteristics of the rider, and influential factors evident in the broader horse culture. Some of these risks can be reduced by the design of thoroughbred facilities such as stud farms, auction venues and racetracks [19], but the horse and rider risk variables remain [2–5]. An appreciation of these risks should prompt researchers across various disciplines to the study of horse related risks for people, including track or equestrian-centre employees, handlers and riders. To address the dearth of psychological equestrian risk literature, it is salient to include already established and evidence-based psychological variables that could potentially mitigate risk of injuries to humans during their interactions with horses.

In the current paper, we specifically use some of Michael Yapko's [20–22] evidence-based ideas of cognitive processes and his emphasis on the effects of one's quality of focus on eventual outcomes as a theoretical anchor. Yapko is best known for combining focused states of awareness (*i.e.*, paying attention under hypnosis) with cognitive therapy to alleviate depression, anxiety and other negative emotional states to assist people in making associations with existing resources or developing new skills [20–22]. He includes attributional style as an influence on the development of cognitive therapy and hypnosis [23]. He has also used focused awareness, such as hypnosis and cognitive therapy, to recalibrate dangerous health behaviours, which essentially recalibrates the cognitive, emotional, physical and behavioural risks for destructive human behaviours [22]. Focus is central to his work because it amplifies awareness that is an integral part of change and this enhances processing new information [20–22]. In essence, he focuses on changing psychological process errors (*i.e.*, association rather the dissociation and being specific as opposed to being global in some circumstances) rather than the content (*i.e.*, basis of the person's story). It is evidence-based and merits consideration in the current context of reducing risks in human–horse interactions, which is a novel arena for the application and extension of his work. There is useful literature on naturalistic decision-making [24,25], in situations that require rapid responses, but this lies beyond the scope of the current article. Indeed, it makes sense to apply established psychological theoretical principles with effective clinical outcomes supported by a body of evidence, to a novel target such as human–horse risks.

Yapko suggests psychological targets, such as context (*i.e.*, situational factors shaping responses), focus (*i.e.*, the direction and quality of one's focus) and cognitive style (*i.e.*, how one assimilates and integrates information), as processes for understanding the mitigation or elevation of the risk for humans who engage in behaviours that may be detrimental to their health. This cognitive and contextual information can be applied to the area of risks arising for employees working in the horse racing industry (e.g., jockeys, track riders, attendants and veterinarians) and others who have contact with horses across the equestrian disciplines.

In the following sections of this paper, we consider whether context, loss of focus, global cognitive style and the application of self as the frame of reference are important and how they can relate to each other, in exacerbating and managing risks.

2. The Relevance of Context in the Changing Profile of Risk

The importance of contextual relevance is evident when some purposeful strategy may work well in one particular area or interaction, yet the same strategy may fail when applied in a second situation when the context does not support its use [20,22]. The effective application of a given strategy would need to account for specific cues perceived by the rider, from the particular context of the immediate human–horse interactions. A strategy from a previous context may or may not work in the current one.

Assessing the contextual cues from humans and horses associated with risky behaviour and then selecting the appropriate (*i.e.*, safe, effective) strategy is paramount. It is possible that, when working with horses, making decisions within a short time-frame may also be a paramount consideration (e.g., quick and efficient decision-making skills) that could help riders respond more effectively and thereby prevent accident or injury. This point is supported by Hausberger *et al.* [26], who indicated that people require different skills for different contexts. They separated the skills necessary for handling horses well from the skills of riding horses well. As such, racehorse attendants' duties, which include animal care, grooming, preparation of the horses, cleaning stables and horse-handling, differ substantially from track riders' and jockeys' occupational requirements [6]. Furthermore, veterinarians' skills are vastly different to the skills of others in the horse industry. Veterinarians assess and treat equine medical problems, such as attending to horses in crushes, examining horses that may be in pain, distressed or in unusual circumstances (e.g., traffic or accidents during transportation in a float), all of which pose different risks to personnel [27]. In summary, different jobs require different skills that include job-specific assessments of human–horse interactions. Arguably, risk mitigation can be progressive, and ranking the skill-sets for different duties relies on re-evaluation of risk across different circumstances, situations and with different horses.

Changes in context are of great relevance in risk management at the human–horse interface. They could simply represent a shift from the horse's home environment to an unknown or less familiar environment (e.g., racecourse, veterinarian facility or competition venue). Cowley *et al.* [6] state that, in a shared track-riding environment, occupational hazards can be especially prevalent because trainers have less influence over the environment than they do in a private context. Equine responses reflect the familiarity, but also predictability, of their current environment. As a prey species, horses are flight animals that are characterised by unpredictability or their instinctive need for safety [13,28]. Three examples are provided in Table 1.

Table 1. Examples from track-work riders and jockeys, equine veterinarians and equestrian competitors that illustrate the relevance of context and purposeful effective strategies that can increase the risk of injuries.

Track-Work Riders and Jockeys	Equine Veterinarians	Equestrian Competitors
Examples of how awareness of the context can be used to recalibrate risk		
A horse in a racecourse environment may be adrenalinised on race day by the atmosphere at the track. The crowd, noise, speakers, barriers, and other horses could prompt a flight response in the horse, particularly with a younger horse just introduced to the new environment. Horses are often kept moving (*i.e.*, walking) to cope with the stressful atmosphere. An unpredictable response from a person in the crowd, such as a flag flying near the horse, could prompt a startle response and flight reaction. The racing attendant may use the focus of the horse to divert its attention from the flag, move the horse away, or habituation to flags could be undertaken prior to race day to prepare the horse.	A horse may not have been exposed to a crush prior to attending a vet's premises. It may be in pain and require treatment. The horse may trial running backwards or sideways, and either kick out or barge over the handler if it fears being put in the crush. Some vets may, with the owner's consent, sedate the horse for safety of the horse and personnel. A contextual alternative could be to use clicker training (*i.e.*, positive reinforcement with food) or exploring the environment with wither scratching, if it is in the training repertoire of established responses. A poorly chosen strategy, such as the handler using a whip for punishment, could produce disastrous, noxious and fear-related results and potentially exacerbate the flight response contributing to possible injuries to those involved.	When a competition horse in the home environment sees an unfamiliar object, such as a camera on a tripod, it may seek to avoid the object and need reassurance, such as calm verbal responses and wither scratching and/or (if under saddle) leg cues to move past the object [27]. The same horse in a competition environment may require time and free exploration to habituate to novel objects. It may also need reassurance before approaching such objects [27]. Thus, the strategy chosen by the rider could vary according to the environmental conditions and potential reactivity or flight response of the horse. Forcing horses past novel objects in the competition atmosphere may exceed their tolerance threshold, increasing the risk of a flight response, and potential for human/animal injury.

3. The Counter-Balance of Focus or Loss of Focus

The second variable to consider as part of a risk-mitigation model relates to loss of human focus. The definition of dissociation is the ability to separate a broad (*i.e.*, global) experience into its component parts and reduce awareness [22]. Essentially, dissociation is a reduction in the direction and quality of focused attention. Yapko [22] emphasises that dissociation is a neutral term. That said, he makes the point that dissociation can be used as a negative or positive process and that it ultimately depends upon the context and specificity of its application [22]. A person's attention may be internally oriented (*i.e.*, they may be thinking about the other tasks, past events or daydreaming), or they may be distracted and diverting their attention to another task in their environment. It is the "attentional drift" [22] at critical moments from the interaction with the horse in specific settings that poses dangers. The awareness or focus of an individual can drift and, when this happens, more automatic responses can emerge with less awareness for recalibrating risk [21,22]. Three examples are shown in Table 2.

Table 2. Examples from track-work riders and jockeys, equine veterinarians and equestrian competitors that illustrate the drifting awareness that increase the risk of injuries.

Track-Work Riders and Jockeys	Equine Veterinarians	Equestrian Competitors
Examples of how the awareness or focus of an individual can drift and, when this happens, more automatic responses can emerge with less awareness for recalibrating risk		
A track rider exercising a horse could be distracted by a discussion with a co-worker riding alongside and not notice a change at the racecourse, such as some new machinery. The horse could spook.	A vet could be distracted while performing an examination and explaining something to an owner. Meanwhile, the horse stands on the vet's foot or kicks out.	A riding competitor is distracted by another horse's behaviour and loses focus on his own horse's reaction. Thus, the rider did not notice the fear building in his horse and not respond early enough to defuse it.

Research shows that inattention has been associated with an increased risk of crashing vehicles, such as cars or trucks [29–31], as well as with risky health behaviours, such as smoking tobacco [22]. Smokers routinely fail to accurately consider that smoking elevates their risk of cancer and cardiovascular diseases [22]. Therefore, risks for injury, death or illness increase when dissociation happens in contexts that are inappropriate and problematic. Similarly, it is feasible that dissociation may be a process that hinders human–horse risk.

Indeed, Hausberger *et al.* [26] reported that observational skills and attentional skills were pivotal in preventing accidents specific to humans and horses. Perhaps contrary to expectation, Hausberger *et al.*'s review [26] concluded that documented accidents did not decrease with the degree of human competence or accumulated experience with horses. Hitchens *et al.* [2] reported that jockeys over 35 years of age had increased rates of falling off horses if they had ridden earlier at the race meeting, suggestive perhaps of fatigue or attentional issues. However, prior evidence reviewed [2,4] indicates other factors such as being less experienced, having less accomplished horses and not wearing a helmet were associated with elevated risk, thus highlighting human experience and horse variables as risks for injuries. There is evidence of a clear decrease in fall rates over time, *i.e.*, with accumulated experience [32].

Thompson and Haigh [5] reported that horse riders "rarely described their own horses as dangerous or unpredictable" because of their experience and familiarity with these animals. However, fractious horse behaviour with a rapid flight response [18] can make horses dangerous in their responses around people [13,27]. Riders often fail to appreciate a change in the direction or the quality of their focus as critical for risk mitigation. Instead, they rely on experience as the predictor, which indeed could be a serious or even fatal error for risk appraisal. The specific context and loss of focus already pose two major risks, but a further issue, such as cognitive style, can also thwart accurate risk reappraisal.

4. A Global Cognitive Style Can Bring Specific Risks

A global cognitive style is a broad style of thinking that focuses on the bigger picture and is over-inclusive [20,21]. It lacks the specificity and filtering processes necessary for some thoughts or perceptions to be accurate, true, integrated and representative [20,21,33]. Again, this is a neutral term that relies on context for its utility. To illustrate this, a global overarching principle for life, say, regarding animal welfare (e.g., "animal abuse is wrong"), may indeed be helpful in protecting animals from harm. A metaphor for a global cognitive style could be seeing the forest but not the trees [20]. A sweeping broad view, such as a global cognitive style cannot be relevant for all circumstances, as it overlooks specific risks. A global cognitive style regarding accident risk could be problematic. In this vein, a study by Thompson and Haigh [9], when they investigated the use or lack of use of helmets, highlighted the global cognition of riders that "accidents happen", "I can control risk" and "it does not feel right". It is highly improbable that each of these statements could be accurate in all situations across horse care, preparation, handling practices, riding, track work, racing, competition venues, veterinary practices or the home environment. So, specific and focused adjustments for contexts and circumstances offer a critical opportunity to reappraise the process of risk assessment and management.

Without the context, focal point and specifics, the opportunity for cognitive errors in assessing risk increases [22]. Specificity is an antidote for a global cognitive error. As specific realistic risks are raised under focused states of awareness, decision-making can be altered to minimise or avoid the risk. When the cognitive error is acknowledged in a focused state, some people recalibrate their behaviours and choices. In his clinical work with people who engage in risky behaviours, Yapko [20–22] has documented the teaching of skills and routine recalibration of risks under focused states of awareness.

Table 3. Examples from track-work riders and jockeys, equine veterinarians and equestrian competitors that illustrate global cognitions that increase the risk of injuries.

Track-Work Riders and Jockeys	Equine Veterinarians	Equestrian Competitors
Examples of human global cognitions that elevate risk		
Global: Everyone runs risks in the workplace. It's just bad luck if you get injured.	Global: Older mares don't need to be scanned in crushes.	Global: Riding horses is no more dangerous than any other sporting activity.
Specific: Some risks in the racing industry can be identified, managed and avoided when safety protocols are followed.	Specific: Even when scanning an experienced broodmare, it would be sensible to reduce the risk of injury by using a crush, especially given it's a veterinary examination that occurs less frequently and calls for extra handling skills.	Specific: Given that a horse's response to fear is to flee, riding horses can elevate the risk of injury and mortality, especially when high speeds and jumping are involved.

A global cognitive style can ultimately be detrimental for the process of differentiation [20–22]. Differentiating and separating certain elements of experiences should be considered a core skill required for mitigating risk. This is especially so, given the cognitive differences between humans and horses, although any discussion of these differences, either real or putative, can be contentious. Humans have the ability to generalise learning across situations; in contrast, equitation scientists generally maintain that horses are context-specific in their approach, do not generalise immediately, and do not possess higher cognitive abilities [34–37]. A rider can train a horse to go forward in a certain place, but a horse, particularly a young one, associates all the cues (e.g., visual) specific to that particular place with the go-forward cue. Therefore, the younger horse may fail to go forward in another situation [35,36]. An example would be when the horse learns to travel through a water obstacle at home, but fails to generalise it to all water obstacles at different venues until all the other concomitant visual details are less relevant than the water itself [35]. It is critical that people who work with horses understand the intricate learning processes of horses and humans, especially given that horses can be context-specific [35,36] with a new skill and the rider may be global [20–22] in his approach. A context-specific horse with a global rider could quickly succumb to an elevated risk of injury. Examples are shown in Table 3.

People who work with horses and who have a global cognitive style, a loss of focus and who fail to consider the context, collectively or separately can have an elevated risk of injury or mortality. As the case builds across this article, the final problematic process is the ability to see risk only through one lens or focal point.

5. Using the Self as a Frame of Reference Hinders Flexibility in Risk Appraisal

The idea that one person's perspective is fact can potentially contribute to the risk of injury. It is important to appreciate that each horse can have a repertoire of current behaviours that vary from the most recently observed [27]. More specifically, a given horse may adopt a certain response in one episode and a different response in the same environment because horses trial behaviours to cope with circumstances and do not always generalise [27,34–36]. If workers in the racing industry, veterinary practice or equestrian disciplines assume that horses will react in the same way as they have done previously, they are embracing a dangerously rigid approach to risk assessment and management.

Yapko [21] defined "using the self as a frame of reference" as perceiving or interpreting information from the person's social learning history without consideration of alternative views. It can indicate a pattern of selective perception. So, being wedded to a particular view that is safety -focused may have an advantage but, plainly, adhering to a view that elevates risk (e.g., "I've ridden without a helmet before and I've been fine") could be problematic [21]. A rigid style discounts the probability of the horse trialing a new behaviour, so riders may be surprised by the behaviour and delayed in their response. A flexible style (using other points of reference) would be more accommodating, allowing riders to respond quickly if they have the skills to cope with such behaviour and correct it. Examples are shown in Table 4.

Table 4. Examples from track-work riders and jockeys, equine veterinarians and equestrian competitors that illustrate self-referencing as a cognitive error that increases the risk of injuries.

Track-Work Riders and Jockeys	Equine Veterinarians	Equestrian Competitors
Examples for the idea of self-referencing as an assessment point		
A horse at home has no history of rearing. On race day, it is exposed to the public address system and rears. The stable attendant is surprised, dragged sideways and delays making a response.	A horse attends the vet's premises for treatment. The owner says the horse has never kicked out before, but in an unfamiliar environment while undergoing veterinary procedures, such as injections, it kicks out as it is insecure and unwell.	A horse spooks at a yellow garbage bin in the warm-up arena at a new competition venue. The horse does not typically shy at home. The rider is perplexed and caught unawares. The rider's self-referencing ideation can heighten the risk of a fall or provide delays in signaling the horse to move forward (e.g., faster with leg speed or longer in the stride) and maintain focus.

6. Conclusions and Future Direction

Drawing on the available literature, our analysis has found that while the four critical concepts for human–horse risk mitigation; namely contextual relevance, inattention or loss of focus, global cognitions and referencing on the basis of past history, are not problematic *per se*, it is the context in which they are applied that is paramount when assessing risk for human–horse interactions. We have provided examples to demonstrate how each of these risk factors relate to the three high at-risk groups identified earlier in this paper; track-work riders and jockeys, equine veterinarians and equestrian competitors.

We posit that to reduce risk, the formation of new associations that realistically appraise the risk in an ongoing manner and prompt actions that are context-specific, focused, and flexible offer a sensible approach [20–22]. In the pragmatic sense, using metaphors and hypnosis to alter the cognitive, emotional and behavioural options of paying attention to horses' cues and behaviours in specific scenarios (e.g., transport loading, racetrack behaviour, or injection as part of veterinary practice) could be used. Establishing a tool-kit with information on how horses learn and respond (e.g., habituation, learning theory; pressure and release, positive reinforcement) could be of use, albeit to illustrate key points, the language could be modified for lay people. Metaphors and hypnosis could emphasise and promote the advantages of acquiring the specific skill-set or to make adjustments in the skill(s) used (e.g., appropriate use of negative reinforcement, such as via the bit as a stop cue; wither scratching for an anxious horse; allowing the horse to assess an aversive object, subject to the arousal state of the horse [38]). Baseline and post-intervention measures could be developed to test for statistical or clinically relevant change (*i.e.*, of practical or applied value in everyday life) [39].

The awareness of risk factors, the association of these risk factors with workers in different parts of the equine industry who are most at-risk, and identifying the implications if these risks translate into actual incidents, are all important steps in reducing equine-related accidents. However, it is important to go beyond these steps and suggest that preliminary solutions that may be developed in the future into a more comprehensive model built on the foundations outlined in this paper. For example, dissociation in a new environment, or even in a familiar environment, can pose risks due to the inattention that arises in the human–horse interaction, specifically distraction from monitoring the horse's responses. Subsequently, the human response will inevitably be delayed. Focused awareness is central to this risk-mitigation model. The human expectation that a horse will respond similarly in all scenarios without the human appropriately assimilating the requirements of the horse or specific situations could elevate the risk of human injury or death. Choosing a poorly matched strategy for the situation could also be disastrous. Combining the contextual, cognitive errors with a rigid self-referenced point of view and a lack of focused attention helps to clarify how risk rapidly emerges, and how others may repeatedly become at risk over time.

Essentially, many workers in the racing industry and people across all equestrian disciplines may be unaware of some of the critical requirements for realistically assessing the risk of injury. This paper offers crucial suggestions on what is important and how to form a pattern of interruption [22] to help to address the risks that equine industry workers and humans in equestrian disciplines currently face.

Acknowledgments: The authors would like to thank Michael Yapko for his comments. We would also like to acknowledge the anonymous reviewers for the valuable feedback on an earlier manuscript.

Author Contributions: Jodi DeAraugo conceived this article. All authors contributed to the writing and approval of the article's content. Jodi DeAraugo is an Honorary Research Affiliate to Federation University.

Conflicts of Interest: The authors declare no conflict of interest.

References

1. Forero Rueda, M.A.; Halley, W.L.; Gilchrist, M.D. Fall and injury incidence rates of jockeys whileracing in Ireland, France and Britain. *Injury* **2010**, *41*, 533–539. [CrossRef] [PubMed]
2. Hitchens, P.L.; Blizzard, C.L.; Jones, G.; Day, L.; Fell, J. Predictors of race-day jockey falls in jumps racing in Australia. *Accid. Anal. Prev.* **2011**, *43*, 840–847. [CrossRef] [PubMed]
3. Lim, J.; Puttaswamy, V.; Gizzi, M.; Christie, L.; Croker, W.; Crowe, P. Pattern of equestrian injuries presenting to a Sydney teaching hospital. *Austr. NZ J. Surg.* **2003**, *73*, 567–571. [CrossRef]
4. Mayberry, J.C.; Pearson, T.E.; Wiger, K.J.; Diggs, B.S.; Mullins, R. Equestrian injury prevention efforts need more attention to novice riders. *J. Trauma.* **2007**, *62*, 735–739. [CrossRef] [PubMed]
5. Thomas, K.E.; Annest, J.L.; Gilchrist, J.; Bixby-Hammett, D.M. Non-fatal horse related injuries treated in emergency departments in the United States, 2001–2003. *Br. J. Sports Med.* **2006**, *40*, 619–626. [CrossRef] [PubMed]

6. Cowley, S.; Bowman, B.; Lawrance, M. Safety in the Victorian Thoroughbred horseracing industry. *J. Occup. Health Saf. Austr. NZ* **2007**, *23*, 421–431.
7. Curry, B.; Hitchens, P.L.; Otahal, P.; Si, L.; Palmer, A.J. Workplace injuries in Thoroughbred racing: An analysis of insurance payments and injuries amongst jockeys in Australia from 2002 to 2010. *Animals* **2015**, *5*, 897–909. [CrossRef] [PubMed]
8. Exadaktylos, A.K.; Eggli, S.; Inden, P.; Zimmermann, H. Hoof kick injuries in unmounted equestrians. Improving accident analysis and prevention by introducing an accident and emergency based relational database. *Emerg. Med. J.* **2002**, *19*, 573–575. [CrossRef] [PubMed]
9. Thompson, K.; Haigh, L. Helmet use amongst equestrians: Harnessing social and attitudinal factors revealed in online forums. *Animals* **2015**, *5*, 576–591.
10. Hawson, L.; McLean, A.; McGreevy, P.D. The roles of equine ethology and applied learning theory in horse-related human injuries. *J. Vet. Behav. Clin. Appl. Res.* **2010**, *5*, 324–338. [CrossRef]
11. Lucas, M.; Day, L.; Fritschi, L. Injuries to Australian veterinarians working with horses. *Vet. Rec.* **2009**, *164*, 207–209. [CrossRef] [PubMed]
12. Riley, C.B.; Liddiard, J.R.; Thompson, K. A cross-sectional study of horse-related injuries in veterinary and animal science students at an Australian university. *Animals* **2015**, *5*, 951–964. [CrossRef] [PubMed]
13. Thompson, K.; McGreevy, P.; McManus, P. A critical review of horse-related risk: A research agenda for safer mounts, riders and equestrian cultures. *Animals* **2015**, 561–575. [CrossRef] [PubMed]
14. Murray, J.K.; Singer, E.R.; Morgan, K.L.; Proudman, C.J.; French, N.P. The risk of a horse-and-rider partnership falling on the crosscountry phase of eventing competitions. *Equine Vet. J.* **2006**, *38*, 158–163. [CrossRef] [PubMed]
15. Thompson, K.; Matthews, C. Inroads into equestrian safety: Rider-reported factors contributing to horse-related accidents and near misses on Australian roads. *Animals* **2015**, *5*, 592–609. [CrossRef] [PubMed]
16. Johns, E.; Farrant, G.; Civil, I. Animal-related injury in an urban New Zealand population. *Injury* **2004**, *35*, 1234–1238. [CrossRef] [PubMed]
17. Fritschi, L.; Day, L.; Shirangi, A.; Robertson, I.; Lucas, M.; Vizard, A. Injury in Australian veterinarians. *Occup. Med-C.* **2006**, *56*, 199–203. [CrossRef] [PubMed]
18. McGreevy, P.D.; Henshall, C.; Starling, M.J.; McLean, A.N.; Boakes, R.A. The importance of safety signals in animal handling and training. *J. Vet. Behav. Clin. Appl. Res.* **2014**, *6*, 382–387. [CrossRef]
19. McManus, P.; Albrecht, G.; Graham, R. Constructing thoroughbred breeding landscapes: Manufactured idylls in the Upper Hunter region of Australia. In *Engineering Earth: The Impacts of Mega-Engineering Projects;* Brunn, S.D., Ed.; Dordrecht: Springer, Berlin, Germany, 2011; pp. 1323–1339.
20. Yapko, M. *Breaking the Patterns of Depression;* Broadway Books: New York, NY, USA, 1997.
21. Yapko, M. *Depression is Contagious;* Free Press: New York, NY, USA, 2009.
22. Yapko, M. *Trancework. An Introduction to the Practice of Clinical Hypnosis*, 4th ed.; Routledge: New York, NY, USA, 2012.
23. Alloy, L.B.; Peterson, C.; Abramson, L.Y.; Seligman, M.E. Attributional style and generality of learned helplessness. *J. Personal. Soc. Psychol.* **1984**, *46*, 681–687. [CrossRef]
24. Klein, G.; Altman Klein, H.; Lande, B.; Bordes, J.; Whitacre, C. Police and military as good strangers. *J. Occup. Organ. Psychol.* **2015**, *88*, 231–250. [CrossRef]
25. Klein, G. Reflections on applications of naturalistic decision making. *J. Occup. Organ. Psychol.* **2015**, *88*, 382–386. [CrossRef]
26. Hausberger, M.; Roche, H.; Henry, S. A review of the human-horse relationship. *Appl. Anim. Behav. Sci.* **2008**, *109*, 1–24. [CrossRef]
27. McGreevy, P.D. Firm but gentle: Learning the importance of handling with care. *J. Vet. Med. Educ.* **2007**, *34*, 539–541. [CrossRef] [PubMed]
28. McGreevy, P.; McLean, A. *Equitation Science;* Wiley-Blackwell: West Sussex, UK, 2010.
29. Asbridge, M.; Brubacher, J.; Chan, H. Cell phone use and traffic crash risk: A culpability analysis. *Int. J. Epidemiol.* **2013**, *42*, 259–267. [CrossRef] [PubMed]
30. Lam, L.T. Distractions and the risk of car crash injury: The effect of drivers' age. *J. Saf. Res.* **2002**, *33*, 411–419. [CrossRef]
31. Zhu, X.; Srinivasan, S. A comprehensive analysis of factors influencing the injury severity of large-truck crashes. *Accid. Anal. Prev.* **2011**, *43*, 49–57. [CrossRef] [PubMed]

32. Hitchens, P.L.; Blizzard, C.L.; Jones, G.; Day, L.M.; Fell, J. The association between jockey experience and race-day falls in flat racing in Australia. *Inj. Prev.* **2012**, *18*, 385–391. [CrossRef] [PubMed]
33. Yapko, M. Hypnotically catalysing experiential learning across treatments for depression: Actions can speak louder than moods. *Int. J. Clin. Exp. Hyp.* **2010**, *58*, 186–201. [CrossRef] [PubMed]
34. Christensen, J. Object habituation in horses: The effect of voluntary *versus* negatively reinforced approach to frightening stimuli. *Equine Vet. J.* **2013**, *45*, 298–301. [CrossRef] [PubMed]
35. McLean, A.; McLean, M. *Academic Horse Training: Equitation Science in Practice*; Australian Equine Behaviour Centre: Clonbinane, Victoria, Australia, 2008.
36. McGreevy, P.; Oddie, C.; Burton, F.; McLean, A. The horse-human dyad: Can we align horse training and handling activities with the equid social ethogram? *Vet. J.* **2009**, *181*, 12–18. [CrossRef] [PubMed]
37. DeAraugo, J.; McLean, A.; McLaren, S.; Caspar, G.; McLean, M.; McGreevy, P. Training methodologies differ with the attachment of humans to horses. *J. Vet. Behav. Clin. Appl. Res.* **2014**, *9*, 235–241. [CrossRef]
38. Starling, M.J.; Branson, N.; Cody, D.; McGreevy, P.D. Conceptualising the impact of arousal and affective states on training outcomes of operant conditioning. *Animals* **2013**, *3*, 300–317. [CrossRef] [PubMed]
39. Kazdin, A.E. The meanings and measurement of clinical significance. *J. Consult. Clin. Psychol.* **1999**, *67*, 332–339. [CrossRef] [PubMed]

Animals **2015**, *5*, 1296-1310; doi:10.3390/ani5040412

OPEN ACCESS

animals

ISSN 2076-2615

www.mdpi.com/journal/animals

Article

Health Problems and Risk Factors Associated with Long Haul Transport of Horses in Australia

Barbara Padalino [1,2,*], Evelyn Hall [1], Sharanne Raidal [3], Pietro Celi [4,5], Peter Knight [6], Leo Jeffcott [1] and Gary Muscatello [1]

[1] The Faculty of Veterinary Science, The University of Sydney, 425 Werombi Road, Camden 2570, NSW, Australia; E-Mails: evelyn.hall@sydney.edu.au (E.H.); leo.jeffcott@sydney.edu.au (L.J.); gary.muscatello@sydney.edu.au (G.M.)

[2] Department of Veterinary Medicine, The University of Bari, Bari 70100, Italy

[3] School of Animal and Veterinary Science, Charles Stuart University, Wagga Wagga 2650, NSW, Australia; E-Mail: sraidal@csu.edu.au

[4] DSM Nutritional Products, Animal Nutrition and Health, Columbia, MD 21045, USA; E-Mail: pietro.celi@dsm.com

[5] Faculty of Veterinary and Agricultural Sciences, The University of Melbourne, Parkville 3010, Australia

[6] Discipline of Biomedical Science, School of Medical Sciences, Sydney Medical School, University of Sydney, Lidcombe 1825, Australia; E-Mail: peter.knight@sydney.edu.au

* Author to whom correspondence should be addressed; E-Mail: barbara.padalino@sydney.edu.au; Tel.: +61-2-9351-1665; Fax: +61-2-9351-1693.

Academic Editor: Kirrilly Thompson

Received: 2 November 2015 / Accepted: 3 December 2015 / Published: 10 December 2015

Simple Summary: Records from road transport of horses from Perth to Sydney over a two year period were analysed to explore the incidence of transport related issues and identify risk factors. Transportation resulted in health problems in 2.8% of the transported horses, and in fatalities in 0.24%. Journey duration and season were risk factors for the development of transport related health problems, while breed, sex and age did not predict disease or injury risk. Overall, this study provides statistics to inform policy development for the equine transport industry and enhance management of the transported horse.

Abstract: Equine transportation is associated with a variety of serious health disorders causing economic losses. However; statistics on horse transport are limited and epidemiological data on transport related diseases are available only for horses transported to abattoirs for slaughter. This study analysed reports of transport related health problems identified by drivers and horse owners for 180 journeys of an Australian horse transport company transporting horses between Perth and Sydney (~4000 km) in 2013–2015. Records showed that 97.2% (1604/1650) of the horses arrived at their destination with no clinical signs of disease or injury. Based on the veterinary reports of the affected horses; the most common issues were respiratory problems (27%); gastrointestinal problems (27%); pyrexia (19%); traumatic injuries (15%); and death (12%). Journey duration and season had a significant effect on the distribution of transport related issues ($p < 0.05$); with a marked increase of the proportion of the most severe problems (*i.e.*, gastrointestinal; respiratory problems and death) in spring and after 20 h in transit. Although not statistically significant; elevated disease rate predictions were seen for stallions/colts; horses aged over 10 years; and Thoroughbreds. Overall; the data demonstrate that long haul transportation is a risk for horse health and welfare and requires appropriate management to minimize transport stress.

Keywords: transport; horse; journey duration; season; risk

1. Introduction

Transport stress in horses is caused by a myriad of stressors (e.g., isolation, confinement, noise, vibration, balance problem) which affect them both mentally and physically, causing behavioural and health problems prior to, during and after travel [1]. Both short and long trips are stressful for horses and require proper management [2]. Longer trips have a greater effect on horse health and require particular attention [3,4], and those longer than 10 h duration may lead to psychological and physical exhaustion and death [5]. Consequently, many animal transport codes include special requirements for longer journeys. For instance, the EU regulation 1/2005 [6] has specific requirements if the transport exceeds 8 h. The Australian Animal Welfare Standards and Guidelines for the Land Transport of Livestock has instead specie-specific maximum journey and minimum rest periods that take into account access to water and food en route [7].

Stress activates hormonal changes in animals, which help them to adapt to the stressful situation. This response is commonly referred as "the flight or fight response", and it is characterized by the activation of the pituitary and adrenal responses and by a release of adrenaline and cortisol. The most common effects of adrenaline are an increase in heart and respiratory rates, and an increase in sweating and defecation [8]. During transportation these hormonal responses are often a result of the horse attempting to adapt to the challenging situation (being transported), but they can affect the horse's immune response, making the horse more susceptible to transport-related diseases [9,10].

Transportation has been associated with physical injuries and heat stress, as well as specific illnesses such as respiratory diseases, colic, laminitis, enterocolitis and rhabdomyolsis [1,3,11]. The most serious, potentially fatal respiratory disease is equine pleuropneumonia, commonly referred to as "travel sickness" or "shipping fever" [12]. The risk of developing this disease increases with journey duration, especially when the duration exceeds 10 h [13]. Predisposing factors for the development of shipping fever include prolonged head elevation [14], poor air quality [15], and pre-existing respiratory diseases [5]. Transport associated dehydration, withholding of food and water, and diet change on arrival have been proposed as risk factors in the development of transport related gastrointestinal disease in horses [1]. Colic during or after transportation is commonly reported, with impaction of large colon most often recognised [1,11]. Enterocolitis caused by *Salmonella spp.* has been also associated with transport stress [16] and can be fatal. While many risk factors for the development of transport related diseases and injuries have been identified, further studies are required to identify additional unrecognised factors, and to determine the relative contribution of different contributing factors to transport related disease and injury. Knowledge of the full range of risk factors related to equine transportation may help to safeguard the welfare and wellbeing of horses.

Surveys on farm animal transportation have been performed in to identify risk factors and explore epidemiological basis of transport related health and welfare issues. For instance, the incidence of mortality during road transport has been calculated for cattle in North America (0.011%) [17], bobby calves in Australia (0.64%) [18], pigs in Europe (0.07%) [19] and in broilers in Brazil (from 0.42% in summer to 0.23% in autumn) [20]. In horses, surveys have been reported only for transport to abattoirs/slaughter plants [21–23]. In these studies, transport related health problems ranged from 7% to 28%. However, large numbers of horses are transported for other commercial activities such as competition and breeding, and for recreational uses. As these animals have a greater economic value than those destined for abattoirs, it is likely that their management and their transport-related health problem incidence will be different.

Millions of horses are moved daily all over the world, with the true global total of horse transport movements so large that it is impossible to estimate [24]. Consequently, there is a gap in our knowledge of the incidence of transport related problems, horse mortality, and risk assessment related to equine commercial transportation. To the authors' knowledge, a survey on commercial equine road transport for any purpose has never been conducted in Australia. As the scientific identification and evaluation of hazards can only be done when the scenario including the animal and the transport environment is defined [25], the records of a horse transport company specialized in long road trips (~4000 km, taking 3.5 days) in Australia were collected and analysed. The objective of the present study was to determine the incidence of transport related injury and illness in horses undertaking commercial long-distance road transportation, and to assess and quantify the relationship between animal (sex, breed, and age), transport conditions (duration, and season) and welfare outcomes measured by the incidence of death, injuries, pyrexia, respiratory and gastrointestinal problems associated with a defined commercial long haul transport in Australia.

2. Experimental Section

2.1. Materials and Methods

Records of all transport movements from April 2013 to April 2015 were obtained from a commercial horse transport company which regularly transports horses between the east and west coast of Australia (~4000 km and at least 3.5 days duration). Care and handling of the animals during transportation was not supervised by the research team. This data set was collected as part of a comprehensive survey on horse transportation approved by the Human Research Ethics Committee of the University of Sydney (2015/308).

2.2. Trip Details

Before booking the trip, each owner had to send to the company the following information: breed, sex, age, body measurement, level of tame, reason for transportation. This information was necessary to allocate the right space to each animal inside the vehicle. As policy the company moved only tamed horses, at least trained to halter and basic commands from ground (e.g., follow and stop at the rope) and advised previous transport experience, the transported animals complied with this policy.

All transportation was performed following a fixed schedule from a collection stable in Sydney. The trip consisted of four stages: Sydney-Melbourne (10 h), Melbourne-Adelaide (8.5 h), Adelaide-Kalgoorlie (24 h) and Kalgoorlie-Perth (6 h). After each stage, horses were given a twelve hour rest period. The total duration was approximately 85 h with approximately 49 h in transit and 36 h for rest stops. The schedule was reversed for Perth-Sydney trips.

At the collection stable and rest points, horses were individually housed in in-walk out rubber lined stables and/or paddocks that were used only for horses in transit.

All animals travelled on the same type of vehicle (Mega Ark Trailers, MAN®, Munich, Germany, Europe) equipped with 15 horse individual stalls, 6 facing backwards and 9 facing forwards. However, since large horses were allocated 1½ stall spaces, the average number of horses transported per trip was 9.1.

The ventilation system comprised venturi vents, louvres and electric fans generating an airflow which the manufacturer verified was compliant with the Australian code of transportation throughout the trailer. When the vehicle was moving fresh air entered through the louvres and was extracted by the venturi vents. The fans were used in extreme heat conditions (> 35 °C–40 °C), and to ensure constant air flow when the truck was stationary (e.g., feeding and watering times, fuel stops).

The horses travelled in individual stalls, restrained by rubber cords which would break under extreme pressure. Foals, ponies, weanlings or un-educated horses were not tied up. Mares and foals travelled in a 3 stall section which allowed them to move around as if in a small box. Horses were fed and watered every 4–5 h, using the stainless steel feed and water bins in each bay; water and food were refreshed regularly en route.

Two drivers were used for all trips for which data were collected. Both were licensed to drive heavy combination vehicles and were experienced horse handlers with many years' experience in commercial horse enterprises.

All journeys complied with the standards and the guidelines for the transport of horses required by the Australian Code of Livestock Land Transportation [26].

2.3. Monitoring of the Animals and Identification of Pathology

The assessment of the fitness for travel of the horses was performed by the drivers and experienced staff members of the company at the collection stable and at each transit stable before loading the animals to continue the journey. The assessment was made in accordance with the Australian Code of Livestock Land Transportation. The condition to be assessed included any signs of colic, raised or lowered rectal temperature, lethargy, diarrhoea, wound or abscesses, lameness (no more or equal to the fourth grade) and body condition score (no less of two) [26]. After the assessment the report was sent to the operation manager, who gave final approval for transport.

During transport, there were two opportunities for monitoring horse health. The first was during the mandatory rest stops which are required after four hours of driving. At these stops, the driver undertook a visual examination of the horses. The second opportunity for monitoring occurred at the transit stables where rectal temperature was recorded, and drinking, feeding and eliminating behaviours monitored. At the rest stop in Adelaide, horses were inspected by an Australian government accredited veterinarian who confirmed that horses were fit to continue their journey, administered a triclabendazole drench and collected a faecal sample in compliance with western Australian quarantine regulations.

As soon as any health problem was identified by the drivers, the company manager was informed and a veterinarian was called for consultation and for treating the affected horse. The company director had a list of veterinarians to call in emergency. The affected horse did not continue the trip if the veterinarian did not evaluate it fit for travel. When health problems were identified by the owners post transport, the transport company manager was informed and he required a veterinarian's report to prove that the problem was related to the previous journey.

2.4. Dataset

The data set included 1650 horses transported from Perth to Sydney (~4000 km) or vice versa for 180 journeys. Horse details (breed, sex, age) and the date of the trip (month and year) were recorded. The data set included reports of problems and issues identified by the drivers and horse owners and sent to the company manager, including the type of problem(s) and where (e.g., location) and when it occurred (*i.e.*, an estimation of the approximate time at which the issues or incidents were first identified). As after the identification of each problem, horses were checked and treated by veterinarians, and after death necroscopy was conducted, their veterinary records were also included in the dataset.

For statistical analysis, the recorded transport related issues and problems were classified according to the time of occurrence in the following categories: pre loading (from the horse's home stable to the company's collection stable in Sydney or Perth); in transit (during the trip or at rest stops); and post-transport (within 3 days after arrival at destination).

Based on the veterinary records, the transport related health problems were also classified into five categories according to clinical signs/body system affected (Table 1).

Table 1. Description of the transport related issues.

Category	Definition
Injuries	Laceration, abrasion, contusion, swelling.
Pyrexia	Rectal temperature >38.5 °C, in the absence of other localising signs.
Gastrointestinal problems	Colic, enterocolitis, large quantity of internal parasites eliminated after triclabendazole treatment.
Respiratory problems	Nasal discharge, coughing, inflammation/infection of the upper or lower respiratory tract, and pneumonia.
Death	Horses found dead or humanely destroyed.

2.5. Statistical Analysis

Descriptive analysis of the dataset was conducted using statulor^{beta} [27]; data were reported as number of injuries or illnesses and as percentages. All further statistical analyses were performed using Gen Stat® Version 14 (VSNi International, Hemel Hempstead, UK). For all statistical analyses, a p value of <0.05 was considered statistically significant.

The details of all the travelled horses were categorized according to sex (mare/filly, gelding, stallion/colt), age (weaning/foal, yearling, 2–5 years, 6–10 years, >10 years), and breed (Arab, Quarter horse, Standardbred, Thoroughbred, Warm Blood). Univariate logistic regression analysis was conducted with development of a transport-related problem as the outcome (1/0: affected/non affected), and sex, breed, and age as the explanatory variables. Wald tests were obtained along with mean predictions of disease rate for each variable.

The date of recorded transport-related issues was categorized into the four Southern hemisphere seasons: winter (June–August), spring (September–November), summer (December–February), autumn (March–May). The time of the recorded transport-related issues (calculated from departure to when the recorded transport related issue or problem was identified) was classified into three categories of journey duration: <20 h, 21 h–40 h, and >40 h. Based on the veterinary records, considering the severity of the clinical signs, the required treatments and the time of recovery, the type of transport-related issues was listed in order of increasing severity as follow: injuries, pyrexia, gastrointestinal problems, respiratory problems, death. Ordinal regression analysis was then conducted to study the association between the type of transport-related issues (outcome) and the journey duration (<20 h, 21 h–40 h, >40 h), and the season of the year (winter, autumn, summer, spring) (factors).

3. Results

3.1. Descriptive Statistics

The general demographics of the population of horses studied is shown in Table 2. Horses were transported for the following reasons: sales-purchase (30%), competition (50%), and breeding (20%).

Table 2. Frequency of the total transported horses by sex, breed and age category.

Variable	Category	Frequency (%)
Sex	Gelding	35.7
	Mare/Filly	49.5
	Stallion/Colt	14.8
Breed	Arab	9.6
	Quarter horse	8.7
	Standardbred	27.5
	Thoroughbred	43.0
	Warm blood	11.2
Age	Weaning/Foal	11.2
	Yearling	12.9
	2–5 yrs	34.7
	6–10 yrs	27.9
	>10 yrs	13.3

Approximately 97 % (1604/1650) of the horses arrived at destination in good health, without any pain, signs of lameness or other pathology and did not develop any diseases post journey.

Only 2.8% (46/1650) were included in the company dataset for a transport related issue at pre-loading, in transit, or post-transit (Figure 1). Of the 46 cases, five cases related to pre-loading events, three were injuries that had occurred before the trip commenced or during transport to the collection stable, and two were cases of colic identified at the departure stable. Of the remaining cases, two horses were injured during loading while resisting loading and 4 injuries happened in transit. All the injuries were minor and the horses were treated topically and continued their journey. Six horses were identified as febrile at rest stops, and another two were identified as febrile upon arrival. No localising signs were identified in any of these courses and all were treated with anti-inflammatory medications. Four horses showed signs of colic at rest stops with another three showing signs of colic post journey. All cases were interpreted as impaction colic; two resolved without treatment (required only monitoring) and 5 were treated medically. Enterocolitis was identified in one horse during transport, and in two horses post transport, with all horses requiring hospitalisation. One horse eliminated a massive quantity of parasites after the anti-parasite treatment. Five horses developed respiratory signs, including nasal discharge, coughing, and pyrexia during the journey, and one developed signs after arrival. The veterinary diagnosis was inflammation of the upper or lower airways, without pneumonia and all cases were treated medically. The specific diagnosis of pneumonia was made on the basis of signs that developed in four horses during transport, and in one horse after arrival. All horses recovered after appropriate medical treatment. There were four transport related deaths, giving an overall death rate of 0.24%. Two occurred during transport, one horse was found dead within 24 h after transportation, and one was humanely destroyed due to enterocolitis post transit. Another horse was found dead two days after transport, and it cannot be confirmed that the death was transport related. Post-mortem examination failed to reveal the cause of death in the four horses that were found dead. If the horse that was found dead two days after transport is included in the statistics, the death rate increases to 0.30%.

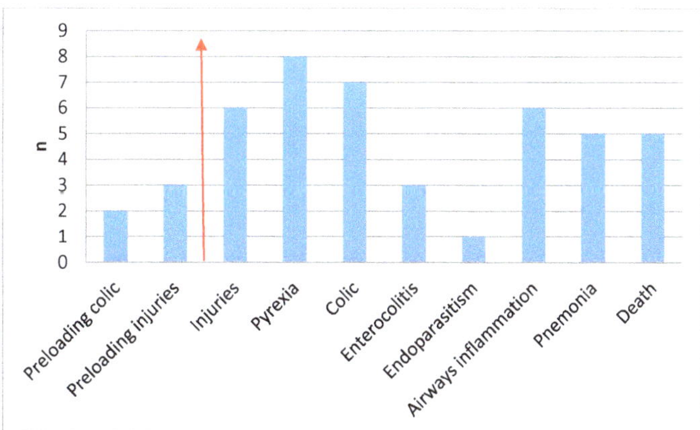

Figure 1. Incidence of transport-related issues as reported by the transport company. The arrow divides issues related to pre-loading from those related to transit and post transit phase.

The incidence of the transport-related issues grouped by category is shown in Table 3.

Table 3. Incidence of transportation issues grouped in 5 major categories according to clinical signs and body system affected.

Category	*n*	Incidence on the Affected Animals (*n* = 41)	Incidence on All Transported Animals (*n* = 1650)
Injuries	6	15%	0.36%
Pyrexia	8	19%	0.48%
Gastrointestinal problems	11	27%	0.66%
Respiratory problems	11	27%	0.66%
Death	5	12%	0.30%

3.2. Logistic Regression

Univariate logistic regression analysis of horses experiencing transport related health issues showed no significant effect of sex, breed, or age (Table 4).

Table 4. Results of the univariate logistic regression analysis with development of a transport-related problem as the outcome (1/0: affected/non affected), with sex, breed, and age as explanatory variable.

Variable	Category	Disease Rate Prediction (%) ± s.e	Estimate ± s.e (%)	OR	Lower 5% CI	Upper 95% CI	*p* Value
Sex	Gelding	7.0 ± 1.5	Ref.				0.611
	Mare/Filly	5.3 ± 1.1	−0.29 ± 0.3	0.74	0.39	1.424	
	Stallion/Colt	7.1 ± 2.4	0.02 ± 0.4	1.02	0.43	2.403	
Breed	Arab	3.3 ± 2.2	Ref.				0.187
	Quarter horse	5.4 ± 3.0	0.14 ± 0.7	1.16	0.25	5.217	
	Standardbred	5.2 ± 1.6	0.56 ± 0.6	1.75	0.50	6.039	
	Thoroughbred	9.2 ± 1.7	0.72 ± 0.6	2.05	0.59	7.141	
	Warm blood	2.5 ± 1.9	0.96 ± 0.6	2.61	0.70	9.743	
Age	Weaning/Foal	1.8 ± 1.0	Ref.				0.523
	Yearling	2.1 ± 1.0	0.51 ± 0.9	1.67	0.26	10.41	
	2–5 yrs	3.2 ± 0.8	−7.20 ± 9.3	0.00	8.12E-12	68858	
	6–10 yrs	3.8 ± 0.9	1.08 ± 0.7	2.94	0.67	12.8	
	>10 yrs	4.8 ± 1.5	−0.10 ± 1.0	0.85	0.11	6.246	

Standard error (s.e), Odds ratio (OR), Confidence Interval (CI).

3.3. Ordinal Regression Analysis

There was a significant association ($p = 0.022$) between type of transport related issues and duration of trip, with a higher probability of a more severe disease after 20 h of transport. Injuries were more likely to occur in the first 20 h of transport (Figure 2).

Table 5 shows odd ratio and confidence interval for each disease occurring in a journey longer than 20 h.

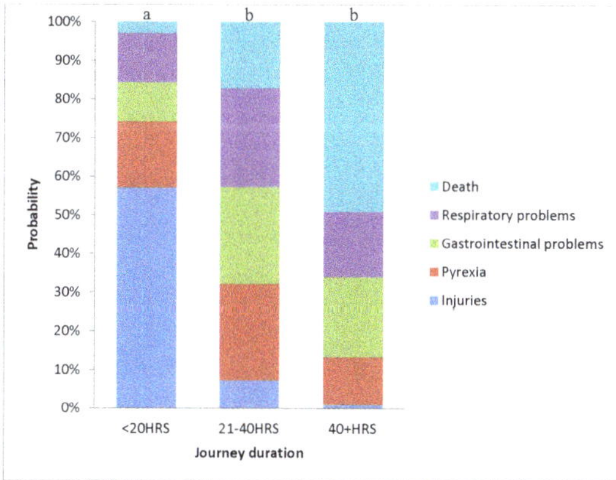

Figure 2. Probability of a more severe transport-related issue to be associated with journey-duration. Bar charts with different letter have a different distribution of transport-related issues: a, b: $p < 0.05$.

Table 5. Estimate, odds ratio (OR) and confidence interval (CI) for each transport-related problem on a journey longer than 20 h.

Transport-Related Problem	Estimate	s.e	*p* Value	OR	Lower 95% CI OR	Upper 95% CI OR
Injury						
Fever	2.56	1.34	0.057	12.91	0.9271	179.8
Colic	3.19	1.3	0.014	24.37	1.91	311.1
Respiratory	2.93	1.32	0.027	18.69	1.399	249.7
Death	4.54	1.52	0.003	93.49	4.783	1827

Season had a significant effect on the distribution of transport-related issues ($p = 0.035$), with a higher probability to have a more severe transport-related issue (gastrointestinal problems, respiratory problems and death) in spring (Figure 3).

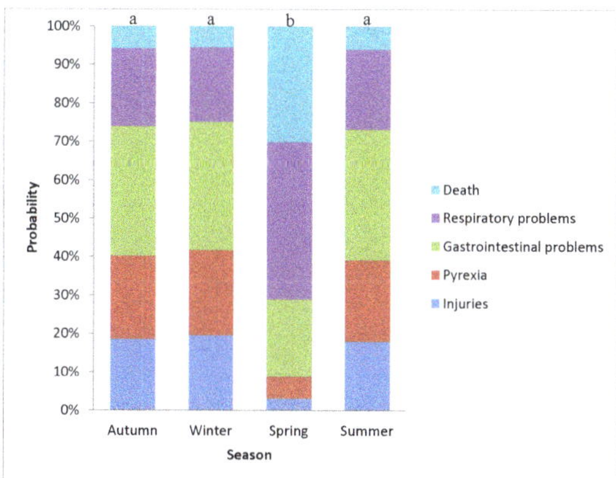

Figure 3. Probability of a more severe transport-related issue to be associated with season. Bar charts with different letter have a different distribution of transport-related issues: a, b: $p < 0.05$

4. Discussion

The present study reports the incidence of transport-related issues and mortality associated with long-haul equine transport by a commercial equine transport company. An overall incidence of transport related injuries or disease of 2.8% was observed in this study, which is much lower than has been reported by horse owners [28,29] and for horses transported to abattoirs for slaughter [21,23]. The data demonstrates that travelling is a risk to equine health and welfare and a correct management of transportation is required for moving horse successfully.

The prevalence of injuries identified in the present study (0.36% of horses) was lower than the rate reported in horses transported to abattoir for slaughter [21,23]. This may reflect differences in the way in which the transport was managed (*i.e.*, individual calculated space), and in the tractability and transport experience of the horses. However, it was also lower than the reported by owners during non-commercial horse transportation [29], which suggests that transport management is a key determinant of the injury rate. The design of the truck (including the floor, suspension and the height of height of the roof) has been identified as a risk factor in the development of injuries and transport related diseases in farm animals [25]. In cattle, the incidence of injuries during long haul transportation has been associated with the years of experience of the drivers [17], and in horses travelling on non-commercial road transport, many incidents are related to poor driving skills, particularly on winding country roads [29]. Vehicle design, road quality and driving that allows horses to keep their balance appears to be important in minimising injuries.

Two injuries occurred during the loading process due to resistance to boarding the truck. The horses' fear of being loaded onto the vehicle manifests through various deleterious and dangerous actions and behaviours (e.g., kicking, rearing) [1]. Handling horses during loading/unloading is therefore considered to be highly dangerous risk for those handling the transported horse [30]. Thus it is important that transport procedures are carried out by experienced horse handlers wearing protective equipment, such as, capped boots and gloves, to minimize the risk of injuries to the horse and the handler during the loading and unloading phase of equine transport.

The body systems most commonly affected by transport are the respiratory and the gastrointestinal systems, and a common clinical sign associated with inflammation to these systems is pyrexia [5]. Pyrexia affected 0.48% of all transported horses in this study. Since early identification of pyrexia prompts investigation and implementation of appropriate therapy and recovery, checking temperature during and after long trips should be seen as best practice when dealing with the transported horse and has been recommended previously [31,32].

Transportation has been associated with the development of airway inflammation and equine pleuro-pneumonia [12]. Poor ventilation inside the truck, forced high head position, and dehydration have been identified as predisposition factors in the development of respiratory diseases associated with transport [3]. The horses in the current study travelled in a vehicle equipped with a forced ventilation system, which should assure good air quality and a comfortable temperature inside the trailer, but horses were not allowed to lower their head beyond the height of the wither. In our study 0.66% of all transported horses developed respiratory problems, and only five developed pneumonia. This rate is less than expected [28], potentially reflecting the importance of a good ventilation system in the transport vehicle. However, there is no evidence to suggest that good ventilation alleviates the need to provide correct watering [33], to minimise the duration of confinement with the head elevated, or to maximize the time available for the horse to physically clear its respiratory tract [34].

Transportation may increase the likelihood of colic for several reasons [35]. Firstly during a stressful situation, preferential perfusion of the brain and the muscles may reduce visceral perfusions (flight and fight response) [8]. Additionally, dehydration during transportation can reduce vascular perfusion of the gut, potentially inducing impaction of the colon [11]. Finally food and water withdrawal, altered diet and/or eating in an unnatural position on route might create change in the pH and gut flora which may influence the chance of colitis. In this study, transport associated gastrointestinal problems were seen in 27% of cases, with enterocolitis seen in 4 out 41 cases, with one requiring euthanasia. Equine enterocolitis, can manifest in sudden death and be associated with over-proliferation of *Salmonella spp*, *Clostridium spp* and *Fusarium spp* in the equine gut [36]. Stress is considered an important predisposing factor for salmonellosis in horses; this pathology has already been associated with transportation, surgery, feed withdrawal, changes in feed, and antimicrobial and anthelminthic therapy [16]. Avoiding prolonged feed and water restriction, abrupt diet changes, or overuse of antibiotics and anthelminthics before, during and after transportation could therefore potentially reduce the incidence of gastrointestinal transport associated illness.

In this study the mortality rate associated with transport events was 0.24% or 0.30%, which is somewhat higher than rates reported for cattle in North America (0.01%) [17] and pigs in Europe (0.07%) [19]. A reason could be that animal transportation is more risky in Australia due to the long journey duration and the climate; the rate described in this study is indeed in line with mortality rates observed in cattle transported in Queensland by rail (an overall mortality rate of 0.10%, ranging from 0.44% in bulls to 0.06% in calves) [37], and in bobby calves (>4 days old) transported by road in Victoria (0.64%) [18]. Thus, moving horses in Australia may require more detailed and specific strategies to cope with extreme distance and weather.

In the current study, two horses were found dead in transit and two died soon after transportation. Even after pathological examination the reasons for these deaths were unknown. It is possible that protracted stress may have contributed to death. Stress is a physiological and endocrionological response that helps individuals to cope with stressors and to survive. However, when an animal fails to adapt, the stress response can lead to death [8]. In horses, transport stress is often followed by the stress of living in unfamiliar environments (e.g., new stall, food, social group), further affecting the horse's health. Consequently, offering similar feed and avoid inserting the recent arrival into a new herd could reduce protracted stress and assist with adaptation to the unfamiliar environment after the journey [1] and potentially reduce the risk of death after transportation events.

Witnessing a death of any animal can have a negative impact on those who have witnessed it, whether they are professionals working in the animal or veterinary industry or members of the general public owning animals [38]. Consequently, minimising equine transport associated mortality rate will have a positive impact on the wellbeing of horses and the mental wellbeing of those dealing with the transported horse on a day to day basis.

No significant effect of sex, age, or breed in the development of transport diseases was found, suggesting that individual horse variability and past experience might be more important in influencing the ability of the horse to cope with the transport event [39]. Elevated, but not statistically significant, prediction rate were seen for males, horses aged over 10 years, and Thoroughbreds, consequently dealing with these categories of horses may require specific transport management strategies to reduce the predicted risk of these horses developing transport related complications.

In agreement with Stefancic and Martin [22], although numbers were small, this current study showed transported related death and respiratory diseases were more likely to occur in spring. This may reflect abrupt temperature increases at a time when many horses still have winter coats, with consequent, impaired thermoregulation in transit. Alternatively, animal behaviour and immune system may be affected by the reproductive hormonal profile of the breeding season [40]. Other factors, such as the occurrence of viral respiratory tract infections or increased pollen or other allergens might contribute to increased risk of respiratory disease at this time. Such speculations warrant further research.

The data presented in the current study, confirm the increased risk of mortality and disease in horses associated with longer transport events, with more severe diseases (e.g., enterocolitis, pleuro-pneumonia) and death more commonly observed after 20 h of transportation. Horses travelling to abattoirs for slaughter, were similarly more likely to die after protracted transportation [22]; and in cattle, higher mortality rates have been associated with trips longer than 36 h in Australia [37], and longer than 30 h in Canada [17]. Better understanding of the increasing risk of severe transport related diseases and death with increasing duration of transport may encourage the adoption of more rigorous preventive strategies, such as a veterinary examination, for horses that are going to be transported for more than 20 h.

The biggest limitation of this study is that the assessment of the horse health before, during and after journeys was not performed by the authors. Consequently some transport-related problems might have been missed by the drivers and the owners, and the incidence of the transport-related diseases could be underestimated. This is particularly likely if horse owners did not associate illness with recent transport, failed to recognise minor or subclinical disease, or failed to report minor or major illness to the company. Pleuro-pneumonia and enterocolitis can manifest up to a week after transport [5], and other effects of transportation stress can take up to one month to manifest after the event [41]. Hence it is possible that owners or agents may have failed to associate disease with transportation, or may have failed to detect mild effects on horse health. The data obtained in this current study is also limited by the lack of environmental parameters measured or recorded during journeys. Extreme hot and cold temperatures have been identified as risk factors in long haul transportation of farm animals [42]. Notwithstanding these limitations, this study is the first carried out on horses undertaking this unique multi-day road trip across one of the harshest continents in the world. It has provided important data for the equine industry on the incidence of health problems associated with long haul transportation in the horse. Preliminary evaluation has identified and suggested some predisposing factors associated with transport related health problems which warrant further evaluation to enhance policy and practices relating to transportation of the horse.

5. Conclusions

Journey duration and season were identified as risk factors contributing to transport related health problems in horses undergoing long distance road transportation. Although the trips were well organized and complied with or exceeded the requirements of the National Code of Practice for the Transportation of Horses, serious diseases still occurred. Moving horses should be considered as a human-related risk to horses and also a horse-related risk to humans [43], so it should be always carried out by professional and experienced horse handlers and drivers, wearing adequate protective equipment, to reduce the risk of injuries and diseases in both horses and humans. Further research to confirm preliminary conclusions based on this data and to recognize other risk factors for the development of equine transport related issues is needed to assist in improvement of the Australian code of horse transportation.

Acknowledgments

The authors thank the equine transport company for providing the data used in this survey.

Author Contributions

Barbara Padalino, Evelyn Hall, Sharanne Raidal, Pietro Celi, Peter Knight, Leo Jeffcott, and Gary Muscatello conceived and designed the experiments; Barbara Padalino and Evelyn Hall analysed the data; Barbara Padalino wrote the paper; Evelyn Hall, Leo Jeffcott, Sharanne Raidal, Peter Knight, and Gary Muscatello edited the paper.

Conflicts of Interest

The authors declare no conflict of interest.

References

1. Padalino, B. Effects of the different transport phases on equine health status, behavior, and welfare: A review. *J. Vet. Behav. Clin. Appl. Res.* **2015**, *10*, 272–282. [CrossRef]
2. Tateo, A.; Padalino, B.; Boccaccio, M.; Maggiolino, A.; Centoducati, P. Transport stress in horses: Effects of two different distances. *J. Vet. Behav. Clin. Appl. Res.* **2012**, *7*, 33–42. [CrossRef]
3. Leadon, D.P.; Hodgson, D.R. Transport of horses. In *The Athletic Horse: Principles and Practice of Equine Sports Medicine*, 2nd ed.; Elsevier-Saunders: St. Louis, MO, USA, 2014; pp. 155–161.
4. Marlin, D.; Meldrum, K.; White, J.; Westen, H.; Parkin, T.; Wood, J.; Broom, D.; Kennedy, M.; Johnson, B.; Coombs, S.; *et al.* Welfare of horses transported long distances in Europe. *Vet. Rec.* **2009**, *165*. [CrossRef]
5. Marlin, D.J. Transport of horses. In *Equine Sports Medicine and Surgery*; Elsevier-Saunders: St. Louis, MO, USA, 2004; pp. 1239–1250.
6. Animal Welfare during Transport. Available online: http://eur-lex.europa.eu/legal-content/EN/TXT/?uri=uriserv:f83007 (accessed on 1 October 2015).
7. De Witte, K. Development of the Australian Animal Welfare Standards and Guidelines for the Land Transport of Livestock: Process and philosophical considerations. *J. Vet. Behav. Clin. Appl. Res.* **2009**, *4*, 148–156. [CrossRef]
8. Moberg, G.P.; Mench, J.A. *The Biology of Animal Stress: Basic Principles and Implications for Animal Welfare*; CABI Publishing: Oxfordshire, UK, 2000; p. 377.
9. Stull, C.L.; Spier, S.J.; Aldridge, B.M.; Blanchard, M.; Stott, J.L. Immunological response to long-term transport stress in mature horses and effects of adaptogenic dietary supplementation as an immunomodulator. *Equine Vet. J.* **2004**, *36*, 583–589. [CrossRef] [PubMed]

10. Stull, C.L.; Morrow, J.; Aldridge, B.A.; Stott, J.L.; McGlone, J.J. Immunophysiological responses of horses to a 12-hour rest during 24 hours of road transport. *Vet. Rec.* **2008**, *162*, 609–614. [CrossRef] [PubMed]
11. Mansmann, R.A.; Woodie, B. Equine transportation problems and some preventives: A review. *J. Equine Vet. Sci.* **1995**, *15*, 141–144. [CrossRef]
12. Copas, V. Diagnosis and treatment of equine pleuropneumonia. *Practice* **2011**, *33*, 155–162. [CrossRef]
13. Austin, S.M.; Foreman, J.H.; Hungerford, L.L. Case-control study of risk factors for development of pleuropneumonia in horses. *J. Am. Vet. Med. Assoc.* **1995**, *207*, 325–328. [PubMed]
14. Raidal, S.L.; Love, D.N.; Bailey, G.D. Inflammation and increased numbers of bacteria in the lower respiratory tract of horses within 6 to 12 hours of confinement with the head elevated. *Aust. Vet. J.* **1995**, *72*, 45–50. [CrossRef] [PubMed]
15. Katayama, Y.; Oikawa, M.; Yoshihara, T. Clinico-pathological effects of atmosferic ammonia exposure on horses. *J. Equine Sci.* **1995**, *6*, 99–104. [CrossRef]
16. Feary, D.J.; Hassel, D.M. Enteritis and colitis in horses. *Vet. Clin. N. Am. Equine Pract.* **2006**, *22*, 437–479. [CrossRef] [PubMed]
17. González, L.A.; Schwartzkopf-Genswein, K.S.; Bryan, M.; Silasi, R.; Brown, F. Relationships between transport conditions and welfare outcomes during commercial long haul transport of cattle in north America. *J. Anim. Sci.* **2012**, *90*, 3640–3651. [CrossRef] [PubMed]
18. Cave, J.G.; Callinan, A.P.L.; Woonton, W.K. Mortalities in bobby calves associated with long distance transport. *Aust. Vet. J.* **2005**, *83*, 82–84. [CrossRef] [PubMed]
19. Averós, X.; Knowles, T.G.; Brown, S.N.; Warriss, P.D.; Gosálvez, L.F. Factors affecting the mortality of weaned piglets during commercial transport between farms. *Vet. Rec.* **2010**, *167*, 815–819. [CrossRef] [PubMed]
20. Vieira, F.M.C.; Silva, I.J.O.; Barbosa Filho, J.A.D.; Vieira, A.M.C.; Broom, D.M. Preslaughter mortality of broilers in relation to lairage and season in a subtropical climate. *Poult. Sci.* **2011**, *90*, 2127–2133. [CrossRef] [PubMed]
21. Marlin, D.; Kettlewell, P.; Parkin, T.; Kennedy, M.; Broom, D.; Wood, J. Welfare and health of horses transported for slaughter within the European Union Part 1: Methodology and descriptive data. *Equine Vet. J.* **2011**, *43*, 78–87. [CrossRef] [PubMed]
22. Stefancic, I.; Martin, D. *Influence of Transport Conditions on Animal Welfare*; ISAH: Warsaw, Poland, 2005; pp. 148–152.
23. Grandin, T.; McGee, K.; Lanier, J.L. Prevalence of severe welfare problems in horses that arrive at slaughter plants. *J. Am. Vet. Med. Assoc.* **1999**, *214*, 1531–1533. [PubMed]
24. Herholz, C.; Füssel, A.E.; Timoney, P.; Schwermer, H.; Bruckner, L.; Leadon, D. Equine travellers to the Olympic Games in Hong Kong 2008: A review of worldwide challenges to equine health, with particular reference to vector-borne diseases. *Equine Vet. J.* **2008**, *40*, 87–95. [CrossRef] [PubMed]
25. Marahrens, M.; Kleinschmidt, N.; Di Nardo, A.; Velarde, A.; Fuentes, C.; Truar, A.; Otero, J.L.; Di Fede, E.; Villa, P.D. Risk assessment in animal welfare—Especially referring to animal transport. *Prev. Vet. Med.* **2011**, *102*, 157–163. [CrossRef] [PubMed]
26. Australian Animal Welfare Standards and Guidelines. Available online: http://www.animalwelfarestandards.net.au/land-transport/ (accessed on 1 October 2015).

27. Descriptive Analysis and Data Visualization. Available online: http://statulator.com/descriptive.html (accessed on 20 September 2015).
28. Padalino, B.; Hall, E.; Raidal, S.L.; Knight, P.; Celi, P.; Jeffcott, L.; Muscatello, G. Survey on horse transportation:issues and practices. In *Postgraduate Conference 2015*; The University of Sydney: Camden, NSW, Australia, 2015; Volume 1, p. 71.
29. Noble, B.; Riley, C.B.; Thompson, K. Incidence and risk factors associated with horse injuries sustained during noncommercial transport. In *International Large Animal Rescue*; ILAR: Adelaide, Australia, 2013; Volume 1, p. 10.
30. Riley, C.B.; Liddiard, J.R.; Thompson, K. A cross-sectional study of horse-related injuries in veterinary and animal science students at an Australian university. *Animals* **2015**, *5*, 951–964. [CrossRef] [PubMed]
31. Raidal, S.L.; Bailey, G.D.; Love, D.N. Effect of transportation on lower respiratory tract contamination and peripheral blood neutrophil function. *Aust. Vet. J.* **1997**, *75*, 433–438. [CrossRef] [PubMed]
32. Racklyeft, D.J.; Raidal, S.; Love, D.N. Towards an understanding of equine pleuropneumonia: Factors relevant for control. *Aust. Vet. J.* **2000**, *78*, 334–338. [CrossRef] [PubMed]
33. Anon. *Practical Guidelines on the Watering of Equine Animals Transported by Road*; World Horse Welfare: London, UK, 2014; p. 36.
34. Stull, C.L.; Rodiek, A.V. Effects of cross-tying horses during 24 h of road transport. *Equine Vet. J.* **2002**, *34*, 550–555. [CrossRef] [PubMed]
35. Tinker, M.K.; White, N.A.; Lessard, P.; Thatcher, C.D.; Pelzer, K.D.; Davis, B.; Carme, D.K. Prospective study of equine colic risk factors. *Equine Vet. J.* **1997**, *29*, 454–458. [CrossRef] [PubMed]
36. Staempfli, H.R.; Townsend, G.G.; Prescot, J.F. Prognostic features and clinical presentation of acute idiopathic enterocolitis in horses. *Can. Vet. J.* **1991**, *32*, 232–237. [PubMed]
37. Jarratt, I.S.; Tobin, J.G.; Saal, R.L. Railed cattle mortalities in Queensland. In *Animal Production in Australia*; Pergamon Press: Sydney, Australia, 1982; pp. 253–256.
38. Foster, S.M.; Maples, E.H. Occupational stress in veterinary support staff. *J. Vet. Med. Educ.* **2014**, *41*, 102–110. [CrossRef] [PubMed]
39. Fazio, E.; Medica, P.; Cravana, C.; Ferlazzo, A. Cortisol response to road transport stress in calm and nervous stallions. *J. Vet. Behav. Clin. Appl. Res.* **2013**, *8*, 231–237. [CrossRef]
40. Styrt, B.; Sugarman, B. Estrogens and infection. *Rev. Infect. Dis.* **1991**, *13*, 1139–1150. [CrossRef] [PubMed]
41. Cregier, S.E.; Gimenez, R. *Non-Commercial Horse Transport: New Standards for Trailer in Canada*; Cregier S: Montague, Canada, 2015; p. 228.
42. Schwartzkopf-Genswein, K.S.; Faucitano, L.; Dadgar, S.; Shand, P.; González, L.A.; Crowe, T.G. Road transport of cattle, swine and poultry in north America and its impact on animal welfare, carcass and meat quality: A review. *Meat Sci.* **2012**, *92*, 227–243. [CrossRef] [PubMed]
43. Thompson, K.; McGreevy, P.; McManus, P. A critical review of horse-related risk: A research agenda for safer mounts, riders and equestrian cultures. *Animals* **2015**, *5*, 561–575. [CrossRef] [PubMed]

Animals **2015**, *5*, 1136-1146; doi:10.3390/ani5040402

OPEN ACCESS

animals

ISSN 2076-2615

www.mdpi.com/journal/animals

Article

Development of a Safety Management Web Tool for Horse Stables

Jarkko Leppälä [1,*], Christina Lunner Kolstrup [2], Stefan Pinzke [2], Risto Rautiainen [1,3], Markku Saastamoinen [4] and Susanna Särkijärvi [4]

[1] Natural Resources Institute Finland (Luke), Latokartanonkaari 9, 00790 Helsinki, Finland

[2] Swedish University of Agricultural Sciences (SLU), 23053 Alnarp, Sweden; E-Mails: christina.kolstrup@slu.se (C.L.K.); stefan.pinzke@slu.se (S.P.)

[3] University of Nebraska Medical Center, Omaha, NE 68198, USA; E-Mail: risto.rautiainen@luke.fi

[4] Natural Resources Institute Finland (Luke), Opistontie 10 a 1, 32100 Ypäjä, Finland; E-Mails: markku.saastamoinen@luke.fi (M.S.); susanna.sarkijarvi@luke.fi (S.S.)

* Author to whom correspondence should be addressed; E-Mail: jarkko.leppala@luke.fi; Tel.: +358-29-532-6326.

Academic Editor: Kirrilly Thompson

Received: 17 August 2015 / Accepted: 4 November 2015 / Published: 12 November 2015

Simple Summary: A new web tool for equine activities, InnoHorse, was developed to support horse stable managers in business, safety, pasture and manure management. The aim of the safety section of the web tool was to raise awareness of safety issues in daily horse stable activities. This section contains a safety checklist, stable safety map and good practices to support human health and horse welfare and to prevent injuries in horse-related activities. Reviews of the literature and statistics, empirical horse stable case studies, expert panel workshops and stakeholder interviews were utilized in designing the web tool.

Abstract: Managing a horse stable involves risks, which can have serious consequences for the stable, employees, clients, visitors and horses. Existing industrial or farm production risk management tools are not directly applicable to horse stables and they need to be adapted for use by managers of different types of stables. As a part of the InnoEquine project, an innovative web tool, InnoHorse, was developed to support horse stable managers in business, safety, pasture and manure management. A literature review, empirical horse stable case studies, expert panel workshops and stakeholder interviews were carried out to support the design. The InnoHorse web tool includes a safety section containing a horse stable safety map, stable safety checklists, and examples of good practices in stable safety, horse handling and rescue planning. This new horse stable safety management tool can also help in organizing work processes in horse stables in general.

Keywords: horse stable; safety; management; web tool

1. Introduction

The equine sector has grown strongly in recent years in many European countries [1,2]. For example, the number of horses has almost doubled in Finland during the past thirty years [3]. In Sweden, in turn, the number of horses per capita is the largest in Europe [1]. The equine sector, with diverse activities, provides an attractive lifestyle and rewarding experiences, but to be successful, good safety management skills and practices are needed. A problem is that safety risks and injuries are high in many horse-related activities. Without awareness of the possible risks and proper knowledge and skills of horsemanship, people engaged in horse-related activities will be exposed to many safety risks that may have serious consequences [4–6]. For example, in Finland, approximately 170 injuries occur per year among horse entrepreneurs and about 300 incidents among persons in other professional sectors working with horses (such as students, farmers, relief workers, veterinarians) (Figure 1). In other professional sectors, most of the injuries involve students and stable workers (Figure 2) [6,7].

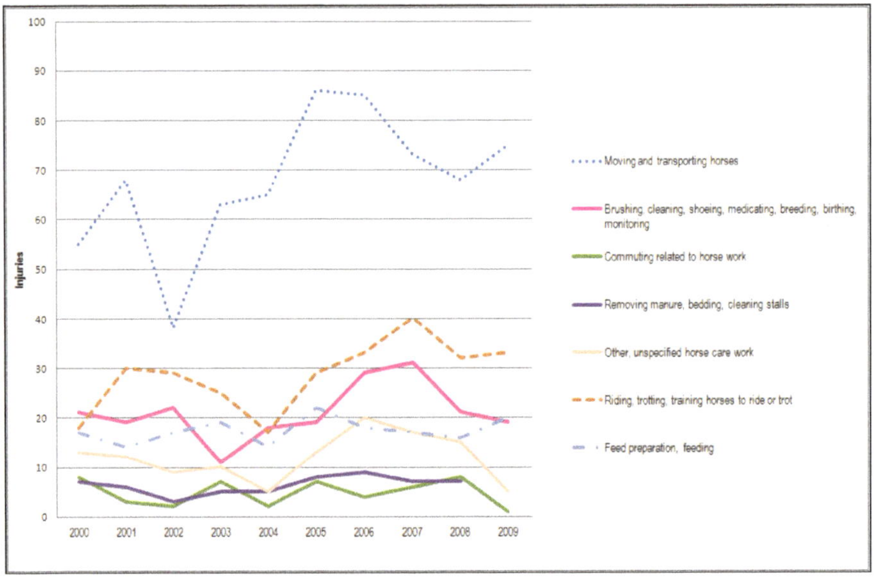

Figure 1. Statistics from Farmers Social Insurance Institution, Finland concerning occupational injuries among equine entrepreneurs in horse-related activities during 1990–2009 [7,8].

Leisure time injury statistics are still largely unknown [6–8]. In Sweden, the exact number of horse-related injuries is uncertain because of underreporting. Nevertheless, in 2012, nearly 12,900 persons went to an emergency centre after being injured in riding accidents or other activities related to horse handling. Nearly nine out of ten injured persons were females and 40% were children younger than 18 years of age. Injuries were more frequent among girls aged 10 to 19 years compared to other age groups [9,10]. According to the statistics of the Finnish Farmers' Social Insurance Institution, Mela [8,11], almost 35% of human injuries in horse activities have been serious incidents that have resulted in over 30 days of sick leave.

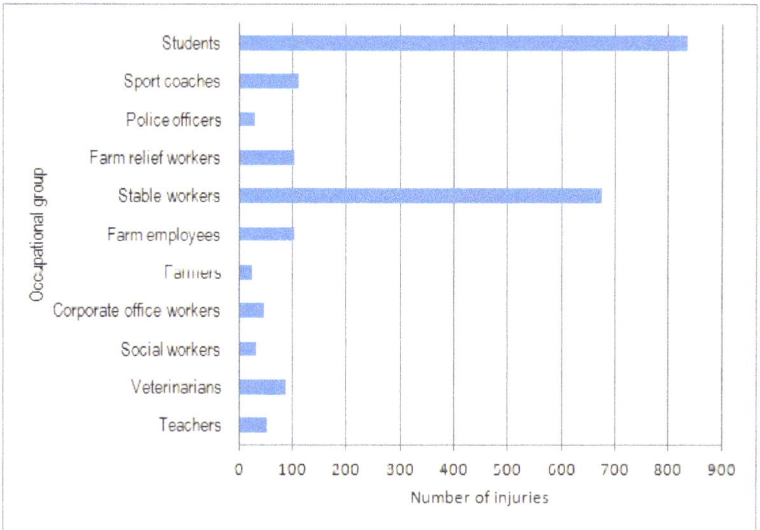

Figure 2. Statutory accident insurance (TVL) statistics from Finland concerning the total numbers of injuries to horse stable workers and other persons related to occupational horse activities during 2003–2010 [12].

Consequently, good risk management tools and safety practices are needed in the equine business and horse activities. A project titled InnoEquine financed by the EU Central Baltic programme was carried out in Finland, Sweden, Latvia and Estonia during 2010–2013. The overall aim of the project was to enhance the competitiveness of equine entrepreneurs in the Baltic region and to promote sustainable management in the equine sector. As one of the results of the InnoEquine project, the InnoHorse safety web tool for horse stables was developed in order to identify new practical solutions for risk and safety management [7].

2. Data and Methods

The InnoEquine project was carried out jointly by MTT Agrifood Research Finland (presently the Natural Resources Institute Finland (Luke), the Swedish University of Agricultural Sciences (SLU) and the Latvian University of Agriculture (LTU). A specific aim of the project was to develop a web tool providing good practices in environmental, human safety and horse welfare activities for the equine sector. This paper focuses on the design and development of the safety web tool in the project. The purpose of this safety web tool was to provide knowledge and practical tools to prevent injury incidents and occupational diseases in the equine sector.

The design, methods and processes of the stable safety web tool, as well as the tasks and timetables are presented in Figure 3 [13]. Basic information related to management activities and safety needs in horse stables in Finland and Sweden were gathered through a customer survey (*N* = 1325) [14]. An expert workshop on horse stable safety was held (at MTT) in December 2012 to assist with the design of the web tool. The workshop identified major risk categories and specific risks associated with horse stable activities and functions. The participants (*n* = 10) included horse sector experts, horse farm managers, farm safety, security, and risk management experts from Finland and Sweden, and Innoequine project representatives. The workshop utilized existing Farm Risk Map [15] tools and procedures as the starting point. Participants used a wallpaper technique, writing their ideas on wall notes, based on their perspectives and experiences. The identified horse stable risks and themes were documented, photographed, grouped, and arranged under redefined risk categories.

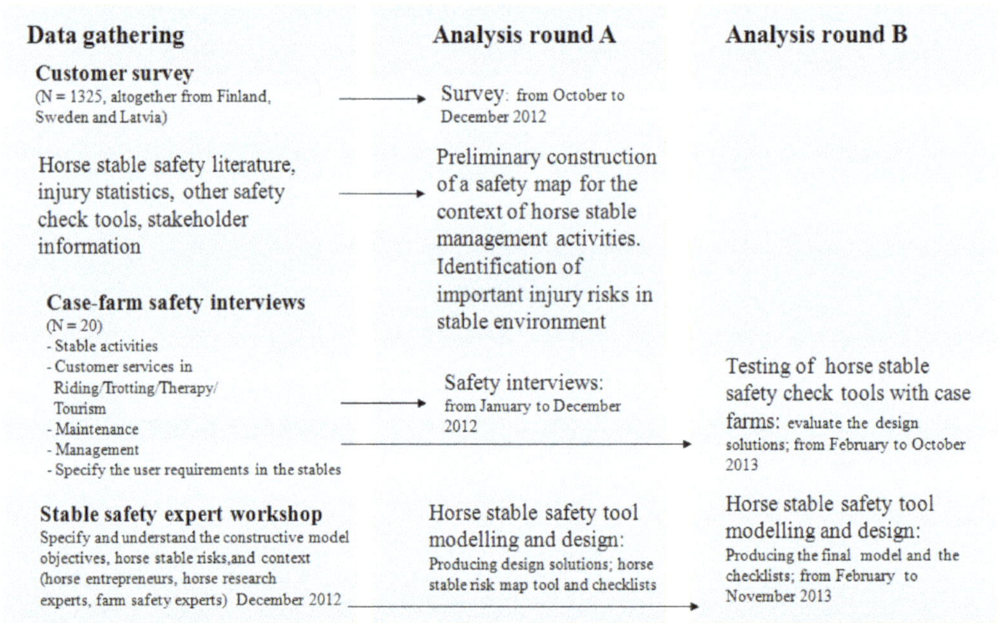

Figure 3. The design, methods, processes, analysis tasks and timetables for preparing the InnoHorse safety web tool [13].

Ten case studies on horse farms were analyzed for the safety section both in Finland and in Sweden. The case farms included enterprises in different size categories and different types of stables. All interviews were documented.

Information on current legislation, standards, and various horse safety and management guidebooks was gathered from the participating countries. These data were analyzed and integrated to produce the background for safety risk identification in horse stable activities. In these cases, we used broader recommendations based on standards, research, International Labour Organization (ILO) recommendations or EU directives [7,16]. The comprehensive collection of photos and documentation to illustrate good practices was gathered through numerous farm visits, equine fairs and equine companies in the three participating countries.

In addition, we conducted literature reviews, analyses of insurance claims involving horse injuries, Internet searches, comparison of various farm safety risk tools, discussions with equine organizations and stakeholders, and interviews and visits to horse farms. Results from these investigations have been reported elsewhere [13,16,17].

All collected data were processed in an iterative development cycle and constructive analysis, which included content analysis and designing the preliminary horse stable safety checklists and risk map. The iterative development cycle process is a standardized method that uses control stages to analyze data before going further in the interactive system design process. The process includes stages like (1) understand and specify context of use; (2) specify user requirements; (3) produce design solutions to meet these requirements; (4) evaluate design against requirements [18,19]. This method is widely used in technical and management sciences. Constructive research method is a problem solving method for construction and testing of models to reach a certain objective in a system or context [20]. The final content of the safety web tool was tested during 2013 prior to making it public online (Figure 3). Feedback and data from the safety web tool were gathered by case farmer phone interviews and by the email. The final content was edited by MTT for the InnoHorse web site in English and Finnish and thereafter also translated into Swedish and Latvian [7,16].

3. Results and Discussion

3.1. The Innohorse Web Tool

In this project, the new InnoHorse web tool was designed to assist in horse stable management practices. The web tool was published by the National Equine Competence Association of Finland (Hippolis). The InnoHorse web site includes management tools for horse stable activities such as safety, manure, pasture, and innovation management. The web tool has been published in English, Finnish, Swedish and Latvian. The layout and information are the same in the different language versions but with minor country-specific differences [7].

All major areas of the web tool include an introduction section followed by good practices applied to the particular horse stable management sections. The horse stable safety management section in InnoHorse provides safety information, safety checklists, safety management practices and a horse stable safety map, which is presented in Figure 4. All tools were designed to improve health, reduce safety risks and prevent injuries among horse stable workers, stable managers, clients, visitors and horses.

3.2. Safety Section of the Web Tool

The section on safety provides information for horse stable managers in the Baltic Sea region related to various aspects of safety and health management. The web pages of the InnoHorse safety section include a stable safety checklists, one-page horse stable safety map, and examples of good practices for stable safety, horse handling and rescue planning.

The safety web site content begins by introducing injury statistics, mainly in Finland, and the characteristics of horse-related injuries in the equine sector. In Finland, injury statistics for all farm owners, including horse farm owners, are maintained by the Farmers' Social Insurance Institution (Mela). Statistics on horse-related injuries in the other participating project member countries are scarce. The Finnish statistics indicate that the risk of injuries is nearly three times higher on horse farms compared to grain farms [8].

Figure 4. The horse stable safety map.

3.3. Safety Checklists and the Horse Stable Safety Map

Systematic risk checklists are practical self-assessment tools for identifying and managing risks in various tasks [15,21,22]. Based on the findings and information from statistics, the stable safety expert workshop, horse farmer interviews and literature search, we developed a stable safety checklist, which addresses potential safety issues in walkways and corridors, built facilities, work ergonomics, equipment and machinery (Figure 4). Personal protective equipment (PPE), fire safety and rescue planning, employee and client safety, as well as some other safety issues such as the transportation of horses and horsemanship skills were also included. The questions in the safety checklists enable the screening of possible risk sources or factors with the potential to cause injuries in and around the stable facilities and around horses. The respondents were asked to estimate whether particular working conditions or activities in a stable were in order. The checklist also includes some guidance or recommendations for reducing potential safety risks. The differences between countries or regions in legislation and safety activities pose a challenge in integrating risk management information in a single safety tool. The horse stable safety map introduces the content of the safety web tool as a one page figure (Figure 4). The idea is the same as in the Farm Risk Map, which was previously designed in Finland by MTT and the Technical Research Centre of Finland (VTT) [15].

3.4. Good Practices

The section on safety management introduces good practices for persons working or visiting horse stables and riding facilities. According to Mela statistics, a large number of injuries and accidents occur when moving and transporting horses (Figure 1). This is why it is important to have spacious and well-lit corridors with sufficiently wide doorways and sliding stable doors for safe passage with the horse, as well as good ventilation and natural light used together with electric lighting to provide a good work environment for horses and people (Figure 5). The Good Practices section was designed in line with the safety checklist questions, providing further information on stable safety management activities. The section contains information, practical tools, illustrative photos and figures, and examples of good safety practices for horse stable safety management. Overall, practical and efficient stable safety tools can be useful management aids for horse stables.

Figure 5. Spacious and bright corridors are important. An example of good safety management practice in horse facilities presented in the InnoHorse web tool. © Christina Lunner Kolstrup.

The stable safety management section in the InnoHorse web tool presents practices and guidelines including safety aspects related to stable work, ergonomics, buildings, equipment and machinery use; all important safety factors in the stable work environment. Musculoskeletal disorders and ergonomic problems are very common in horse stables [23]. For example, traditional hand tools are not always adapted for the users, which increases the risk of musculoskeletal disorders in the upper extremities and lower back. Bent shafts in some hand tools help create a more upright posture for the back (Figure 6).

Good handles provide better grip and lightweight tools reduce the workload. Ergonomic tools reduce the workload and the risk of upper limb and lower back musculoskeletal problems. The cleaning of stalls, manure transport and feeding of horses are typical routines. They are time consuming as well as physically demanding work tasks in a stable that require special attention. Feeding and the handling of feeds takes about five to seven minutes per horse per day, and the cleaning of stalls (mucking, replacement of bedding materials) takes approximately 10 minutes per horse daily if no machinery is used [24,25]. Good working clothes, proper equipment and the use of personal protective equipment (PPE), combined with good working conditions, form a good basis for an improved safety culture in horse stables.

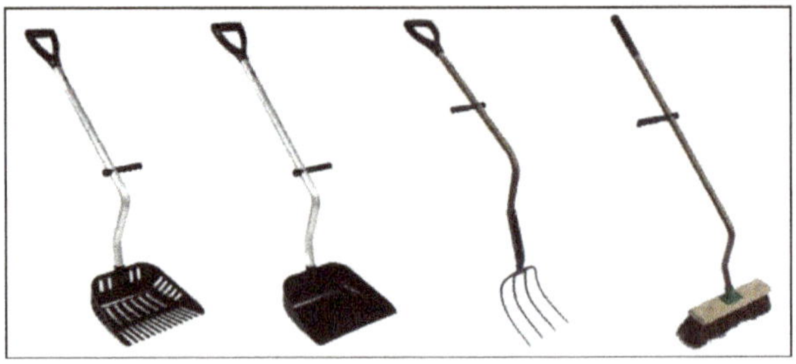

Figure 6. Ergonomically designed long-shafted tools. Reproduced with permission by Lite-Lift Ltd. (http://www.lite-lift.com/, 23 June 2015).

3.5. Case Study Interviews

The horse stable safety map content was tested with the case farmers. All case farmers were also horse stable managers. The case farmers thought that the horse stable safety map was a comprehensive safety management list for the stable management. They could not find any missing safety management areas that should be added to the horse stable safety map. The structure of the map worked well also during the actual stable visit and stable safety check. It is possible that the case farmers were more interested in safety management than stable managers on average. Their stables were in good condition and well managed. Yet some minor shortcomings were found in almost all stables during the safety check. Most of the shortcomings concerned slippery areas (winter time) on corridors, uneven walkways, lack of lights in the stable and the stable yard, and the farmer's own welfare. After the safety check the case farmers made several safety improvements in their stable and safety management practices. They had started to use rubber mats on the corridors, improve lighting and ergonomics in the stalls, and they started to think more about their own welfare. They tried to find free time to rest or have a vacation.

According to the interviewed horse stable managers, the most challenging task is to improve people's safety skills and to get them to behave in a safe manner. Thus, the web tool includes management information on activities such as fire safety and rescue skills, customer and worker safety management, and other behavioral safety issues such as examples of good horsemanship and the importance of rules in stable safety. It is known that poor safety habits are easy to adopt in organizations, so an initial and essential management task is to show and train workers in safe working habits and communicate why safety is important [26]. Thus, the implementation of good safety practices needs to be easy and understandable, and every worker needs to be trained beginning from the first day in a stable. The stable manager's own self-commitment to safe behavior in stable work is also important as a role model for employees and consumers.

4. Conclusions

The differences between countries in legislation and safety activities pose a challenge in designing safety or environmental management guidelines. For this reason, some recommendations are provided on a general level. However, some standards, research studies or directives may help stable managers in acquiring more practical information. Another challenge is human safety behaviour in the horse stable environment. Without good management, poor safety habits may spread in the organization. Thus, the application of good safety practices needs to be as easy as possible and every worker needs to be trained beginning from the first day in the stable. The stable manager's own commitment to safe behaviour in stable work is also important.

The InnoHorse safety web tool aims to help in organizing and managing safety activities in horse stables and facilities. This tool provides a practical context model for identifying risks in horse stable activities. It contains physical and behavioral risks, which are listed in a compact horse stable safety map. This holistic approach provides a new comprehensive model for risk identification and risk management for the equine sector. The horse stable safety map and other safety check tools in the Innohorse web tool may not solve all the safety problems in horse stables, but hopefully they help some horse stable managers to improve their stable safety management. It is intended to provide tools for the equine sector to inspire, motivate and encourage people to act and behave more safely around horses in order to prevent horse-related injuries.

Acknowledgments

The InnoEquine project was led by MTT Agrifood Research Finland (currently the Natural Resources Institute Finland, Luke) and carried out in collaboration with the Swedish University of Agricultural Sciences (SLU) and Latvia's University of Agriculture. The horse stable safety management section in the web tool was designed by MTT and SLU. The InnoEquine project was funded by the European Union Central Baltic Interreg IVA Programme. We would like to thank all horse stables owners and employees who participated in the InnoEquine project. Some of the Finnish or Swedish references may be hard to find. All the references are available on request from the authors.

Author Contributions

Jarkko Leppälä was the main author of this article and designed the horse stable safety map, injury figures and most parts of the InnoHorse safety section. He also conducted several interviews on horse farms in Finland and arranged the workshop on the horse stable safety map content. Christina Lunner Kolstrup co-authored the article and conducted interviews on horse farms in Sweden, worked with the Swedish statistics, good safety practices and horse stable management, and participated in the workshop on the horse stable safety map. Associate Stefan Pinzke studied horse stable ergonomics and good safety practices and participated in the workshop on the horse stable safety map. Risto Rautiainen worked with the injury statistics, good safety practices and participated in the workshop on the horse stable safety map. Markku Saastamoinen took part in editing the safety checklist and web tool and wrote the general introduction parts on the horse industry in Finland, as well as commenting on the article results and conclusions. Susanna Särkijärvi participated in collecting information for the web tool and editing the safety checklists and the InnoHorse web tool in general.

Conflicts of Interest

The authors declare no conflicts of interest.

References

1. Swedish University of Agricultural Sciences. EU Equus 2001—The Horse Industry in the European Union. 2001. Available online: http://www.horse-web.net/docs/EU_Equus_2011.pdf (accessed on 2 June 2015).
2. European Pari Mutuel Association. The Economic and Social Contribution of Horse Racing in Europe. 2009. Available online: http://www.parimutuel-europe.org/ Download/ EPMA_Economic_impact_of_horse_racing_03Sept09.pdf (accessed on 2 June 2015).
3. The Finnish Trotting and Breeding Association (Suomen Hippos). Available online: http://www.hippos.fi/ (accessed on 2 June 2015).
4. Swedish Board of Agriculture (SJV). *Hästhållning i Sverige 2010 (Horse Keeping in Sweden 2010)*; Report 1; Swedish Board of Agriculture: Jonkoping, Sweden, 2012.
5. Thompson, K.; McGreevy, P.; McManus, P. A critical review of horse-related risk: A research agenda for safer mounts, riders and equestrian cultures. *Animals* **2015**, *5*, 561–575. [CrossRef] [PubMed]
6. Riley, C.B.; Liddiard, J.R.; Thompson, K. A Cross-Sectional Study of Horse-Related Injuries in Veterinary and Animal Science Students at an Australian University. *Animals* **2015**, *5*, 951–964. [CrossRef] [PubMed]
7. The National Equine Competence Association (Hippolis). InnoHorse—Web Tool. Available online: www.hippolis.fi/innohorse/home/ (accessed on 2 June 2015).
8. Farmers' Social Insurance Institution (Mela). *MATA—Vahinkotilasto (Finnish Statistics on Farmer Injuries)*; Farmers' Social Insurance: Espoo, Finland, 2012.
9. Swedish National Board of Health and Welfare. *Injury Data Base (IDB)*; Swedish National Board of Health and Welfare: Stockholm, Sweden, 2015.
10. Swedish Civil Contingencies Agency. Olyckor Med Häst (Accidents with Horses). FAKTA 6 December 2013. Available online: http://www.dinsakerhet.se/Global/pdf/ Fritid_resor/ Olyckor%20med%20h%C3%A4st.pdf?epslanguage=sv (accessed on 2 June 2015).
11. Leppälä, J.; Myllymäki, M.; Saastamoinen, M.; Rautiainen, R. Target group safety intervention and safety check tests on horse farms. In Proceedings of the Nordic Meeting on Agricultural Occupational Health and Safety (NMAOHS), Porvoo, Finland, 25–27 August 2014; p. 18.
12. Tapaturmavakuutuslaitosten liitto TVL. *Frequencies of Accidents at Work. Statutory Accident Insurance*; Tapaturmavakuutuslaitosten liitto TVL: Helsinki, Finland, 2012.

13. Leppälä, J.; Kolstrup, C.L.; Pinzke, S.; Rautiainen, R.; Saastamoinen, M. Web tool for horse stable safety practices and management. In Proceedings of the Workshop Presentation and Paper in Equimeeting Seminar, Le Lion d'Angers, France, 6–7 October 2014.
14. Swedish University of Agricultural Sciences. How to Be Innovative in the Equine Sector? In *A Guidebook on Practices and Inspiration from Central-Baltic*; Report 2013:33; Korpa, V., Ed.; SLU: Alnarp, Sweden, 2013.
15. Leppälä, J.; Murtonen, M.; Kauranen, I. Farm risk map: A contextual tool for risk identification and sustainable management on farms. *Risk Manag.* **2012**, *14*, 42–59. [CrossRef]
16. Pinzke, S.; Järvinen, M.; Kettunen, K.; Lunner-Kolstrup, C.; Leppälä, J.; Löfqvist, L.; Myllymäki, M.; Rautiainen, R.; Reilas, T.; Saastamoinen, M.; *et al. Säkerhet i Häststallet (Horse Stable Safety)*; LTJ-Report 2013:32; SLU, Work Science, Business Economics & Environmental Psychology: Alnarp, Sweden, 2013; p. 75.
17. Lunner-Kolstrup, C.; Pinzke, S.; Löfqvist, L.; Järvinen, M.; Korpa, V.; Paula, L.; Kursītis, A. *Current Status of the Equine Sector in the Central Baltic Region (Finland, Latvia and Sweden)*; LTJ-Report 2013:17; SLU, Work Science, Business Economics & Environmental Psychology: Alnarp, Sweden, 2013; p. 61.
18. International Organization for Standardization. *Human-Centered Design Processes for Interactive Systems*; ISO 13407:1999; International Organization for Standardization: Geneva, Switzerland, 1999.
19. International Organization for Standardization. *Ergonomics of Human-System Interaction*; ISO 9241-210:2010; International Organization for Standardization: Geneva, Switzerland, 2010.
20. Kasanen, E.; Lukka, K.; Siitonen, A. The constructive approach in management accounting research. *J. Manag. Account. Res.* **2003**, *5*, 243–284.
21. Committee of Sponsoring Organizations of the Treadway Commission. *Enterprise Risk Management—Integrated Framework*; Executive Summary 2004; Committee of Sponsoring Organizations of the Treadway Commission (COSO): City, Country, 2004.
22. Juran, J.; Godfrey, A. *Juran's Quality Handbook*; McGraw-Hill: New York, NY, USA, 1998.
23. Löfqvist, L. Physical Workload and Musculoskeletal Symptoms in the Human-Horse Work Environment. Ph.D. Thesis, Swedish University of Agricultural Sciences, Alnarp, Sweden, 2010.
24. Korkiamäki, H.; Lahin, P.; Laaksonen, K. Hevostallien työmenetelmät, case-tutkimus kuudella tallilla. In *Horse Stable Work and Estimated Working Hours*; Maatalousticdote 9, Report in Finnish; TTS Institute: Rajamäki, Finland, 1998.
25. Laitinen, I. Manure Handling with Machinery in the Horse Stables. Master's Thesis, Department of Agricultural Engineering, University of Helsinki, Helsinki, Finland, 2007.
26. Reason, J. *Managing the Risks of Organizational Accidents*; Ashgate Publishing Ltd.: Aldershot, UK, 1997.

Animals **2015**, *5*, 1072-1091; doi:10.3390/ani5040399

OPEN ACCESS

animals

ISSN 2076-2615

www.mdpi.com/journal/animals

Article

Jump Horse Safety: Reconciling Public Debate and Australian Thoroughbred Jump Racing Data, 2012–2014

Karen Ruse *, Aidan Davison and Kerry Bridle

School of Land and Food, Discipline of Geography and Spatial Sciences, University of Tasmania, PB 78, Hobart, Tasmania 7001, Australia; E-Mails: Aidan.Davison@utas.edu.au (A.D.); Kerry.Bridle@utas.edu.au (K.B.)

* Author to whom correspondence should be addressed; E-Mail: Karen.Ruse@utas.edu.au; Tel.: +61-2-6286-4750.

Academic Editor: Kirrilly Thompson

Received: 17 August 2015 / Accepted: 16 October 2015 / Published: 22 October 2015

Simple Summary: This paper documents the dynamics of Australian thoroughbred jump racing in the 2012, 2013, and 2014 seasons with the aim of informing debate about risks to horses and the future of this activity. We conclude that the safety of Australian jump racing has improved in recent years but that steeplechases are considerably riskier for horses than hurdle races.

Abstract: Thoroughbred jump racing sits in the spotlight of contemporary welfare and ethical debates about horse racing. In Australia, jump racing comprises hurdle and steeplechase races and has ceased in all but two states, Victoria and South Australia. This paper documents the size, geography, composition, and dynamics of Australian jump racing for the 2012, 2013, and 2014 seasons with a focus on debate about risks to horses. We found that the majority of Australian jump racing is regional, based in Victoria, and involves a small group of experienced trainers and jockeys. Australian jump horses are on average 6.4 years of age. The jump career of the majority of horses involves participating in three or less hurdle races and over one season. Almost one quarter of Australian jump horses race only once. There were ten horse fatalities in races over the study period, with an overall fatality rate of 5.1 fatalities per 1000 horses starting in a jump race (0.51%). There was significant disparity between the fatality rate for hurdles, 0.75 fatalities per 1000 starts (0.075%) and steeplechases, 14 fatalities per 1000 starts (1.4%). Safety initiatives introduced by regulators in 2010 appear to have significantly decreased risks to horses in hurdles but have had little or no effect in steeplechases. Our discussion considers these data in light of public controversy, political debate, and industry regulation related to jump horse safety.

Keywords: thoroughbred; horse-racing; steeplechase; hurdles; risk; safety; animal-human relationships; media; public debate

1. Introduction

Thoroughbred jump racing sits in the spotlight of contemporary welfare and ethical debates about horse racing [1,2]. This activity encompasses hurdling, steeplechasing, point-to-point, and mountain racing and is controversial primarily because it involves higher risk of accident or fatality to both horse and rider compared with flat racing [3–5]. Debates about jump racing have been played out via mass media [6,7] and have relevance in ongoing cultural renegotiation of the meanings, norms, and governance of human-animal relationships in modern societies [8,9]. These debates fuse questions about defensible human-animal relations with those about the conduct of human entertainment, sport, and gambling [10]. In particular, falls and fatalities in jump racing are widely reported in the media and result in negative public opinion and criticism of racing and of the racing industry's approach to equine welfare [2]. Arguably, public concern about horse welfare has contributed to the decline of jump racing in many parts of the world, although economic factors have also been significant [11,12]. Jump racing continues, however, in 18 countries across four continents [11]. While jump racing comprises only a small proportion of horse races and related gambling turnover and prize money, this activity has political, cultural, and economic importance in particular countries, regions and towns [2,11].

Australia provides an important context in which to explore the decline and resilience of jump racing (where it is commonly known as jumps racing). Jump racing is no longer conducted in four of Australia's six states. However, hurdles and steeplechases remain a feature of thoroughbred racing in Victoria and, to a lesser extent, South Australia. Total annual prize money in Australian jump racing is around $2.7 million (Australian dollars used throughout), with average prize money per race equivalent to that of flat racing [13]. Australian jump racing is a familiar and divisive subject of passionate advocacy and critique in conventional and social media and has been a notable electoral issue in Victorian and South Australian state politics over the past decade [14,15]. In addition to criticism from established animal welfare organisations, such as the Humane Society and Royal Society for the Prevention of Cruelty to Animals (RSPCA), Australian jump racing has been the focus of protest, activism, and lobbying including from groups with a sole focus on horseracing, such as the Coalition for the Protection of Racehorses.

Table 1 outlines key features of Australian jump racing. Australian jump racing occurs in regional, as well as metropolitan settings in autumn and winter when tracks are softer. Thoroughbreds are specifically bred for jump racing in some countries, but not in Australia where jump horses are drawn from the flat racing population. Australian jump racing is conducted under state-based Local Rules of Racing. There are some differences between the two states: for example, the minimum distance of a hurdle race in South Australia is 2800 m and in Victoria it is 3200 m. In both states, the hurdles in a hurdle race are a maximum of one metre high with horses carrying a minimum weight of 64 kg (jockey and saddle, plus handicap weight). In a steeplechase, horses jump over fences which are at least 1.15 m high. Hurdle races usually involve fewer jumps over shorter distances than steeplechases. In Australia, horses must undergo qualifying trials to be eligible to compete in a hurdle race. Horses may then progress through a sequence from maiden to novice to open hurdle races, before then being eligible to qualify for steeplechase races [16,17].

Table 1. Key features of Australian jump racing.

Locations of races	Restricted to 15 South Australia and Victorian race courses, predominantly non-Metropolitan.
Type of races	Two thirds hurdle races, one-third steeplechases.
Regulatory bodies	Administered under state-based Local Rules of Racing by Thoroughbred Racing South Australia (TRSA) and Racing Victoria (RVL).
Industry bodies	Australian Jumping Racing Association (AJRA).
Race regulations	Specify minimum weight (64 kg); course condition rating; height, number and placement of obstacles; maximum field size; use of whips; horse boots; *etc.*
Industry review	Seven safety performance reviews by regulatory bodies since 1994.
Race review	In addition to the TRSA and VRL Stewards Committee review, TRSA and VRL Jump Review Panels review each horse's jump at each obstacle in each race and may refer horse or jockey to undergo further training. The Panel includes a former jump jockey who can provide individual coaching if needed.
Qualification	Horses, trainers and jockeys must undergo qualification training and trials, overseen by VRL and TRSA, in order to compete in a maiden hurdle race. Horses that progress to open hurdle races are then eligible to qualify for steeplechase races. Mandatory trainer and jockey skills workshops held annually.
Veterinary inspections	Of each horse, before and after each race.
Race season	March to September. Races are scheduled at approximately fortnightly intervals.
Number of races	Less than 100 per season.
Horses	Thoroughbreds, drawn from flat racing population, must be at least three years old, and may race in both flat and jump races during the jump race season.
Tracks	Left handed turf tracks, no steep downhill runs to finishing-lines.
Race field	Field sizes are small, with less than 8 horses on average in a race. Low fields are not uncommon (<5 starters).
Race start	Starting gates used at commencement of races.
Race speed	Races are run on slow tracks (heavy conditions), with heavier weights carried (>64 kg).
Hurdle obstacles	Hurdles are padded panels, maximum 1 metre in height, with standardised design.
Steeple obstacles	Steeples are a mix of brush top panels and live hedges not less than 1.15 m in height, depending on race course, with height and width specified by regulator. No water jumps or drops.

During the 2008 and 2009 Australian jump racing seasons, the highly visible deaths of 14 horses in jump races across Victoria and South Australia inflamed criticism by welfare and activist groups, heightened public concern and prompted the prospect of its banning in Victoria [18–20]. In late 2008, a group of activist organisations led by Animals Australia presented a submission to the Victorian parliament calling for a ban on jump racing [21]. Their submission reported that 13.1 out of every 1000 horses (1.31%) starting in a jump race (hereafter referred to as "starts") died [21]. Their submission also summarised longer term fatality rates from 1989 to 2004 based on a 2006 study in the Equine Veterinary Journal documenting the risk of a fatality in Australian jump racing as almost 19 times that in flat racing [22]. This study found that catastrophic limb failure, the predominant cause of horseracing deaths, was approximately 18 times greater for Australian jump racing than flat racing, with cranial or vertebral injury 120 times greater and sudden death 3.5 times greater [22]. Following a further eight deaths in the 2009 jump racing season, Racing Victoria (RVL), the principal authority governing thoroughbred racing in Victoria, commissioned a review of jump racing. The fall and fatality rates for the 2009 season were reported by this review as being 50.8 per 1000 starts and 12.7 per 1000 starts, respectively, the highest recorded during the 2005 to 2009 seasons [23]. These results were despite six previous reviews of horse and jockey safety in Australian jump racing since 1994, with the recommendations of the last of these, the 2008 Jones Report, implemented prior to the 2009 season [21,23]. In November 2009, following their latest review, RVL announced a two year transition plan to phase out jump racing after 2010. However, in January 2010, RVL handed jump racing a tentative reprieve, allowing the sport to continue in 2010 subject to stringent safety conditions and standards including a reduction in fatality rate to approximately half of that for 2009 (*i.e.*, 6.5 per 1000 starts) and a reduction in fall rate to 30 per 1000 starts [24]. The RVL Chairman warned that if the new conditions were not met, jump racing would cease.

In September 2010, following an improvement in the safety of hurdle racing, RVL gave hurdle racing the go ahead for a three year program subject to meeting a key performance indicator (KPI) of a horse fatality rate of not more than 6.5 deaths per 1000 starts (0.65%), measured as a rolling three year average [25]. At the same time, RVL declared the performance of steeplechasing unsatisfactory and requested further measures to improve its safety [25]. However, in October 2010, without the introduction of further measures, RVL agreed to a steeplechase program for 2011, and determined that its future beyond 2011 be subject to a fatality KPI of 6.5 deaths per 1000 starts (0.65%), measured as a rolling two year average [26]. These performance targets were accompanied by a raft of measures to improve horse and rider safety, including the ability of a jockey to withdraw a horse during a race because it is fatigued and out of contention and a danger to itself or the jockey. These initiatives were closely followed in November 2010 by a change of government in Victoria that saw a vocal advocate of jump racing, Dr. Denis Napthine, a veterinarian, installed as Premier and Minister of Racing [27]. With government support, the period 2011 to 2014 saw jump racing authorities in Victoria increase prize money and invest in jump race safety and training infrastructure improvements. By November 2011, RVL discontinued the safety KPIs on the basis that both hurdling and steeplechasing had met their KPIs relating to fatalities for the past two years, undertaking to monitor the safety performance of jump racing on an ongoing basis and to undertake reviews as required [28].

In the context of industry efforts to address concerns about risks to horses in Australian jump racing, advocates argue that horses love to race and jump, that jump racing extends a horse's career and that many of these animals would be slaughtered if not for jump racing [29,30]. Opponents such as the RSPCA argue that horses have evolved to avoid rather than jump obstacles, that the heightened prospect of injury or death to jump horses is an unacceptable focus of human entertainment, and that the risk of being injured or killed in jump racing is not an acceptable alternative to the slaughterhouse [31–34]. The Humane Society describes jump racing as institutionalised cruelty [35].

Recent debate about Australian jump racing has taken place in the absence of sufficient robust or current data, with opponents continuing to rely on Boden *et al.*'s 2006 study [32,34–36].

The composition and dynamics of the cohorts of jockeys, trainers and horses involved in Australian jump racing, the ages and career trajectories of jump horses, the geography of this activity, and annual changes in activity have not been reliably documented nor related to data about horse safety. In response, this paper compiles, synthesises, and analyses data collected by Racing Australia (RA, formerly Racing Information Services Australia), for the 2012, 2013, and 2014 jump race seasons in Victoria and South Australia. In addition to informing public and policy debate, the paper contributes to international understanding of the dynamics of Australian jump racing in the context of changes in the horse racing industry more generally, including changing attitudes on questions about the use of animals in public entertainment.

2. Methodology

Jump race data (hurdle and steeplechase) were obtained from RA for the 2012, 2013, and 2014 racing seasons from the Results area of the RA website. Although publicly available, these data are recorded on a race-by-race basis and are not aggregated, nor are composite trends identified. Data retrieved included, race name, location, date, time, distance and course condition rating, as well the name of each horse, jockey, and trainer involved in each race. Seasonal data on total starts for the period 2007 to 2010 were obtained from the Australian Racing Fact Book. Every thoroughbred race in Australia is reviewed by an official Stewards Panel which monitors racing conduct and injuries to horses. For jump races, a Jump Review Panel (JRP) subsequently reviews how well each horse jumped each obstacle. Stewards' reports were obtained from RA and JRP reports from RVL. Although a similar panel reviews South Australian jump races these reports are not publicly available.

Data were entered into a Microsoft Excel ™ (2013) database and ordered in the form of a "start", or an individual horse leaving a starting gate in a jump race. Data were organised by racing season, which extends from March to September of a calendar year. Annual data thus relate to a single jump race season, except for data in Figure 4 which has been standardised to an annual racing year (1 July to 30 June) to allow for comparison with datasets from other studies. Horses listed in a jump race may be "scratched" before a race due to a variety of reasons, for example, disqualification by a veterinarian. Scratched horses are not included in the database. The database records individual horse performances and race placings and lists intra-race incidents in each race from the official Stewards reports, including, falls, run outs, lost rider, brought down, and failed to finish (a term used in official race reports to describe a horse withdrawn during a race at the discretion of the jockey). Jump Review Panel Reports about horse performance at each jump were matched to the official Stewards Reports for each horse, including details of fatal falls and other race incidents. Our database therefore provides a comprehensive picture of individual horse performances as well as a means to aggregate Australian jump race information over the study period. Our initial purpose was to uniquely identify jump horses, their trainers, and jockeys in order to describe the size, scope, and location of Australian jump racing and risks to horses over this period.

Starts were summed by horse by the state they raced in. The number of individual horses participating over this period was calculated by aggregating starts against horses' names and uniquely identifying each horse based on the RA horse search. The home state of each horse and trainer was identified by matching horses to trainers and identifying the trainer's place of residence from the addresses shown for official qualified trainers. The information generated included starts per state, as well as the number of starts by each trainer. Our analysis also identified horses that only raced in hurdles, horses that competed in hurdles and steeplechases, and those who competed only in steeplechases.

Horse falls and fatality rates were calculated by dividing the number of falls and fatalities by the total number of starts in all races, for each season and in the overall sample. Only race fatalities were included in the analysis; training and trial fatalities were not considered. The average number of starts in each race was calculated by dividing the total number of jump race starts by the total number of races. A trainer operating both in a partnership and also in their own name was counted as two separate entries.

3. Results

3.1. Location of Australian Jump Racing

The study encompassed 257 jump races, comprising 171 hurdle races and 86 steeplechases, conducted in Victoria and South Australia over the 2012, 2013, and 2014 jump race seasons. This represents less than 1.5% of total thoroughbred races in South Australia and Victoria over this period [37]. The majority of jump races (67%) were held in Victoria. Only 15 of the 386 racing clubs in Australia conducted jump racing; five in South Australia and 10 in Victoria [37]. The distribution of clubs is shown in Figure 1.

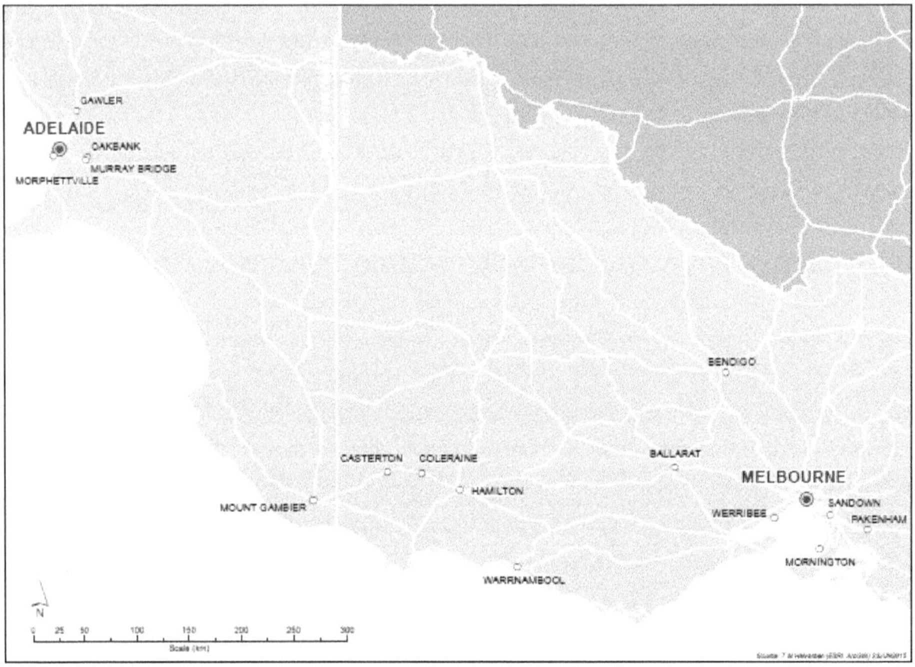

Figure 1. Location of Australian jump racing clubs, South Australia and Victoria. Clubs hosting jump races over the period 2012 to 2014.

Australian jump racing is concentrated in western Victorian country (rural) racing clubs (Figure 1). Warrnambool Racing Club hosted 28.4% of jump races over the study period, more than any other location, providing the hub for the Australian jump racing industry. The Warrnambool Racing Club's May carnival attracts around 30,000 people and includes Australia's longest jump race, the Grand Annual Steeplechase (total prize money t.p.m. $250,000), over 5.5 km and 33 jumps (4). Only two metropolitan clubs held jump races, although they accounted for 26% of races over the study period. The South Australian Jockey Club at Morphettville, Adelaide, held 44% of South Australian jump races, and 10.1% of all jump races. The Melbourne Racing Club at Sandown, Melbourne, hosted 21.2% of Victoria's jump races, and 16.3% of all jump races, including the Grand National Hurdle (t.p.m. $200,000) and Grand National Steeple (t.p.m. $250,000).

3.2. Participants in Australian Jump Racing

3.2.1. Horses

Over the 2012, 2013, and 2014 seasons, 438 individual horses participated in 1970 jump race starts. In keeping with the proportion of hurdle to steeplechase races, over two thirds of jump horses (302, 69%) competed only in hurdle races; just under one quarter (99, 23%) competed in both forms of Australian jump racing; and 37 (8%) competed only in steeplechases.

Figure 2 shows the number of starts per horse over the study period. More than half of horses (55%) competed three times or less, with almost one quarter (22%) competing in only one race. Another quarter raced between four and 10 times. Less than 10% of horses competed more than 10 times, with one horse racing 32 times. The median number of starts per horse was three, the first quartile was two starts and the fourth quartile was six starts. The range was one to 32 starts.

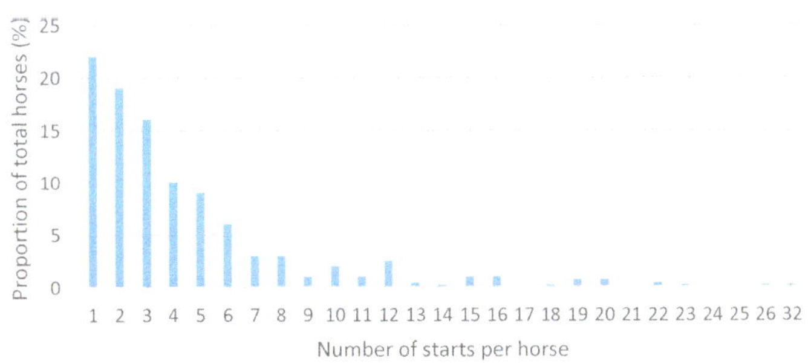

Figure 2. Proportion of horses (%) by number of starts per horse over the 2012, 2013, and 2014 seasons, South Australia and Victoria combined.

Annual turnover in the jump horse cohort was analysed by identifying individual horses that ran in two of the three race seasons, either in successive seasons or in 2012 and 2014, or in all three seasons (Table 2). Of the 2012 jump horse cohort, 37% raced in 2013 and 21% raced in 2014. Only 7% of the 2012 cohort raced in 2014 but not in 2013, and only 14% raced in all three seasons. Of the 2013 cohort, 29% raced in 2014.

Table 2. Australian jump horse annual turnover from the 2012 to the 2013 and 2014 seasons.

Number of Horses 2012	Number of Horses 2013	Number of Horses 2014	Number of Horses 2012 & 2013	Number of Horses 2012 & 2014	Number of Horses 2012, 2013 & 2014	Number of Horses 2012 & 2014 but not 2013
176	209	195	65 (37%)	40 (21%)	27 (14%)	13 (7%)

Number of horses that jump by season(s) of participation. All % figures indicate a proportion of the 2012 jump horse cohort. Horses are counted in the cohort of each season in which they competed.

The average age of horses in the sample was 6.4 years, as of 1 March in the seasons in which they raced in the 2012–2014 time period. Median age was six years, the first quartile median age was five years and the fourth quartile median age was seven years. (Figure 3). Consistent with the regulatory requirement that a horse be at least three years old to begin jump training and racing, and the time required for this training, just eight horses (2%) in our sample were aged three years in their first season of jump racing. Twenty five horses (7%) were aged 10 years or more. The range was three to 12 years.

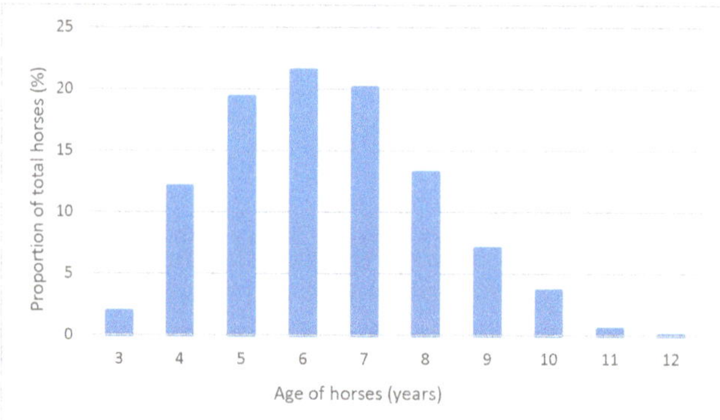

Figure 3. Age profile of jump horses, as a proportion of all horses jumping during the 2012, 2013, and 2014 seasons. The age census date is the beginning of the jump racing season (1 March) of each calendar year in which a horse competed. The total horse pool is the sum of the 2012, 2013, and 2014 horse cohorts, and not the number of individual horses over this period.

3.2.2. Trainers

In 2014, there were 4027 registered race horse trainers in Australia, with 914 in Victoria and 304 in South Australia (5). Only 145 trainers, less than 4% of the total, started a horse in a jump race between 2012 and 2014 (Table 3). Three quarters (76%) of these trainers were based in Victoria (110), with 31 (21%) based in South Australia, one in New South Wales and three in New Zealand. Victorian trainers account for the majority of starts in jump racing. A small number of prominent Victorian trainers account for a disproportionately large share of activity. Table 3 lists the five most prolific trainers in Victoria and South Australia, who accounted for 34% and 6%, respectively, of jump starts in the study period. Five trainers (two Victorian, two South Australian, and one New Zealander) accounted for half of the starts in South Australia, and the majority of horses in South Australian jump races were trained in Victoria. The most prolific Victorian trainer accounted for 20% of all starts in South Australia.

Table 3. Trainer participation by location in Australian jump racing, 2012–2014.

Trainer Rank	Number of Starts	Proportion of Total Starts (%)	Proportion of Victorian Starts (%)	Proportion of South Australian Starts (%)	Number of Horses Trained
Victoria					
1	231	11.7	9.2	19.9	34
2	136	6.9	7.7	4.4	21
3	135	6.9	7.3	5.5	24
4	89	4.5	5.1	2.8	25
5	79	4.0	4.8	1.5	18
South Australia					
1	67	3.5	1.0	12.1	15
2	32	1.6	0.1	6.6	3
3	10	0.5	0.2	1.5	6
4	7	0.3	n/a	1.5	3
5	7	0.3	n/a	1.5	3
New South Wales					
1	4	0.4	0.5	0	4
New Zealand					
1	33	1.7	0.6	5.2	12
2	3	0.15	n/a	0.4	2
3	1	n/a	n/a	0.2	1

Top trainers are ranked by number of starts. Only the 5 top trainers for Victoria and South Australia are listed.

3.2.3. Jockeys

Relative to horses (438) and trainers (145), the cohort of jump jockeys in the study period was small, at 51. Around 30 jockeys rode in all three Australian jump racing seasons, with the majority moving frequently between South Australia and Victoria. Ten jockeys accounted for 62.1% of starts, with three jockeys accounting for 25.3% of starts. The most prolific jockey rode in almost four out of every five (77%) jump races over this period.

3.3. Horse Falls and Fatalities

Table 4 shows the fall, fatality, and finish rates for jump horses in the study period. The overall fatality rate was 5.1 per 1000 starts (0.51%). The overall fall rate was 33 per 1000 starts (3.3%). Around 10% of all jump horse starters were retired before the race finished, falling into the category "failed to finish". Other reasons for not completing a race include "lost rider" (*i.e.*, jockey falling off) (1.6%), "brought down" (*i.e.*, horse brought down by another horse's fall) (0.35%) and, in one case, "run out" (*i.e.*, horse jumped out of the race course). Overall, 85% of horses starting a jump race completed the race.

A significant difference in risk profile was evident between steeplechase and hurdle races during the study period (Table 4). Of 10 horse fatalities, nine occurred in steeplechases, with the single hurdle fatality occurring on the flat at the start of a race, rather than over a hurdle. This disparity is increased when the larger proportion of hurdle races is taken into account, with one fatality in 1328 hurdle starts compared to nine fatalities in 642 steeplechase starts. The fatality rate for steeplechases was 14.0 per 1000 starts, more than double the KPI; and the rate of 0.75 per 1,000 starts for hurdles, was almost an order of magnitude below the KPI. This disparity is less marked when considering the rate of falls in hurdles (29 per 1000 starts) and steeplechases (40 per 1000 starts). However, the proportion of falls that result in fatalities in steeplechases (35%) is more than an order of magnitude greater than for hurdles (2.6%), with over one third of steeplechase falls proving fatal.

Table 4. Hurdle and steeplechase horse falls, fatalities and finishes, Victoria and South Australia, for the 2012, 2013, and 2014 seasons.

Race Type	Starts	Finishes	Deaths	Fatality Rate (Deaths per 1000 Starts)	Falls	Fall Rate (% of Starts)	Fatalities as Proportion of Falls (%)	FF *	BD **	RO ***	LR ****
Hurdle	1328	1135	1	0.75	39	2.9	2.6	128	7	1	18
Steeplechase	642	537	9	14	26	4.0	35	65			14
Total	1970	1672	10	5.1	65	3.3	15	193	7	1	32

* Failed to Finish (FF): horse withdrawn during race as fatigued and uncompetitive at discretion of jockey; ** Brought Down (BD): horse brought down during race by another horse; *** Run Out (RO): horse leaves track during race; **** Lost Rider (LR): jockey falls from horse during race.

We compared our study period against longer term trends using publicly available data for Victoria (comparable data were not available for South Australia). The resulting data (Figure 4) indicate that annual Victorian fatality rates during the study period are lower than any other consecutive three year period since 1986. These rates show considerable variability, with the lowest annual fatality rate in our study period, 5.5 deaths per 1000 starts in 2011–2012 and 2012–2013; the lowest annual rate being 4.1 deaths per 1000 starts (0.41%) in 1999–2000 (and 4.5 deaths per 1000 starts (0.45%) in 1997–1998.

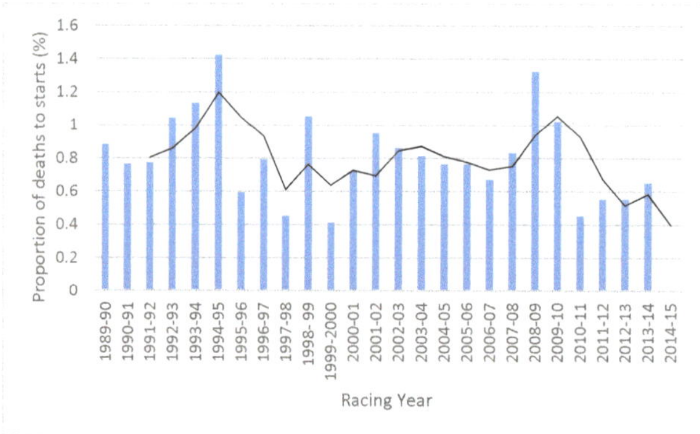

Figure 4. Horse fatality rates in Victorian jump racing, 1986 to 2015. Source data: Boden *et al.* 2006 [22], Australian Racing Fact Book 2013 [37], and the Animals Australia submission to the Victorian Parliament 2008 [21]. A racing year is defined as the period 1 July to 30 June in the following year. Fatality rates for the period 2011–2014 rates were estimated by identifying the date of the fatality and aggregating deaths over the racing year. Data for 2004–2005 and 2005–2006 are average values sourced from the 2008 Animals Australia submission to Victorian MPs as individual fatalities could not be located for this period [21].The line represents the moving three year average.

4. Discussion

Current public debate about Australian jump racing is taking place in the absence of comprehensive and reliable data and is characterised by competing claims by supporters and opponents. We use the data presented above to assess a range of contentious issues related to risks to horses in Australian jump racing.

4.1. Risks to Horses in Hurdles and Steeplechases

There is currently no horse fatality rate target mandated in Australian jump racing [16,28]. However, our data show that the fatality rate across the 2012, 2013, and 2014 jump racing seasons in Victoria and South Australia combined, 5.1 deaths per 1000 starts (0.51%), is below the Victorian industry KPI of 6.5 deaths per 1000 starts (0.65%) set in 2010 (and discontinued in 2011). The fatality rate observed in any one season of the study period is below half of that in the 2008, 2009, and 2011 seasons, and no more than three quarters of the rate in 2010 (Figure 4). This indicates that, despite the removal of mandated KPIs, safety initiatives introduced by RVL in 2010, and also adopted by South Australia, have had some success in reducing fatalities. These initiatives included improving the placement of obstacles, improving schooling and trialing facilities, and assessing the suitability of venues to conduct steeplechase races [12,26,28,38]. In particular, our data show that a new rule in 2010, allowing jockeys to retire a horse during a race if fatigued and out of contention, was invoked for about 10% of starts. Given that fatigued horses are more likely to pose a risk to themselves, their rider, or other horses, frequent application of this rule is likely to have been a significant contributor to improved horse and human safety in the study period [39–41]. Our finding that just 10 jockeys accounted for almost two thirds of jump race starts in the study period, raises the question also of whether improved safety performance is linked to the presence of a small, highly-experienced cohort of jump jockeys. Previous research in the United Kingdom supports jockey experience and consistent pairing of horse and jockey as key variables in reducing the risk of falls [42,43]. While the pool of Australian jump trainers is almost three times that of jockeys, this activity is also concentrated within a small, highly-experienced group, with five Victorian trainers accounting for one third of starts in Victoria and South Australia (Table 3).

To regulators and advocates of Australian jump racing, recent reductions in horse deaths indicate a laudable and sustainable improvement in horse welfare [44,45]. However, our data also indicate that horses continue to die in jump races. In addition to the risk of catastrophic injuries associated with falls over jumps [22,46], it has been suggested that an older horse population and longer race distances may contribute to a higher intrinsic risk in jump racing, compared to flat racing [47]. To opponents of Australian jump racing, then, jump racing cannot be run safely or humanely [31–34,48] with unjustifiable and avoidable horse deaths during jump races each season being predictable; "you are just waiting for it (a fall) to happen" [33]. Reflecting this, each horse death during the study period attracted media attention, with activist groups calling for an immediate ban on Australian jump racing, raising petitions and stating their intention to mount public protests at every jump race trial and race meeting in Victoria [48,49]. Horse injury and death in training and trials were not included in industry reporting of Australian jump racing fatality rates in our study period. A complete picture of horse fatalities in Australian jump racing would need to include training and trials as well as races. In 2015, RVL strengthened their Local Rules of Racing to require the death of a horse in training (including trials) to be reported by the trainer [17]. While our data show that industry responses to concern about horse welfare in Australian jump racing have reduced risks over the period 2012–2014, considering hurdle and steeplechase races separately reveals a disparity rarely raised in debate about Australian jump racing. No horse died while jumping a hurdle in a race in the study period, despite the fact that there were 39 falls during hurdle races. The rate of 0.75 fatalities per 1000 starts for the study period is considerably lower than that previously documented for hurdle races (6.3 fatalities per 1000 starts), suggesting that the current design and placing of hurdles is effective in reducing risks to horses during falls [22]. Steeplechase races, however, were riskier for horses, with nine of 26 falls over fences resulting in a fatality, despite the introduction of significant safety measures prior to the study period. The steeplechase fatality rate in the study period of 14.0 fatalities per 1000 starts (1.40%) is comparable to that documented by Bailey *et al.* of 14.3 [50], and higher than that documented by Bourke of 11.0 [51]. While the higher obstacles of steeplechases are likely to contribute to increased fall rates, the presence of more obstacles over longer distances, compared to hurdle races, enhances the risks of the onset of fatigue and the related risks of falling. In the United Kingdom, the risks of falling and/or injury were associated with greater race speed [40,41,46,52], race distance [53], fence type and location [40,41,53], going or track condition [54], number of runners [53,54], and experience of the horse [40] or jockey [41,54]. Fall risk increased in races with over thirty runners [42,43], greater race speed, especially in the second half of the race, races run at a faster pace [39–41,53] and longer races [42]. Over 90% of falls in UK jump racing were associated with horses colliding with an obstacle [40].

While steeplechases comprised only one third of jump races during the study period, and around 30 races in any one season, each of the 10 horse deaths occasioned in these races generated considerable negative media attention, energised anti-racing activists, and increased political pressure on the thoroughbred racing industry more generally [31–34]. However neither the RSPCA (South Australia and Victoria) nor the Coalition for the Protection of Racehorses distinguish between the inherent risk of hurdle races and steeplechases. While the Coalition for the Protection of Racehorses drew attention to a peak in hurdle deaths in 2009, they continue to combine data for steeplechase and hurdles in their critique, claiming that the statistics are so variable from year to year that there is no clear cut answer to whether or not steeplechases are riskier than hurdle races [33].

Supporters of Australian jump racing have not sought to respond to negative publicity about jump racing by drawing attention to the recent improved safety record of hurdle racing, the dominant form of Australian jump racing. The lack of any fatality over a hurdle in the study period, and an overall hurdle race fatality rate close to that of flat racing could potentially be used to question the assumption of critics that jump racing is inevitably far riskier than flat racing. Certainly, steeplechase races are among the most prestigious in Australian jump racing and include the two most lucrative jump races (in terms of prize money), with the average prize pool of steeplechases and hurdles in 2015 in Victoria being $64,000 and $35,000, respectively [29]. In the context of the status of steeplechase racing in Australian jump racing, future research to understand why safety interventions that appear to have successfully decreased the risk of hurdle races have not realised a similar reduction in risk for steeplechases seems warranted. In particular, the risk associated with specific race tracks is an area worthy of further study given the work of Williams *et al.* (2013) about risk factors for falls and fatalities associated with individual race courses [4,5,39]. Objective risk analysis and better understanding of sources of risk that result in falls and fatalities may ultimately result in better targeted risk mitigation strategies with a consequent reduction in falls, injuries and fatalities to both horses and jockeys [55]. A multi-disciplinary agenda for research that could reduce accident, injury, and death through risk mitigation strategies, as suggested by Thompson, McGreevy, and McManus, appears warranted [56].

Conflicting perceptions of acceptable risk, and the diverse value-judgments on which these may be based, are at the heart of debate about Australian jump racing. Typically those horse sports where there is a higher risk of injury or death to the horse, for example, bull fighting, rodeo, chuck wagon racing, and the cross country component of eventing, attract adverse media commentary, controversy, and public debate about horse deaths and the use of horses for public entertainment [6,57–59]. Horse deaths in jump races are highly public and often involve spectacular falls, providing activist groups with graphic photographs. The widespread use of such images on social media increases the visibility and awareness of the risks of Australian jump racing. Increased visibility is an advantage in promoting improved welfare outcomes, and a larger volume of protective legislation is generated in the case of animals with a high level of visibility. As O'Sullivan argues, "the community needs to know and like an animal for that animal to have a chance of receiving effective legal protection" [60]. Equally, the highly public and inherent risk involved in Australian jump racing may well be integral to the attraction this activity holds to those who view it as "the thrill of the chase" and an entertainment comprised of courageous horses, hardened trainers, fearless riders, and controlled danger in "the greatest show on earth" [61,62] .

4.2. Does Jump Racing Extend a Horse's Racing Career?

Unlike those in the UK and Ireland, Australian jump horses are not selectively bred for jumping, and are usually former flat racing horses bred to run longer distances. The progression from flat to jump racing in Australia is the basis of claims from advocates that jump racing extends a horse's racing career, giving them "a new lease on life"; not just by extending their tenure within the industry, but also by renewing a horse's desire for racing [63,64]. In Australia, rules of racing mandate that a horse has to be at least three years old before it can commence jumping. Given the requirements to qualify a horse to jump, it is likely that the majority of Australian jump horses will be at least four years old before racing over obstacles. Our data show the median age of jump horses is six years, with a range of three to 13 years and almost one quarter aged eight and above (Figure 3).

The anti-jump racing organisation, the Coalition for the Protection of Racehorses, has argued that Australian race horses are separated from their mothers at about six months of age to commence preparation for sale and training and have an "average career of ... less than three years," after which "the majority will be killed" [33,34]. If this claim has merit, our finding that the median age of a jump horse is six years, with almost one quarter aged eight and above, indicates that Australian jump racing may well significantly extend the tenure of some horses within the racing industry. Given that jump racing accounted for only 1.5% of all thoroughbred racing in Victoria and South Australia the study period, this career option will be offered only to a small minority of race horses.

Our data about the age profile of jump horses needs to be understood, however, in the context of a high rate of annual turnover in the jump horse cohort. Despite the investment in training and qualifying a jump horse, 55% of horses started in three or fewer races across the 2012, 2013, and 2014 seasons, with 22% starting in only one race (Table 2). Only 37% of the 2012 cohort jumped in 2013, while only 29% of the 2013 cohort raced in 2014. Less than 5% of the 2012 cohort participated in all three seasons. The career of the majority of jump horses thus involves a small number of hurdle races in a single season of racing. This finding is consistent with the pattern of jump racing in New Zealand where jump horses have fewer starts than flat racing thoroughbred horses and represent an older horse population [65]. Many flat racing careers are also of limited duration and may also consist of a limited number of starts [66]. This suggests that any career extension enabled by Australian jump racing is short lived for the majority of horses. It also suggests that, given a median jump horse age of six years, the flat careers of jump horses either started later or progressed for longer than claimed by Coalition for the Protection of Racehorses. It is also clear that Australian jump racing is sustained by high levels of new horse entries each season. This high rate of turnover raises questions about horse pathways in and out of Australian jump racing. These questions are relevant to public debate about the drivers of horse breeding in Australian racing [11,67].

4.3. The Future of Jump Racing

Much opposition to Australian jump racing has called for the abolition of this activity, a prospect that seemed imminent in Victoria in 2010 [20]. In response to these calls, many advocates point to the vital role of jump racing within the Australian racing industry as a whole and within specific regional economies and cultural identities [7,45,68,69]. Recent media commentary suggests that Australian jump racing has enjoyed a resurgence following the election of the pro-jump Napthine State Government in 2011 [45,69]. Our study shows no evidence of either significant decline or significant growth in Australian jump racing over the period 2012 to 2014. However, in 2015, the South Australian Jockey Club expressed a wish to phase out jump racing at Morphettville. The Minster for Racing, Leon Bignell, stated his desire to end the activity; a parliamentary committee has been set up to investigate the future for jump racing; and a member of State Parliament, the Greens MLA, Tammy Franks, has introduced a bill to ban the activity [14,70].

The extent of public controversy about Australian jump racing might suggest that this activity is widespread and substantial. Our data, however, indicates that geographically, and in terms of the human participant base, this activity is highly concentrated. Three race tracks—Morphettville (Adelaide), Sandown (Melbourne) and Warrnambool (South-West Victoria)—accounted for over half (55%) of this activity, with South-West Victoria its vital heartland. Three trainers in Victoria trained one in four of all jump horses in Australia, including almost one in three in South Australia. Three jockeys accounted for over one quarter of all jump race starts (25.3%), with each participating in a majority of races. These data indicate that the presence of jump racing in South Australia is highly dependent upon participation from the Victorian racing industry. Any reduction or banning of jump racing in South Australia will have some impact on the Victorian industry. The banning of jump racing in Victoria would likely mean the end of this activity in Australia. While an end to Australian jump racing might affect only a relatively small number of industry livelihoods, racing clubs, and regional communities, the extent of this impact on these individuals, groups and places will be profound. It is not surprising, therefore, that the defence of Australian jump racing is vociferous, particularly in regional Southwest Victoria [68]. The Warrnambool Racing Club's May carnival, for example, attracts an audience equivalent to the entire population of this regional centre [68]. This event is regarded locally as a major boost for regional tourism and local businesses. "After all, the three day carnival is said to be worth $15 million to the town (Warrnambool). Pubs and restaurants swell in an annual blur of heartiness and hangovers. 'If we lose it the town is screwed,' says one well-placed observer" (p. 20, [71]). Similarly, in South Australia, the iconic Oakbank Easter racing carnival is reputably worth $13 million to the local economy and draws crowds around 70,000 to the two days of jump racing [70].

It is too early to assess the implications for Australian jump racing of a change of government in Victoria in late-2014, although the new Minister for Racing has said that the future of jump racing rests in the hands of RVL [72]. It is noteworthy that the 2015 Victorian jump racing season is being conducted entirely at regional race courses, and not at Sandown in Melbourne, a venue that accounted for 16.1% of all jump races between 2012 and 2014 [73]. In light of on-going public debate about Australian jump racing, longitudinal, industry-wide and composite data, such as that presented here, is vital for informed discussion and effective regulation. We note that while the data on which this paper is based are publicly available, and are also highly fragmented and dispersed, requiring considerable effort to assemble and integrate. We recommend that racing authorities consider forms of data collection, recording and archiving that are more amenable to analysis of industry-wide and long-term trends. Reasons underlying short-lived thoroughbred jump racing careers also deserve further research, particularly given the significantly greater investment in time required to train and qualify a jump horse in Australia.

5. Conclusions

We conclude that the safety of Australian jump racing over the 2012, 2013, and 2014 seasons has improved significantly from the 2008 and 2009 seasons that provoked strong public opposition and led regulators to canvass the banning of this activity. However, it is not clear that there has been any improvement in horse safety in steeplechasing, with this activity accounting for nine of the 10 horse deaths in the study period. The risks to horses in hurdle racing during the study period, in contrast, were close to those documented for flat racing. While the average age of Australian jump horses indicates that jump racing may extend their tenure within the racing industry, the jump racing career of a majority of horses comprises no more than three races conducted in one season. Jump racing in South Australia is substantially dependent upon Victorian involvement and, overall, Australian jump racing relies upon significant participation from a very small number of trainers and jockeys.

Acknowledgments

The authors acknowledge the support of the Australian Research Council (DP130104933 project "Caring for Thoroughbreds: addressing social, economic and welfare issues in international horse racing") and three anonymous reviewers for their guidance.

Author Contributions

Karen Ruse conducted data collection and analysis, literature review, and contributed to the argument and writing of the paper. Aidan Davison and Kerry Bridle contributed to the argument, analysis and writing of the paper.

Conflicts of Interest

The authors declare no conflict of interest.

References

1. Evans, D.; McGreevy, P. An investigation of racing performance and whip use by jockeys in thoroughbred races. *PLoS ONE* **2011**. [CrossRef] [PubMed]
2. McManus, P.; Montoya, D. Towards new understandings of human-animal relationships in sport: A study of Australian jumps racing. *Soc. Cultur. Geogr.* **2012**, *13*, 399–420. [CrossRef]
3. Hitchens, P.; Blizzard, C.L.; Jones, G.; Day, L.; Fell, J. Enduring sport: The incidence of race day jockey falls in Australia, 2002–2006. *Med. J. Austr.* **2009**, *190*, 83–86.
4. Williams, J.M.; Marks, F.; Mata, F.; Parkin, T. A case control study to investigate risk factors associated with horse falls in steeplechase races at Cheltenham racetrack. *Compar. Exerc. Physiol.* **2013**, *9*, 59–64. [CrossRef]
5. Williams, J.M.; Marlin, D.M.; Langley, N.; Parkin, T.D.; Randle, H. The Grand National; A review of factors associated with non-completion and horse-falls, 1990–2012. *Compar. Exerc. Physiol.* **2013**, *9*, 131–146. [CrossRef]
6. Daniels, J. Should the Grand National be Banned? 2014. Available online: http://horsetalk.co.nz/2014/should-the-grand-national-be-banned/LZexzz3JbMISd97 (accessed on 26 August 2015).
7. Alexander, M. Premier Annointed Patron Saint of Jumps Racing. 2013. Available online: http://www.standard.net.au/story/1474965/premier-anointed-patron-saint-of-jumps-racing/ (accessed on 26 August 2015).
8. Yeo, J.-H.; Neo, H. Monkey business: Human-animal conflicts in urban Singapore. *Soc. Cultur. Geogr.* **2010**, *11*, 681–699. [CrossRef]
9. Mattson, D.J.; Clark, S.G. The discourses of incidents: Cougars on Mt Elder and in Sabine canyon, Arizona. *Policy Sci.* **2012**, *45*, 315–343. [CrossRef]
10. McManus, P.; Graham, R.; Ruse, K. The construction of human-animal relationships. National jumps day 2013, Te Rapa, Hamilton, NZ. *NZ Geogr.* **2013**, *70*, 190–200. [CrossRef]
11. McManus, P.; Albrecht, G.; Graham, R. *The Global Horseracing Industry. Social, Economic, Environmental and Ethical Perspectives*; Routledge: Abingdon, VA, USA; New York, NY, USA, 2013.
12. Vamplew, W. Horses, history and heritage; A comment on the state of going. *Austr. Sport. Soc. Histor. Stud.* **2008**, *23*, vii–x.
13. International Federation of Horse Racing Authorities. Annual Report, 2013. 2013. Available online: http://www.horseracinfed.com/Default.asp?section=Resources&area=2 (accessed on 26 August 2015).
14. Franks, T. Animal welfare (Jumps Racing) Amendment Bill. 2015. Available online: http://www.tammyfranks.org.au/?s=jumps+racing (accessed on 25 August 2015).
15. Law Society of South Australia. Jumps Racing Submission. 2011. Available online: http://.lawsocietysa.asn.au/LSSA/media/Archived_Submissions.aspx?Websitekey=bf611791-fae7-49b2-884d-cf15580b554e (accessed on 25 August 2015).

16. Racing Victoria. Jumps Racing Qualification. 2015. Available online: http://www.rv.racing.com/racing-and-integrity/rules-of-racing (accessed on 26 August 2015).
17. Racing Victoria. Rules of Racing. 2015. Available online: http://www.rvracing.com/racing-and-integrity/rules-of-racing (accessed on 25 August 2015).
18. Smith, P. Jumps Racing is Dead, Now We Have to Muster the Courage to Bury it. 2009. Available online: http://www.theaustralian.com.au/archive/news/jumps-racing-is-dead/story-e6frg7mo-1225710546646 (accessed on 25 August 2015).
19. Racing Victoria. David Jones to Conduct Jumps Review. 2008. Available online: http://racing.co/news/2008-04-09/david-jones-to-conduct-jumps-review (accessed on 26 August 2015).
20. Victorian Jumps Races Suspended after More Deaths. 2009. Available online: http://www.smh.com.au/news/sport/horseracing/victorian-jumps-races-suspended-after-more-deaths/2009/05/07/1241289301661.html (accessed on 25 August 2015).
21. Animals Australia. Submission to State of Victoria, Members of Parliament, Dec 2008. Available online: http://www.animalsaustralia.org/documents/Jumps-Racing-Submission-to-Victorian-MPs-Dec-2008.pdf (accessed on 26 August 2015).
22. Boden, L.; Anderson, J.A.; Charles, K.L.; Morgan, J.M.; Parkin, T.D.H.; Slocombe, R.F.; Clarke, A.F. Risk of fatality and causes of death of thoroughbred horses associated with racing in Victoria, Australia: 1989–2004. *Equine Vet. J.* **2006**, *38*, 312–318. [CrossRef] [PubMed]
23. Racing Victoria. Jumps Racing Review 2009 Racing Victoria Limited. 2009. Available online: http://www.rspca.vic.org/issues-take-action/jumps-racing/RSPCA-2009-Racing-Victoria-Limited-review.pdf (accessed on 6 August 2015).
24. Racing Victoria. Jumps Racing Decision. 2010. Available online: http://www.racing.com/news/2010-01-2/jumps-racing-decision-january-21 (accessed on 25 August 2015).
25. Racing Victoria. Board Decision on Hurdle and Steeplechase Racing. 2010. Available online: http://www.racing.com/news/2010-09-02/board-decision-on-hurdle-and-steeplechase-racing (accessed on 25 August 2015).
26. Racing Victoria. One Year Steeplechase Racing Program Approved. 2010. Available online: http://www.racing.com/news/2010-10-07/one-year-steeplechase-racing-program-approved (accessed on 26 August 2015).
27. Fortnum, S. Racing Minister Pledges Support for Jumps Racing. 2011. Available online: http://www.races.com.au/2011/04/09/racing-minister-pledges-support-for-jumps-racing (accessed on 26 August 2015).
28. Racing Victoria. Jumps Racing Program Released. Available online: http://www.racing.com/news/2011-11-22/jumps-racing-program-released (accessed on 26 August 2015).
29. Barnes, J. Brief History of Jumps Racing in Australia. 2015. Available online: http://barnesphotography.com.au/jumps11/page1.htm (accessed on 25 August 2015).
30. Bazley, N. Jumps Ban. 2009. Available online: http://www.abc.net.au/btn/story/s2573586.htm (accessed on 30 July 2015).

31. RSPCA Victoria. 2014 Another Horse Death Marks Another Day in Jumps Racing: When Will It End? 2015. Available online: http://rspcavic.org/documents/media/Media%/releases/2014/RSPCA-Media-Release-Another-horse-death-marks-another-day-in-jumps-racing.pdf (accessed on 26 August 2015).
32. RSPCA South Australia. Jumps Racing. 2015. Available online: http://www.rspca.org/the-issues/jumps-racing/ (accessed on 26 August 2015).
33. Coalition for the Protection of Racehorses. Jumps Racing. 2015. Available online: http://www.horseracingkills.com/the-issues/jumps-racing (accessed on 26 August 2015).
34. Coalition for the Protection of Racehorses. Deathwatch Report 2014. 2014. Available online: http://www.horseracingkills.com/the-issues/jumps-racing (accessed on 26 August 2015).
35. The Humane Society. Jump Racing. 2015. Available online: http://www.hsi.org.au/go/to/567/jumps-racing.html#.VZ8rG10ViUk (accessed on 26 August 2015).
36. RSPCA Victoria. Jumps Racing. 2015. Available online: http://www.rspcavic.org/issues-take-action/jumps-racing/ (accessed on 26 August 2015).
37. Australian Racing Board. Australian Racing Fact Book 2013/14. 2014. Available online: http://www.australianracingboard.com.au/fact_book.aspx (accessed on 26 August 2015).
38. Racing Victoria. RVL Implements Further Jumps Policies. 2008. Available online: http://racing.com/news/2008-08-22/rvl-implements-further-jumps-policies (accessed on 26 August 2015).
39. Williams, J.M.; Marks, C.F.; Parkin, T.D.M.; DaMata, F. An epidemiological review of factors that increase the risk of horse falls in steeplechase races at Cheltenham. *Compar. Exerc. Physiol.* **2013**, *9*, 59–64. [CrossRef]
40. Pinchbeck, G.L.; Clegg, P.D.; Proudman, C.J.; Morgan, K.L.; French, N.P. Case control study to investigate factors for horse falls in hurdle racing in England and Wales. *Vet. Rec.* **2003**, *152*, 583–587. [CrossRef] [PubMed]
41. Pinchbeck, G.L.; Clegg, P.D.; Proudman, C.J.; Morgan, K.L.; French, N.P. Whip use and race progress are associated with horse falls in hurdle and steeplechase racing in the UK. *Equine Vet. J.* **2004**, *36*, 384–389. [CrossRef] [PubMed]
42. Williams, J.M.; Aldous, N.; Da Mata, F. Factors associated with horse falls in British steeplechase racing from 2008 to 2011. In Proceedings of the 8th International Society of Equitation Science; International Society of Equitation Science: Edinburgh, UK, 2012; p. 22.
43. Williams, J.M.; Smith, K. Factors associated with horse falls in British class 1 steeplechases 1999 to 2011. In Proceedings of the 8th International Society of Equitation Science; International Society of Equitation Science: Edinburgh, UK, 2012; p. 143.
44. Lynch, M. Bool Brings Hope and Relief to Jumping Fraternity. 2014. Available online: http://www.smh.com.au/sport/horseracing/bool-brings-hope-and-belief-to-jumping-fraternity-2014 0503-zr3r3.html (accessed on 25 August 2015).
45. Auld, T.; Fawkes, A. Jumps Racing Back on Track. 2012. Available online: http://www.standard.net.au/story/370326/jumps-racing-back-on-track/ (accessed on 25 August 2015).
46. Pinchbeck, G.L.; Clegg, P.; Proudman, C.; Riggs, C.; Singer, E.R.; Webbon, P.M.; Morgan, K.L. Race and course level risk factors for fatal distal limb fracture in racing Thoroughbreds. *Equine Vet. J.* **2004**, *36*, 521–526.

47. Parkin, T. Epidemiology of training and racing injuries. *Equine Vet. J.* **2007**, *39*, 466–469. [CrossRef] [PubMed]
48. Demo for Try Pickle. Available online: http://www.facebook.com/banjumpsracing/event/4456227 6530/ (accessed on 1 September 2015).
49. Jumps Racing: The Week That Never Was. 2014. Available online: http://www.banjumpsracing. com/2014/05 (accessed on 5 April 2015).
50. Bailey, C.J.; Hodgson, D.R.; Bourke, J.M. Flat, hurdle and steeple racing; Risk factors for musculoskeletal injury. *Equine Vet. J.* **1998**, *30*, 498–503. [CrossRef] [PubMed]
51. Bourke, J.M. Fatalities on racecourses in Victoria. A seven year study. In Proceedings of the 10th International Conference of Racing Analysts and Veterinarians; R & W Publications: Elma, NY, USA, 1995; pp. 265–268.
52. Pinchbeck, G.L.; Clegg, P.D.; Proudman, C.J.; Morgan, K.L.; French, N.P. A prospective cohort study to investigate factors for horse falls in UK hurdle and steeplechase racing. *Equine Vet. J.* **2004**, *36*, 595–601. [CrossRef] [PubMed]
53. Pinchbeck, G.L.; Clegg, P.D.; Proudman, C.J.; Morgan, K.L.; Wood, J.L.; French, N.P. Factors and sources of variation in horse falls in steeplechase racing in the UK. *Prev. Vet. Med.* **2002**, *55*, 179–192. [CrossRef]
54. Parkin, T.D.H.; Clegg, P.D.; French, N.P.; Proudman, C.J.; Riggs, C.M.; Singer, E.R.; Webbon, P.M.; Morgan, K.L. Risks for fatal lateral condylar fracture of the third metacarpus/metatarsus in UK racing. *Equine Vet. J.* **2004**, *37*, 192–199. [CrossRef]
55. Curry, B.A.; Hitchens, P.L.; Otarhal, P.; Si, L.; Palmer, A.J. Workplace injuries in thoroughbred racing: An analysis of insurance payments and injuries amongst jockeys in Australia from 2002 to 2010. *Animals* **2015**, *5*, 897–909. [CrossRef]
56. Thompson, K.; McGreevy, P.; McManus, P. A critical review of horse-related risk: A research agenda for safer mounts, riders and equestrian cultures. *Animals* **2015**, *5*, 561–575. [CrossRef]
57. O'Brien, D. Eventing in Crisis. 2015. Available online: http://www.horsetalk.co.nz/saferdie/ 131-eventingincrisis.shtml#ixzz3Xue7tgCh (accessed on 2 July 2015).
58. Grand National Deaths at Aintree Fuel Criticism from Animal Rights Groups. Available online: http://www.london24.com/sport/other/grand_national_2013_horse_deaths_at_aintree_fuel_ criticism_from_animal_rights_groups (accessed on 5 April 2015).
59. O'Sullivan, S. *Animals, Equality and Democracy*; Palgrave MacMillan: Basingstoke, UK, 2011.
60. McGinley, S. The Glorious Spectacle of Jumps Racing. 2013. Available online: http://www.theroar.com.au/2013/06/05/the-glorious-spectacle-of-jumps-racing (accessed on 1 September 2015).
61. Adams, J. *Over the Hurdles*; Melbourne Books: Melbourne, Australia, 2012.
62. Best, G. Gotta Take Care Rewards Trainers Judgement with Galleywood. 2014. Available online: http://www.smh.com.au/sport/horseracing/gotta-take-care-rewards-trainers-judgment-with-galleywood-20140430-zr219.html (accessed on 25 August 2015).
63. Best, G. Let the Fun Times Roll in the Brierly. 2012. Available online: http://www.standard.net.au/ story/71258/let-the-fun-times-roll-in-brierly/ (accessed on 25 August 2015).

64. Bolwell, C.F.; Rogers, C.W.; Gee, E.K.; Rosanowski, S.M. Descriptive statistics and the pattern of horse racing in New Zealand: Part One Thoroughbred racing. *Animal Prod. Sci.* **2014**. [CrossRef]
65. Velie, B.D.; Wade, C.M.; Hamilton, N.A. Profiling the careers of thoroughbred horses racing in Australia betwen 2000 and 2010. *Equine Vet. J.* **2013**, *45*, 182–186. [CrossRef] [PubMed]
66. Geelen, R. The internet age of misinformation. *Breed. Rac.* **2014**, *117*, 52–54.
67. Auld, T. Bumper Results for May Racing Carnival. 2015. Available online: http://www.standard.net.au/story/3074625/bumper-results-for-may-racing-carnival/ (accessed on 25 August 2015).
68. Auld, T. Napthine Hails Valediction Win a Success for Jumps Racing. 2014. Available online: http://www.standard.net.au/story/2213088/napthine-hails-valediction-win-a-success-for-jumps-racing/ (accessed on 25 August 2015).
69. Stephens, T. Motion to Establish a Joint Committee on Jumps Racing. 2015. Available online: http://www.terrystephens.com.au/news/Speeches/ID/164/Motion-To-Establish-a-Joint-Committee-on-Jumps-Racing (accessed on 1 September 2015).
70. Carlyon, C. *D-Day for Jumps. After a Brutal Backlash, the Carnival could be Over*; Herald Sun: Melbourne, Australia, 2009; p. 20.
71. Paluka Visit a Jump Ahead. 2015. Available online: http://www.standard.net.au/story/3145927/pakula-visit-a-jump-ahead/ (accessed on 25 August 2015).
72. Auld, T. Prizemoney Boost as Races Transferred to Warrnambool. 2015. Available online: http://www.standard.net.au/story/3051460/prizemoney-boost-as-races-transferred-to-warrnambool/ (accessed on 25 August 2015).
73. Racing Victoria to Transfer Grand Nationals from Sandown. 2015. Available online: http://www.smh.com.au/sport/horseracing/racing-victoria-to-transfer-grand-nationals-from-sandown-20150427-1muins.html (accessed on 19 October 2015).

Animals **2015**, *5*, 965-977; doi:10.3390/ani5040394

OPEN ACCESS

animals

ISSN 2076-2615

www.mdpi.com/journal/animals

Article

Reducing Respiratory Health Risks to Horses and Workers: A Comparison of Two Stall Bedding Materials

Markku Saastamoinen [1,*], Susanna Särkijärvi [1] and Seppo Hyyppä [2]

[1] Natural Resources Institute Finland (Luke), Opistontie 10 a 1, 32100 Ypäjä, Finland; E-Mail: susanna.sarkijarvi@luke.fi

[2] Ypäjä Equine College, Opistontie 9, 32100 Ypäjä, Finland; E-Mail: Seppo.hyyppa@hevosopisto.fi

* Author to whom correspondence should be addressed; E-Mail: markku.saastamoinen@luke.fi; Tel.: +358-29-532-6509.

Academic Editor: Kirrilly Thompson

Received: 14 July 2015 / Accepted: 29 September 2015 / Published: 8 October 2015

Simple Summary: In this study, the effect of wood shavings and peat was examined on stable air quality and health of horses and stable workers. The ammonia level in the boxes in which peat was used as bedding was non-existent or very low. The respiratory symptoms in horses increased regardless of the bedding material at the beginning of the study. The health status of the horses on peat bedding returned to the initial level in the end of the trial but horses in stalls bedded with wood shavings continued to be symptomatic. The hooves of the horses in stalls with peat bedding had a better moisture content. The results suggest that peat is a better bedding material for horses and people working or visiting horse stables than wood shavings.

Abstract: Stable air quality and the choice of bedding material are an important health issue both in horses and people working or visiting horse stables. Risks of impaired respiratory health are those that can especially be avoided by improving air quality in the stable. The choice of bedding material is particularly important in cold climate conditions; where horses are kept most of the day and year indoors throughout their life. This study examined the effect of two bedding materials; wood shavings and peat; on stable air quality and health of horses. Ammonia and dust levels were also measured to assess conditions in the stable. Ammonia was not detected or was at very low levels (<0.25 ppm) in the boxes in which peat was used as bedding; but its concentration was clearly higher (1.5–7.0 ppm) in stalls with wood shavings as bedding. Personal measurements of workers revealed quite high ammonia exposure (5.9 ppm_{8h}) in the boxes in which wood shavings were used; but no exposure was observed in stalls bedded with peat. The respiratory symptoms in horses increased regardless of the bedding material at the beginning of the study. The health status of the horses in the peat bedding group returned to the initial level in the end of the trial but horses bedded with wood shavings continued to be symptomatic. The hooves of the horses with peat bedding had a better moisture content than those of the horses bedded with wood shavings. The results suggest that peat is a better bedding material for horses than wood shavings regarding the health of both horses and stable workers.

Keywords: bedding material; respiratory health; peat; wood shaving

---

1. Introduction

The stalls of horses are bedded to absorb urine, moisture, and gases and to increase the comfort, health, and well-being of the horses. In addition, a large number of people are engaged in the horse industry as trainers, riders, stable workers, farriers, and other roles working or visiting stables for many hours daily, and being subjected to the influences of the stable environment. Most of them are young people, for example riding school pupils.

Horses in northern climatic conditions (e.g., in the Nordic countries and Canada) are exercised outdoors usually for 1 to 2 h and spend, consequently, the major part of the day (often up to 23 h) indoors. Because of this, stable air quality is of considerable importance. Furthermore, training and racing in cold weather conditions expose the vulnerable respiratory system to health problems, increasing inflammatory cells in the lungs [1]. As a consequence, respiratory disorders are common problems, and respiratory allergy is commonly diagnosed as a condition affecting the equine lung. When the condition becomes protracted it is referred to as chronic obstructive pulmonary disease (COPD) or heaves (or RAO, recurrent airway obstruction), an animal model of asthma. Anecdotal reports suggest that the condition is rare in climates where animals are outside all year around but is common in climates where horses are stabled indoors [2]. Clinical signs in horses with this chronic lung disease include poor athletic performance, chronic couching, purulent nasal discharge, and ultimately difficulties in breathing [2,3].

People working in and visiting horse stables may also be exposed to the effects of the stable air. Causes of chronic airway disease both in horses and humans usually involve exposure to excessive concentrations of airborne dust, molds, viruses, bacteria, spores, aeroallergens, and endotoxins which mostly originate from bedding and feed [4–7]. Furthermore, the inhalation of gaseous irritants such as ammonia may initiate airway obstruction and exacerbate or prolong the clinical signs of COPD in affected horses [2] as well as humans [3].

The effect of bedding material on the quality of stable air, both on stable dust and the ammonia concentration, is significant [8]. The various forms of bedding in a stable and even the differences in beddings between boxes within a stable [8] influence the stable dust and gas loads, and consequently the risk to airway health of both horses and humans.

Currently, several materials are available for the bedding of boxes in horse stables. The most common bedding materials are wood shavings, saw dust, straw, and peat. Many other materials are also used as bedding, including processed (pelleted) wooden materials and (pelleted, chopped) straw from different plants. In addition, shredded or cut paper and some plant materials (by-products), as well as woodchips are used. Each of these has individual properties, including advantages and disadvantages [9]. Stall mats are also available, but they are usually used together with bedding because of the binding of urine. It has also been reported that horses prefer bedding material for lying down as compared with areas without it [10].

Factors considered when selecting bedding are its availability, cost, cleanness (free from dust and foreign material) and its effect on stable air quality. The bedding material should also be easy and light to handle, to avoid excessive physical exertion in stable workers. From the point of view of horses' health and well-being, good bedding creates a layer of insulation between the horse and cold floor, pads the hard surface, prevents bruised knees, elbows, hocks and hips, and keeps the horse clean. The bedding material also affects the behavior of horses [11–13], for example the incidence of stereotypic behaviour. In addition, good bedding material has a better potential to be re-used e.g., in farming and horticulture [14,15].

The objective of this study was to examine the effects of two different bedding materials, wood shavings and peat, on the health of horses. This issue was evaluated on the basis of respiratory and overall health and quality of hooves, and by measuring stable air quality.

2. Materials and Methods

The experiment was conducted in the facilities of MTT Agrifood Research Finland (MTT, currently Natural Resources Institute Finland Luke) in the south western part of Finland (latitude 60°) under autumn/early winter (October to December) climatic conditions. The duration of the experiment was 84 days. Twelve Finnhorse brood mares (four of which were pregnant) aged between 5 and 17 years were housed in box stables in individual stalls (3 m × 3 m), divided into separate sections of the stable according to the bedding material (peat; wood shavings). The stable sections were of the same size and had an identical mechanical ventilation system. The horses were held on pasture from the beginning of June to the middle of September.

The two bedding materials were selected because they are the most common materials used for bedding in Finland. They both have a low content of harmful components when manufactured, selected and stored properly. Peat is favoured as a bedding material because of its good properties in soil improvement and good composting ability, as well as its superior capacity to bind ammonia and fluid [13,16,17] compared to other materials. Both bedding materials were manufactured for use as beddings in horse stalls; peat by Vapo Ltd. (Jyväskylä, Finland) and wood shavings by Joutsenon purupaali Ltd. (Joutseno, Finland).

The horses were exercised daily in paddocks in groups for four hours, and for one hour by riding or driving during the course of the experiment. The stalls were manually cleaned by the same person between 8 and 12 am when the horses were in outdoor paddocks. All feces and wet material were removed and new bedding material was added. The depth of the bedding was about 10 cm. All removed and added bedding materials were measured by their volume.

The horses were individually fed according to their needs three times per day (morning, noon, evening) with silage/haylage (DM 26.6%–6.9%) and pelleted compounded feed (DM 88%) (Suomen Rehu Ltd., Turku, Finland) to minimize the release of airborne particles from the feeds. The diet was balanced for protein (nitrogen) intake to avoid nitrogen lost in urine and, thus, to minimize the ammonia in the stable. The forage was produced by MTT and its fermentation and hygienic quality fulfilled the criteria of good quality haylage and silage [18]. The forage was placed on the floor.

Outdoor temperatures and weather conditions were recorded daily at 8:00 am. The average outdoor temperatures in October, November and December, respectively, were −1.7 °C (−10 to 2 °C), −3.8 °C (−14 to 3 °C) and −6.6 °C (−20 to 0 °C). According to the statistics of the Finnish Meteorological Institute the temperatures in December were quite normal, but in November the daily temperatures were highly variable, and in October the temperatures were exceptionally low.

The stable temperatures and humidity, as well as ammonia and carbon dioxide levels and amount of dust, were measured daily in both stable sections. Methane (day 0) and hydrogen sulphide (days 0 and 42) contents were measured, but because of undetectable values the measuring was not continued. Gases concentrations were measured at a height of 120 cm from the ground using an Accuro gas detection pump which draws air through sampling tubes (Dräger Safety AG, Lübeck, Germany). The measurements were carried out at 6:00 am in three boxes of each stable section; from the middle of the box at the level of the muzzle of the horse. Dust was continuously collected into dust cases that were fitted in empty boxes in both stable sections at the level of 40 cm from the ground.

Exposure of the stable workers to ammonia was evaluated with personal measurements using sampling tubes attached to the lapel of the person (Dräger Safety AG, Germany) in the middle and at the end of the experiment during the cleaning of the stalls. The measurement result was converted to correspond to an exposure period of eight hours (HTP_{8h}) [19]. HTP value is the concentration that is harmful to people.

A respiratory endoscopic examination was performed three times during the study (days 0, 42 and 84), including examination of the ethmoidal region, pharyngeal openings of guttural pouches, soft palate, larynx, and trachea (symptoms = 1; no symptoms = 0). Tracheobronchial aspirates were drawn during the endoscopy and cytological and bacteriological (neutrophil cells) evaluation was carried out. The classification of the neutrophil cells in bronchoalveolar smear samples was as follows: none or some single cells (−); single cells and few small pool of cells (+); several large pools of cells (++); abundant pools of cells (+++); and an extreme abundance of cells (++++).

Blood analyses, fecal analyses and hoof quality evaluation were used as measures of health and well-being of the horses. These samples were taken with the same interval as the endoscopic examination. In addition, rectal body temperature was measured and, heart rate (with stethoscope) and respiration rate via auscultation were recorded by a veterinarian researcher. Blood samples were collected from the jugular vein, and the blood analysis consisted of hemoglobin, haematocrit, serum urea, iron, protein, and differential cell count. Bacteriology, parasites, and the pH of faeces were determined. The quality of hooves was assessed from the dry matter content of hoof horn. The hoof horn samples were collected from the hooves of front legs when the horses were in shoeing. All samples were analyzed in the clinical laboratory of MTT.

The experimental design was a randomized block design with repeated measurements. After the first endoscopy, the horses were formed into pairs based on their symptomatic similarity. The two horses of each pair were then randomly allotted to different bedding material groups (peat bedding or wood shaving bedding). The procedure was repeated until all horses were divided in the two groups. The information from the first endoscopy was excluded from the data because it was included in the animal pair-variable in the model. The data (samples from horses) were analyzed using the MIXED procedure of the SAS system with the following statistical model: $Y_{ijk} = \mu + p_i + b_j + (p \times b)_{ij} + t_k + (p \times t)_{jk} + (b \times t)_{jk} + e_{ijk}$, where Y_{ijk} is the observation, μ is the overall mean, p_i is the random effect of ith animal pair ($i = 1 \ldots 6$), b_j is the fixed effect of jth bedding material ($j = 1 \ldots 2$), t_k is the fixed effect of the time period ($k = 2$ or 3), and e_{ijk} is the normally distributed error with a mean of 0 and variance δ^2. Terms $(p \times b)_{ij}$, $(p \times t)_{jk}$ and $(b \times t)_{jk}$ are compound effects of factors. The best fitting covariance structure for repeated measurements was selected on the basis of the Akaike information criterion. The differences were tested with Tukey's test. Categorical variables (neutrophil cells in tracheal mucus) and 0/1-variables were not tested statistically, but were presented descriptively, because of the small number of observations and their subjective scoring making them less informative.

In animal handling and sample collection, the European Union recommendation directives (1999/575/EU) and national animal welfare and ethical legislation set by the Ministry of Agriculture and Forestry of Finland were followed carefully. The experimental procedures were evaluated and approved by The Animal Care Committee of MTT before the study was started. The endoscopic examination was carried out and all samples from the horses were collected by a veterinarian researcher.

3. Results and Discussion

3.1. Air Quality

The average temperatures in the stable sections (peat bedding *vs.* wood shaving bedding) in October, November and December, respectively, were 9.2 *vs.* 10.3 °C, 9.4 *vs.* 9.4 °C, and 8.0 *vs.* 8.0 °C. These temperatures are within the target indoor temperature range (8–12 °C) in horse stables in Finland [20]. The average moisture content of the stable air (peat *vs.* wood shavings) was 54.3% *vs.* 54.6%, 56.0% *vs.* 57.6% and 53.0% *vs.* 58.6% in the corresponding months. During the lowest outdoor temperatures the moisture contents were naturally at the lowest levels (38% to 44%). The moisture of the stable air originates from horses' respiration, urine, feces and drinking and washing water. Excessively high temperatures and moisture may increase the release of ammonia from the bedding [21].

The bedding material numerically influenced the ammonia content of the stable air (Table 1). Measurements we made early in the morning before any other activity in the stables. Thus, the ammonia concentrations represent the situation at its worst after the night. The ammonia level in the middle of the boxes in which peat was used as bedding was non-existent or very low (<0.25 ppm). However, the ammonia concentration in the stalls with wood shavings as bedding was numerically (6–8 times) higher (1.5–7.0 ppm) and at the highest close to levels (10 ppm) considered harmful [22]. The ammonia levels observed in the present study were lower than recently reported gas levels in the morning in stables with bedding consisting of pine wood shaving [23], but under warmer conditions (summer, in North Dakota, US). Ammonia in the stable originates from urine. The urinary production depends the diet (N-intake) and water intake. Both urinary production and N-losses increase with increasing N-intake [24]. In the present study the diet was individually balanced for protein (N) intake, and excretion of N was not obviously very high. There were no differences in ammonia concentrations in the stalls of horses on peat bedding (0–0.25 ppm) because of the superior ammonia absorption capacity of peat. Concerning the horses on wood shavings bedding the ammonia content varied between the horses (stalls) and measuring dates from 1.5 to 7.0 ppm.

Table 1. Gas concentrations in the stable air.

	Wood Shaving Bedding	Peat Bedding
	Day 0	
Ammonia	0.5 ppm	0 ppm
Carbon dioxide	650 ppm	500 ppm
Hydrogen sulphide	0 ppm	0 ppm
Methane	0 ppm	0 ppm
	Day 42	
Ammonia	1.5–7.0 ppm	0–0.25 ppm
Carbon dioxide	500 ppm	700 ppm
Hydrogen sulphide	0 ppm	0 ppm
	Day 84	
Ammonia	4.0–7.0 ppm	0–0.25 ppm
Carbon dioxide	700 ppm	600 ppm

The amount of dust collected was small for both bedding types, and no major differences were observed in dust measurements between the bedding materials. Both bedding materials were specially manufactured for use in horse stalls. The carbon dioxide values were lower than the upper acceptable limit values for horses (3000 ppm) and for humans (1000 ppm) [3]. Carbon dioxide levels were similar in both stable sections.

The bedding materials also affected the environment of the people working in the stables. Personal measurements of the ammonium exposure of the workers revealed was higher (5.9 ppm_{8h}) in the boxes in which wood shavings were used. No exposure was observed in stalls bedded with peat. This is important to consider, because workers can spend a considerable amount of time each day in the stables. In this study it took about 13 min to clean one box. According to previous studies, feeding and handling of feed take about 5 to 7 min per horse per day [25,26], and cleaning the stalls (mucking out, replacement of bedding materials) takes approximately 10 min per horse daily if no machinery is used [25–27]. The upper limit of HTP_{8h} is, however, as high as 20 ppm [19].

Studies on peat as a bedding material are scarce. Airaksinen *et al.* [17] and Nikama *et al.* [15] have reported a superior ability of peat bedding to bind ammonia, which is based on its low pH value. The pH value of the peat for bedding (Vapo Ltd.) used in the present study was 4.0. The pH value of wood shavings used here was not available, but according to a study the pH of wood shavings is higher (pH 5.5) than that of peat [28]. Peat tended to create a better stable environment than pelleted sawdust due to higher absorption of ammonia and lower levels volatile organic compounds [3], but no differences between the bedding materials were observed regarding the amount of dust. However, depending on its origin, peat has been shown to vary widely in dustiness and hygienic quality [17].

According to several other studies, the type of the bedding material has a considerable effect on stable dust, ammonia, bacteria, and endotoxin concentrations in horse stalls [4,7,23,29–31].

Fleming *et al.* [7] observed that the gaseous ammonia concentration was lowest when straw pellets were used. The order based on ammonia concentrations among the studied bedding materials in their study was straw pellets, linen, hemp, wood shavings, paper cuttings, and wheat straw. In a study by Garlipp *et al.* [31] ammonia emissions from wood shavings were considerably lower than from straw. In some studies mucking out and handling of bedding materials influenced the dust and gas (ammonia) emissions in the stable [7,30,32]. In the present study, the ammonia content of the stable air was higher when wood shavings were used, and during mucking exposure of ammonia occurred only in stalls bedded with wood shavings, resulting from the superior ammonia binding capacity of peat.

Pelleting of the bedding material reduces the generation of airborne particles by the bedding material [33]. Fleming *et al.* [30] found the lowest particle generation with straw pellets. In their two studies [7,30] they concluded that straw pellets may promote an improvement in the stable climate in relation to airborne particle formation, ammonia binding and ammonia transformation. Pelleted newspaper also appears to have a good potential as a bedding material for horses [33].

In one study [34] the researchers observed that the generation of airborne particles in straw, wood shavings, flax, and hemp can be reduced with a separation technology. They also found that the generation of particles increased during the storage of the bedding.

Proper ventilation is important to remove moisture, gases, and dust and other particles from the stable. However, in many cases the ventilation of stables does not provide adequate exchange of fresh air. Thus, the quality and properties of the bedding material are of considerable importance.

3.2. Horse Health

The first endoscopic examination at the beginning of the experiment revealed that 4 of the 12 horses had respiratory symptoms (+ ... +++). Thus, moving the horses from pasture to indoor housing in the middle of September (two weeks before beginning of the study) appeared to expose the horses to respiratory disease because of the air quality of the stable. In a Swedish study the highest dust measurements were observed in winter when the stable doors were closed [3]. Slightly increased airborne bacteria levels were also observed in stables in September compared to other seasons in that study. In the present study the frequency of respiratory symptoms increased in both groups during the first half of the study period, but then decreased in the horses bedded with peat such that the number of horses with symptoms in this group was the same at the beginning and end of the experiment. In the horses bedded with wood shavings the number of symptomatic horses remained larger than at the beginning, being twice of that compared to the horses with peat bedding (Table 2). Thus, the peat bedding seemed to be a better bedding choice than wood shavings regarding the health of respiratory tract.

The number of neutrophil cells did not differ between the groups (data not shown). The tracheobronchial aspirates obtained during endoscopy contained either scarce or moderate numbers of neutrophils (peat bedding: − ... ++; wood shavings bedding: 0 ... +++). An elevated number of neutrophils or the detection of Curschman's spirals is suggested to correlate with CODP symptoms [35]. One of the horses bedded with wood shavings had a high neutrophil percentage and also spirals in its sample at the second and third samplings, and was therefore diagnosed as a CODP horse at that time.

Table 2. Symptomatic horses based on endoscopy examination.

Wood Shaving Bedding Horse	**Day 0**	**Day 42**	**Day 84**
1	1	1	1
2	0	0	0
3	0	0	1
4	1	1	1
5	0	1	0
6	0	1	1
Horses with symptoms	2	4	4
Total symptomatic during the experiment			10
Peat bedding Horse	Day 0	Day 42	Day 84
7	0	1	0
8	1	1	1
9	0	1	0
10	0	1	1
11	0	0	0
12	1	0	0
Horses with symptoms	2	4	2
Total symptomatic during the experiment			8

There were no statistically significant differences in blood parameters of the horses between the peat bedding and wood shaving bedding groups (data not shown). The parameters in all horses were within the range of reference values. The bedding did not affect the respiration or heart rate of the horses (data not shown). The microflora of feces was also unaffected by the bedding, which is in agreement with Tanner *et al.* [4] and Hübinette [36]. The pH value was somewhat lower ($p = 0.01$) in the horses with peat bedding (6.78) than in those bedded with wood shavings (7.08) at the end of the experiment (day 84), and the colour of their feces was darker, which was obviously a result of observed eating peat in small amounts. Hübinette [36] found no effect of bedding material (wood shavings or peat) on the faecal pH.

The moisture content of the hoof horn at the end of the study was higher ($p < 0.05$) in the horses bedded with peat (32.6%) compared to the horses with wood shavings as bedding (30.5%). In the middle of the study the difference was not statistically significant (peat 34.2%, wood shavings 33.2%). Dryness of the hooves can cause problems when the natural elasticity and toughness is lost [37]. The weakening of the hoof mechanism can lead to hoof cracks and impose an additional strain on the legs. Tanner *et al.* [38] found that the hooves were dryer and more caked when phone book paper was used as a bedding material than in horses bedded on sawdust.

The results supported by literature [17,30] suggest that bedding materials have the potential to affect stable air conditions and animal health and welfare. However, in some studies, no differences have been observed. For example Tanner *et al.* [38] found no difference in the respiratory health of horses when bedded with either sawdust or (shredded and milled) phonebook paper. However, the choice of bedding material is especially important in cold climatic conditions, which forces horses to be kept indoors for a large part of the year.

Although there appears to be clear differences in the properties and influences between various bedding materials and types, the quality and origin of a particular bedding is important regarding the airborne dust concentrations originating from the bedding material [17,30,39], and stable owners and managers should thus also pay particular attention to this issue when selecting bedding materials.

Horses appear to have individual preferences for bedding material, and no significant overall preference for example for either wood shavings or straw was observed [12]. Werhan *et al.* [13] also found individual differences, but the horses generally preferred straw. The authors concluded that on the basis of the longest time being occupied, straw seems to support the welfare of the horse better than wood shavings or straw pellets.

3.3. Consumption of the Bedding Materials

The consumption of bedding materials differed considerably. The consumption of peat was 59% of that of wood shavings (by volume), obviously due to its superior ability to bind liquids. This affects the cost of bedding as well as the need for storage for both bedding material and manure, thus influencing the construction costs of the facilities.

3.4. General Discussion

An issue of increasing importance is the influence of manure on environment, which means that the amount of manure produced should be minimal and that it should be easily used as a fertilizer or in soil improvement [14,15], or even as a source of energy, for example in methane production [40]. Rapid composting and a good ability to bind and transfer nitrogen are important properties of peat bedding requested by farmers and other users of manure. Poeplau *et al.* [41] reported positive trends in organic carbon storage in Swedish agricultural soils due to increased horse industry and horse manure use in agriculture during the past two decades. It is also important that the bedding material can be easily handled in the stable, which is influenced, for example, by how it is packed, or how much it is consumed daily.

Peat is a good alternative for bedding material in those countries where it is produced for agricultural/horticultural or energy use, for example in the Nordic and Baltic countries, Russia, Poland *etc.* However, it is important to consider that agricultural peat soils should be managed sustainably and that cultural and socio-economical aspects of peatlands are taken into account [42]. In many other countries, such as in The Netherlands and Germany, the percentage of remaining pristine mires is small [43].

The battery of methods used to monitor indoor air quality and animal health in the stable was limited in the present study. In addition, outdoor exercise of the horses makes this issue complicated to investigate, regardless of the methods applied. Horses also individually differ in their sensitivity to exposure to environmental factors, *i.e.*, their genetic predisposition for example, to RAO [44]. On the other hand, many studies regarding airway health have been field studies without a controlled or standardised environment, e.g., examining bedding and feeding practices and outdoor exercise. The findings of the present study suggest that further research with a large sample size would be warranted in order to gain a better understanding on the effect of bedding materials on stable air quality and health and wellbeing of horses.

4. Conclusions

The results suggest that choice of bedding material is of large importance regarding stable air quality, at least in terms of the ammonia level. Both horses and people working in stables are exposed to ammonia if the ability of the bedding material to bind gases and fluids is poor. This may predispose both horses and humans to airway diseases. Based on the results, peat is superior to wood shavings regarding the ability to bind ammonia and reduce ammonia concentration and the risk of ammonium exposure of horses and stable workers. It seems also that horses on peat bedding may have better airway health. The moisture content of the hooves of the horses on peat bedding was higher compared to those bedded on wood shavings. The findings suggest that further research with a larger sample size is warranted.

Author Contributions

Markku Saastamoinen designed the experiment with Susanna Särkijärvi, who also analyzed the data. Seppo Hyyppä took the samples and carried out the endoscopic examination and analyzed materials concerned. Markku Saastamoinen wrote the paper.

Conflicts of Interest

The authors declare no conflict of interest. The study was partly financed by Vapo Ltd. having no role in the design of the study; in the collection, analysis, or interpretation of data; in writing of the manuscript, and in the decision to publish the results.

References

1. Riihimäki, M.; Raine, A.; Elfman, L.; Pringle, J. Markers of respiratory inflammation in horses in relation to seasonal changes in air quality in a conventional racing stable. *Can. J. Vet. Res.* **2008**, *72*, 432–439. [PubMed]
2. Derksen, F.J. Chronic Obstructive Pulmonary Disease. In *Equine Respiratory Disorders*; Beech, J., Ed.; Lea & Febiger: Malvern, PA, USA, 1991; pp. 223–235.
3. Elfman, L.; Wålinder, R.; Riihimäki, M.; Pringle, J. Air quality in horse stables. In *Chemistry, Emission Control, Radioactive Pollution and Indoor Air Quality*; Mazzeo, D., Ed.; Intech: Rijeka, Croatia, 2011; pp. 655–680.
4. Tanner, M.K.; Swinker, A.M.; Beard, M.L.; Cosma, G.N.; Traub-Dargatz, J.L.; Martinez, A.B.; Olenhock, S.A. Effect of phone book paper *versus* sawdust and straw bedding on the presence of airborne gram-negative bacteria, fungi and endotoxin in the horse stalls. *J. Equine Vet. Sci.* **1998**, *18*, 457–461. [CrossRef]
5. Tutluoglu, B.; Atis, S.; Anakkaja, A.N.; Altug, E.; Tosun, G.A.; Yaman, M. Sensitization to horse hair, symptoms and lung function in grooms. *Clin. Exp. Allergy* **2002**, *49*, 1–8. [CrossRef]

6. Berndt, A.; Derksen, F.J.; Robinson, N.E. Endotoxin concentrations within the breathing zone of horses are higher in stables than on pasture. *Vet. J.* **2010**, *183*, 54–57. [CrossRef] [PubMed]
7. Fleming, K.; Hessel, E.F.; van den Weghe, H.F.A. Generation of airborne particles from different bedding materials used for horse keeping. *J. Equine Vet. Sci.* **2008**, *28*, 408–418. [CrossRef]
8. Clements, J.M.; Pirie, R.S. Respirable dust concentrations in equine stables. Part 2: The benefits of soaking hay and optimizing the environment in a neighbouring stable. *Res. Vet. Sci.* **2007**, *83*, 263–268.
9. Saastamoinen, M. Bedding Choices for Horses—An Overview. In Housing and Management of Horses in Nordic and Baltic Climate, Proceedings of the NJF Seminar 437 Housing and Management of Horses in Nordic and Baltic Climate, Reykjavik, Iceland, 6–7 June 2011; NJF: Reykjavik, Iceland, 2011; pp. 33–36.
10. Hunter, L.; Houpt, K.A. Bedding material preferences of ponies. *Equine Athl.* **1991**, *4*, 17–19.
11. McGreevy, P.D.; Cripps, P.J.; French, N.P.; Green, L.E.; Nicol, C.J. Management factors associated with stereotypic and redirected behaviour in the Thoroughbred horse. *Equine Vet. J.* **1995**, *27*, 86–91. [CrossRef] [PubMed]
12. Mills, D.S.; Eckley, S.; Cooper, J.J. Thoroughbred bedding preferences, associated behaviour differences and their implications for equine welfare. *J. Anim. Sci.* **2000**, *70*, 95–106.
13. Werhahn, H.; Hassel, E.F.; Bachhausen, I.; van den Weghe, H.F.A. Effects of different bedding materials on the behaviour of horses housed in single stalls. *J. Equine Vet. Sci.* **2010**, *8*, 425–431. [CrossRef]
14. Airaksinen, S. Bedding and Manure Management in Horse Stables. Its Effect on Stable Air Quality, Paddock Hygiene and the Compostability and Utilization of Manure. Ph.D. Thesis, University of Kuopio, Kuopio, Finland, 2006; p. 52.
15. Nikama, J.; Keskinen, R.; Närvänen, A.; Uusi-Kämppä, J.; Särkijärvi, S.; Myllymäki, M.; Saastamoinen, M. The role of bedding material in recycling the nutrients of horse manure. In Equi-Meeting Infrastructures Horses and Equestrian Facilities, Proceedings of the le Lion d' Angers, France, 6–7 October 2014; Haras National du Lion: France, 2014; pp. 34–37. Available online: http://urn.fi/URN:NBN:fi.fe201501081077 (accessed on 18 August 2015).
16. Greatorex, J.M.; Karlsson, S.; Tersmeden, M.; Steineck, S. Composting Horse Manure. In *Influence of Bedding Material on Ammonia Emissions*; Swedish Institute for Agricultural and Environmental Engineering: Uppsala, Sweden, 1998; p. 13.
17. Airaksinen, S.; Heiskanen, M.-L.; Heinonen-Tanski, H.; Laitinen, S.; Linnainmaa, M.; Rautiala, S. Variety in dustiness and hygiene quality of peat bedding. *Ann. Agric. Environ. Med.* **2005**, *12*, 53–59. [PubMed]
18. Müller, C. Fermentation patterns of small-bale silage and haylage produced as a feed for horses. *Grass Forage Sci.* **2005**, *60*, 109–118. [CrossRef]
19. Kemian työsuojeluneuvottelukunta. HTP-values. In *Työsuojelusäädöksiä 3*; Ministry of Social Affairs and Health: Tampere, Finland, 2002; p. 57.
20. Saastamoinen, M.; Jansson, H. Hevosen elinympäristö. In *Hevosen Ruokinta ja Hoito*; Saastamoinen, M., Teräväinen, H., Eds.; ProAgria Maaseutukeskusten Liitto: Porvoo, Finland, 2007; pp. 126–138. (In Finnish)

21. Pratt, E.V.; Rose, S.P.; Keeling, A.A. Effect of ambient temperature on losses of volatile nitrogen compunds from stored laying hen manure. *Bioresour. Technol.* **2002**, *84*, 203–205. [CrossRef]
22. Zeitler-Feicht, M.H.; Bohnet, W.; Düe, M.; Esser, E.; Pollmann, U. Positionspapier zu den Leitlinien zur Beurteilung von Pferdehaltungen unter Tierschutzgesichtspunkten. Available online: http://www.tierschutz-tvt.de/positionspapierpferdehaltung.pdf (accessed on 18 August 2015).
23. Borhan, S.; Rahman, S.; Hammer, C. Water absorption capacity of flax and pine horse beddings and gaseous concentrations in bedded stalls. *J. Equine Vet. Sci.* **2014**, *34*, 611–618. [CrossRef]
24. Frape, D. *Equine Nutrition and Feeding*, 2nd ed.; Blackwell Science: Oxford, UK, 1998; p. 564.
25. Korkiamäki, H.; Lahin, P.; Laaksonen, K. *Horse Stable Work and Estimated Working Hours*; TTS-Institure: Rajamäki, Finland, 1998; p. 6.
26. Laitinen, I. Manure Handling with Machinery in the Horse Stables. Master's. Thesis, Department of Agricultural Engineering, University of Helsinki, Helsinki, Finland, 2007; p. 53.
27. Kontioniemi, K. Vacation Replacement Service's Service Plan for Horse Farms. Bachelor's Thesis, HAMK University of Applied Sciences, Hämeenlinna, Finland, 2012; p. 28.
28. N'Dayegamiye, A.; Isfan, D. Chemical and biological changes in compost of wood savings, sawdust and peat moss. *Can. J. Soil Sci.* **1991**, *71*, 475–484. [CrossRef]
29. Woods, P.S.; Robinson, N.E.; Swanson, M.C.; Reed, C.E.; Broadstone, R.V.; Derksen, F.J. Airborne dust and aeroallergen concentration in a horse stable under two different management systems. *Equine Vet. J.* **1993**, *25*, 208–213. [CrossRef] [PubMed]
30. Fleming, K.; Hessel, E.F.; van den Weghe, H.F.A. Evaluation of factors influencing the generation of ammonia in different bedding materials used for horse keeping. *J. Equine Vet. Sci.* **2008**, *28*, 223–231. [CrossRef]
31. Garlipp, F.; Hessel, E.F.; van den Weghe, H.F.A. Characteristics of gas generation (NH_3, CH_4, N_2O, CO_2, H_2O) from horse manure added to different bedding materials used in deep litter bedding systems. *J. Equine Vet. Sci.* **2011**, *31*, 383–395. [CrossRef]
32. Sadegh, M.; Wouters, I.M.; Houben, R.; Heederik, D.J.J. Exposure to inhable dust, endotoxins, beta (1–3) glucans and airborne microorganisms in horse stables. *Ann. Occup. Hyg.* **2009**, *53*, 595–603.
33. Ward, P.L.; Wohlt, J.E.; Zajac, P.K.; Cooper, K.R. Chemical and physical properties of processed newspaper compared to wheat straw and wood shavings as animal bedding. *J. Dairy Sci.* **2000**, *83*, 359–367. [CrossRef]
34. Garlipp, F.; Hessel, E.F.; van der Hurk, M.; Timmermann, M.F.; van den Weghe, H.F.A. The influence of a particle separation technology on the generation of airborne particles from different roughages and bedding materials used for horses. *J. Equine Vet. Sci.* **2010**, *30*, 545–559. [CrossRef]
35. Clarke, A.F. Chronic pulmonary disease—A multifaceted disease complex in the horse. *Ir. Vet. J.* **1987**, *41*, 258–264.
36. Hübinette, L. *Effects of Peat and Wood Shavings as Bedding on the Faecal Microflora of Horses*; Swedish Agricultural University: Uppsala, Sweden, 2010; p. 21.
37. Douglas, J. Mechanical aspects of the equine hoof wall. *Equine Athl.* **1994**, *7*, 15–20.

38. Tanner, M.K.; Swinker, A.M.; Traub-Dargatz, J.L.; Stiffler, L.A.; McCue, P.M.; van der Wall, D.K.; Johnson, D.E.; Vap, L.M. Respiratory and environmental effects of recycled phone book paper *versus* sawdust and bedding for horses. *J. Equine Vet. Sci.* **1998**, *18*, 468–475. [CrossRef]
39. Vandeput, S.; Istasse, L.; Nicks, B.; Lekeux, P. Airborne dust and aeroallergen concentrations in different sources of feed and bedding for horses. *Vet. Q.* **1997**, *19*, 154–158. [CrossRef] [PubMed]
40. Wartell, B.A.; Krumins, V.; Alt, J.; Kang, K.; Schwab, B.J.; Fennell, D.E. Methane production from horse manure and stall waste with softwood bedding. *Bioresour. Technol.* **2012**, *112*, 42–50. [CrossRef] [PubMed]
41. Poeplau, C. Positive trends in organic carbon storage in Swedish agricultural soils due to unexpected socio-economic drivers. *Biogeosciences* **2015**, *12*, 3241–3251. [CrossRef]
42. Takko, A.; Vasander, H. Socioeconomic Aspects of Commercial Uses of Peatlands in Finland. In Wise Use of Peatlands, Proceedings of the 12th International Peat Conference, Tampere, Finland, 6–11 June 2004; Päivinen, J., Ed.; Finnish Peatland Society: Tampere, Finland, 2004; pp. 1313–1319.
43. Leupold, S. Recreation—A Complementary after-use Option for Cutway Peatlands. In Wise Use of Peatlands, Proceedings of the 12th International Peat Conference, Tampere, Finland, 6–11 June 2004; Päivinen, J., Ed.; Finnish Peatland Society: Tampere, Finland, 2004; pp. 1190–1193.
44. Marti, E.; Gerber, H.; Essich, G.; Oulehla, J.; Lazary, S. The genetic basis of equine allergic diseases. 1. Chronic hypersensitivity bronchitis. *Equine Vet. J.* **1991**, *23*, 457–460. [CrossRef] [PubMed]

Animals **2015**, *5*, 951-964; doi:10.3390/ani5040392

OPEN ACCESS

animals

ISSN 2076-2615

www.mdpi.com/journal/animals

Article

A Cross-Sectional Study of Horse-Related Injuries in Veterinary and Animal Science Students at an Australian University

Christopher B. Riley [1,†,*], Jessica R. Liddiard [2,†] and Kirrilly Thompson [3,†]

[1] Institute of Veterinary, Animal and Biomedical Sciences, Massey University, Tennent Drive, Palmerston North, 4442, New Zealand

[2] School of Animal and Veterinary Sciences, University of Adelaide, Roseworthy, SA 5371, Australia; E-Mail: jessicaliddiard23@gmail.com

[3] Appleton Institute, Central Queensland University, Adelaide, SA 5034, Australia; E-Mail: kirrilly.thompson@cqu.edu.au

† These authors contributed equally to this work.

* Author to whom correspondence should be addressed; E-Mail: c.b.riley@massey.ac.nz; Tel.: +64-6-356-9099 (ext. 85164); Fax: +64-6-350-5636.

Academic Editor: Clive J. C. Phillips

Received: 30 July 2015 / Accepted: 18 September 2015 / Published: 24 September 2015

Abstract: Specific estimates of the risk of horse-related injury (HRI) to university students enrolled in veterinary and animal sciences have not been reported. This study aimed to determine the risk of student HRI during their university education, the nature and management of such injuries. A retrospective questionnaire solicited demographic information, data on students' equine experience prior to and during their educational programs, and on HRI during their program of study. Of 260 respondents, 22 (8.5%) reported HRI (27 incidents). Including concurrent injuries the most commonly injured body parts were the foot or ankle (nine of 32 injures), the upper leg or knee (eight of 32), and hands (three of 32). Trampling and being kicked by a hind limb were each associated with 30.4% of HRI, and 13% with being bitten. Bruising (91.3% of respondents) and an open wound (17.4%) were most commonly described. No treatment occurred for 60.9% of incidents; professional medical treatment was not sought for the remainder. Most incidents (56.5%) occurred during program-related work experience placements. Although injury rates and severity were modest, a proactive approach to injury prevention and reporting is recommended for students required to handle horses as part of their education. Student accident and injury data should be monitored to ensure effective evaluation of risk-reduction initiatives. The risk and nature of university student horse-related injury (HRI) was studied. Of 260 students, 22 (8.5%) reported HRI (27 incidents). Including multiple injuries, reports described involvement of the foot or ankle (nine of 32 injures), upper leg or knee (eight of 32), and hands (three of 32). Trampling (30.4%) and being kicked (30.4%) accounted for most HRI. The injuries were usually bruising (91.3%) or an open wound (17.4%). Most (60.9%) injuries were untreated; professional medical treatment was not sought for the rest. Most incidents (56.5%) occurred during program-related off-campus work experiences. A proactive approach to injury prevention is recommended for students handling horses.

Keywords: horse-related; injury; accident; student; education; veterinary

---

1. Introduction

Horses are powerful and frequently unpredictable animals, capable of moving at high speeds and of generating great force with a single kick [1,2]. Due to the combination of behavioral characteristics, nature of responses to adverse stimuli, speed, power, and size of these animals, people that interact with horses professionally or recreationally are at risk of severe or fatal injuries. Studies of equine-related injury have been conducted for people engaged in a diverse range of equestrian activities, but the majority of these have investigated the risk of injury to riders [3,4]. In comparison, research on occupational injury by horses within the veterinary profession and other equine industries is limited [2,5–7]. A cross sectional survey of Australian veterinarians (the Health Risks of Australian Veterinarians study; HRAV) found that most large animal (65% equine and/or food animal), and mixed animal (59%; large and companion animals) veterinarians had suffered chronic musculoskeletal or severe acute injuries, placing them at the highest risk of significant injuries [2,5,6]. Significant injuries were classified as an incident resulting in hospital admission, or having a substantial detrimental effect on the ability to work. Of the serious injuries reported, 29% were directly equine associated [2], of which 70% occurred despite safety precautions that were claimed to have been in place, indicating that the latter may have been insufficient or incorrectly applied [5]. Such findings are not unique to Australian veterinarians, and studies in other countries have found similar concerning statistics [8,9].

Currently protocols are in place in many universities with equine educational programs to ensure the safety of students and staff. These include testing the suitability of horses for teaching, hazard evaluation, and incident reporting systems, though few protocols have been published [10,11]. More information on risk factors and prevalence is needed to develop specific and effective evidence-based recommendations for minimizing horse-related injuries (HRI) [10,12]. Reports of occupational HRI for experienced veterinarians provide useful information, but may not be generally applicable to university students enrolled in programs requiring exposure to horses, as the latter have more variable levels of equine experience and interest. At the University of Adelaide, students enrolled in the animal (AnSci) and veterinary science (VS) programs may elect (AnSci and VS) or are required (VS) to obtain practical experience with horses as part of their educational programs. The authors hypothesized that injury rates for students at the University of Adelaide differed from those that of reports for practicing veterinarians and others involved in the equine industry from similar cultural backgrounds. The objectives of this study were to determine the prevalence of injury for university students exposed to horses during structured and unstructured learning activities associated with their program of education, and to identify particular risk factors. The authors found low rates of injury, but a lack of action taken to see medical evaluation in response to these injuries raises concerns about risk culture in Australian students, and its possible carry over into professional life [13].

2. Experimental Section

The study followed the recommended assessment procedures for studies with low ethical risk approved by the Human Research Ethics Committee of the University of Adelaide.

2.1. Background Information Collection

In order to clarify the background conditions underpinning the experience of the survey participants, current safety protocols for on campus and extramural (equine work experience) learning activities involving horses were reviewed. The campus Health and Safety Officer (HSO) was also asked to provide a summary of the number of HRI formally reported by students during the five years preceding the survey

2.2. Survey Data Collection

Students were approached and directly invited to participate in the paper-based survey within the daily class schedule. This occurred over a 29-day sampling period. A response rate of 60% from each year cohort of students within each educational program was anticipated. At the conclusion of the initial survey participation request period, a final attempt was made to acquire results from non-responding students by emailing the survey to each of the cohorts with a description of the study to be undertaken. No identifying information was recorded on the survey or by other means.

Students enrolled in the undergraduate Animal Science (AnSci), Veterinary Bioscience (VetBio), or Masters by coursework Doctor of Veterinary Medicine (DVM) programs at the University of Adelaide (South Australia) for the 2012 academic year were considered eligible for the study (n = 461). An open source sample calculator was used to determine an appropriate target for the number of respondents from this study population to obtain a survey error (type I) no higher than a 5% [14]. It was calculated that to be within a 95% confidence interval 214 respondents were required, and to be within a 99% confidence interval 279 respondents were required [14].

An anonymous retrospective questionnaire was created consisting of 33 questions in three sections: (1) background demographic information, (2) qualitative and quantitative data regarding attitudes and experiences of students prior to and during their program of education, and (3) incidence of horse-related injury, reporting and management. Background demographic information collected included gender, age, educational program and year level, hours of horse exposure in practical equine courses completed on-campus, and length of time spent at equine work experience placements outside of the university campus. Student attitudes and experiences were solicited and semi-quantitative data was obtained using rating scale questions in which the scale range was from 0 to 100. These questions surveyed the students' interest and prior contact with horses, and perceptions of safety procedures.

Finally, the incident report section of the survey was used to identify injury incidence, type, mechanism, severity, and the various contributing factors as perceived by the students. To ensure consistent interpretation of quantitative responses, where a range value was provided the mean of the range was utilized for the purpose of analysis. For example if a respondent commented that they had 10–20 hours of contact with horses per month the mean value of 15 was utilized. Scaled questions offered a semi quantitative range of responses from 0 (most negative or greatest magnitude of disagreement) to 100 (most positive or greatest magnitude of agreement), and results were interpreted at a minimum interval of five units. Where only a year value was given in response to how much time the respondent had spent with horses prior to the program, the respondent was conservatively estimated to have had two hours of contact with horses for each day of the time indicated to generate a final approximation for analysis. Alternatively, where insufficient information was provided to make such an assumption (e.g., "a lot"), the response was neither transcribed nor included in the data analysis.

2.3. Data Analysis

Associations between categorical variables including program type, program year, gender and the occurrence of injury were evaluated by the Chi-squared test (considered significant at $p < 0.05$); where the expected values were <5, Fishers exact test was performed. The distribution of values for continuous variables (age in months, duration of equine experience before enrollment, and equine contact time before and during enrollment) was evaluated by the Shapiro-Wilk test; none were normally distributed. The effects of these continuous variables on the occurrence of injury were analyzed using the Kruskal-Wallis test; differences were considered significant at $p < 0.05$.

3. Results

3.1. Safety Protocols-On Campus Safety and Teaching Horse Assessment

The review of safety protocols in use at the time of the survey found that horses acquired for live animal teaching on the university campus were identified by the Teaching Services Technical Manager, and examined by a registered staff equine veterinarian for health, temperament, and ease of handling. Suitable horses were then transported to the campus for two to four weeks of handling and further evaluation of temperament by equine teaching staff. Horses that failed to adapt to the teaching environment, or showed evidence of recurrent intemperate behavior during teaching activities, were removed from the teaching herd. Qualified staff supervised all equine-related teaching activities on-campus. Formal qualifications held by the staff ranged from Australian equine industry certification in equine handling and training, to graduation from an AnSci or veterinary degree program. The staff to student ratio for classes with horses was ~1:6. All sessions involving horses required students to read and understand standard operating procedures for basic handling and restraint, and the veterinary technical procedures developed for the learning activities. Equine related teaching was delivered in purpose-built yards, stalls, and crushes. All adverse incidents observed by staff or reported by students were noted and forwarded to the campus HSO.

3.2. Safety Protocols-Extramural Equine Safety

A designated staff member was assigned to coordinate student placement (work experience) agreements with host-providers and manage extramural placements from the campus. Placement hosts were required to provide relevant experiences (animal husbandry for AnSci and VetBio students; veterinary clinical training for DVM students), to provide a local safety induction to students, and to adhere to Australian legal requirements for work place health and safety. Feedback on extramural placements was sought from hosts and students, and responded to directly by the coordinator.

3.3. Demographic and Background Data

Of 481 eligible AnSci and VS students, 260 (54.1%) returned surveys (Table 1). The resulting calculated sampling error was 4.1% at the 95% confidence level [14]. Female students constituted 219 of 264 (83%) of respondents (compared to 81% in the eligible AnSci and VS student population), and males constituted 45 of 264 (17%) respondents (compared to 19% in the eligible population). The median age of the participants was 21.5 years (range 17.8 to 42.5). For background questions not related to describing an incident, response rates per question ranged from 78 to 100%. Questions pertaining to an incident (22 students; 23 of 27 incidents described) had a 100% response rate except for the date of the incident (74% response rate).

Table 1. The number of respondents, demographic data, survey response rate for each year level and program.

	Animal Science			**Veterinary Bioscience**			**Doctor of Veterinary Medicine**		**Total**
Year of Program	1	2	3	1	2	3	1	2	
2012 Enrolments	119	48	54	42	65	64	53	36	481
Number Surveyed	33	26	23	29	32	45	46	26	260
% of Class Surveyed	27.7	54.2	42.6	61.9	49.2	70.3	86.8	72.2	54.1

Structured university courses with an equine practical component were compulsory and the duration of on-campus equine contact was constant within each cohort (Figure 1). On-campus class contact time with horses of AnSci and VetBio cohorts were approximately equivalent, whilst DVM students had a significantly greater amount of on-campus class equine contact time in comparison ($p < 0.05$; Figure 1). The time spent on extramural placements (a compulsory requirement of the VetBio after the first year and the DVM programs) varied among veterinary students (Figure 1). A minimum two-week extramural placement to gain experience in equine handling and husbandry was required of the VetBio program respondents, and a minimum five-week extramural placement for equine clinical experience was required of the DVM program respondents. Comparatively, there was no compulsory requirement for an equine based extramural placement in the AnSci program; approved placements were elective.

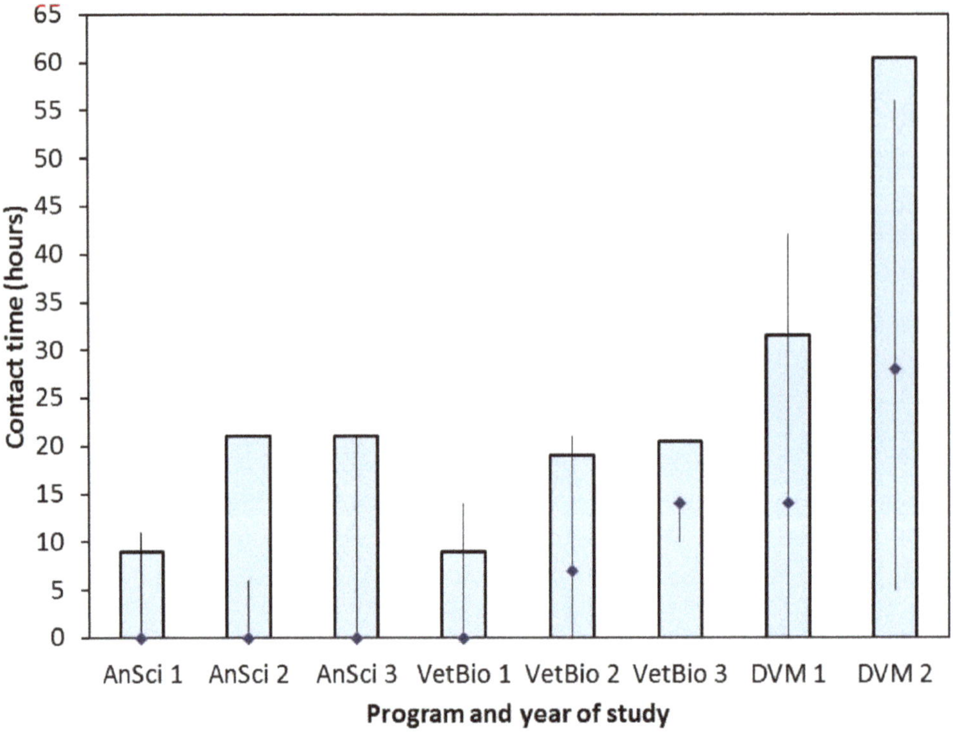

Figure 1. Contact time spent with horses during on-campus instruction (blue column) and extramural equine work experience placements (diamond = median; vertical bars indicate range) for Animal Science (AnSci), Veterinary Bioscience (VetBio) and Doctor of Veterinary Medicine (DVM) students at the University of Adelaide. Note, on-campus contact time was prescribed by the courses in which students were enrolled.

Sixty percent (156 of 260) of respondents reported contact with horses prior to beginning their university program. The approximate duration of this experience varied widely (mean 340 $\pm$ 811 days). Only 41 (16%) of students indicated that working with horses strongly influenced their decision to enroll in their current program (indicated by a score >70/100). These values did not differ significantly between AnSci and VS students, or among year levels within programs. A significant difference between injured and non-injured students in the duration of equine experience before enrollment was not identified.

3.4. Injury Data

Of the 260 respondents, 8.5% (22) reported a total of 27 incidents resulting in injury (Table 2). Only details of 23 incidents (one per student in 21 cases, and one reported the details of the two most recent incidents) were provided for analysis. Only one incident was reported to the campus HSO during 2010 to 2012; HSO records were not available for the three years prior to 2010. Most injured students were female (90.9%; 20/22), and 9.1% (2/22) were male (Table 2). There was no statistically significant gender association with the risk of injury. Students that had experienced a horse-related injury were found to have a median age of 22 years, which was not significantly different than non-injured respondents (21.5 years).

There was a significant association between student contact time (non-linear) with horses during programmed educational activities, and an increased risk of injury (p = 0.016) that could not be differentiated from associations with the program of study.

Table 2. The injury response rates and relative risk for each year level and program.

	Animal Science			**Veterinary Bioscience**			**Doctor of Veterinary Medicine**		**Total**
Year of program	1	2	3	1	2	3	1	2	
Number of injured students	2	0	0	0	2	7	5	6	22
Number of incidents	2	0	0	0	2	7	5	7	23
Relative risk by year	0.06	0	0	0.07	0.16	0.02	0.11	0.04	
Relative risk by program		0.02			0.14 [1]			0.12 [2]	

[1] The year in which the injury occurred was not reported in one case; [2] The year in which the injuries occurred were not reported in four cases.

Of the 23 incidents described, AnSci students reported two (8.7%), the VetBio students reported nine (39.1%), and the DVM students reported 12 (52.2%) (Table 2). The AnSci students were less likely to sustain HRI than VetBio and DVM (p = 0.013), and the two AnSci incidents described involved riding and colliding with a gate during their extramural placement on farms. The DVM students were associated with a higher risk of HRI than the AnSci students (p = 0.038); the difference between VetBio and DVM students was not statistically significant. In only 17 incidents did injured students report the year in which the incident occurred, and the relative risk was highest in the second year of the VetBio and first year of the DVM programs (Table 2).

The anatomic locations and nature of reported injuries sustained are summarized in Table 3. Six respondents reported more than one type of tissue injury due to their HRI. The most common mechanisms of injury were a horse trampling the foot or ankle, being kicked by a hind limb, or being bitten. The most commonly reported types of injuries were bruises or an open wound; three students reported concurrent bruising and a laceration.

Most injured students, (60.9%; 14) did not treat the injury and remaining injuries (39.1%; nine) were self-treated. There were no incidents for which first aid treatment was sought, a doctor was seen, or a student admitted to hospital. None of the injured respondents reported requiring time off from university study or work following injury and only one student (4.3%) with a muscular lower back injury of unreported duration, indicated that she had not fully recovered from the incident. A median of 14 days (range 0 to 60 days) was reported for the remaining students to fully recover from their injuries.

Table 3. Description of the injuries sustained by students including bodily location, mechanism of injury, and nature of injury.

Factor		Number of Injuries	Percentage of Incidents
Affected Bodily Location	Foot/Ankle	9	39.1
	Upper Leg/Knee	8	34.8
	Hand	4	13.0
	Lower Leg (Calf/Dorsal Tibia)	2	4.3
	Head	1	4.3
	Face	1	4.3
	Forearm	1	4.3
	Lower Back	1	4.3
	Other	1	4.3
	Eye	0	0
	Upper Arm/Shoulder	0	0
	Neck	0	0
	Spine	0	0
	Pelvis	0	0
	Total injuries reported	28	-
Mechanism of Injury	Trampled	8	30.4
	Kicked by hind limb	7	30.4
	Bitten	3	13.0
	Fall while riding	2	8.7
	Struck by forelimb & Crushed by body	1	4.3
	Knocked over	1	4.3
	Struck by horse's head	1	4.3
	Total mechanisms reported	23	-
Nature of Injury	Bruising/Soft Tissue Injury	20	91.3
	Open Wound	4	17.4
	Muscle or Tendon Injury	2	8.7
	Rope Burn	2	8.7
	Non-Specified Internal Injury	1	4.3
	Crushed Tissue Injury	1	4.3
	Muscular Strain	1	4.3
	Fracture	0	0
	Intracranial injury	0	0
	Total injuries reported	31	-

Incidents occurred most frequently during extramural placements to obtain experience in equine husbandry (56.5%; 13 of 23 incidents). All injuries to second-year VetBio students and first-year AnSci students occurred during an extramural placement (Table 2), whereas for the DVM and other VetBio program years injuries occurred in approximately equivalent numbers on and off campus. No injuries associated with extramural placements at veterinary practice facilities were reported to the HSO. Nine adverse horse-related incidents (39.1%; nine of 23 incidents) occurred on the premises of the university during teaching activities, and one (4.3%) at an equine event unrelated to an educational program. Specifically, nine (39.1%) HSI occurred in handling yards or a fenced enclosure, eight (34.8%)

in a stable or barn, two (8.7%) in an open paddock, two in an arena (8.7%), one (4.3%) in a horse float (trailer), and one in a non-specified location.

The activities or actions being undertaken by the students at the time of injury included standing near a horse (26.1%; six incidents), performing a handling or husbandry procedure (21.7%; five cases; e.g., moving horses or rounding up livestock on horseback), or performing a non-invasive physical examination procedure (17.4%; four incidents; e.g., auscultation of the chest). Walking near a horse, working with the horses' limbs or feet, and catching a horse in a paddock were each associated with two incidents. Leading a horse in an outside location and trailering (floating) horses were each associated with a single incident. The activities for three incidents were not described. One or more factors were believed by respondents to be associated with their injury including resistance by the horse to handling (12 students; 52.2%), inexperience on their part (nine students; 39.1%), and seven admitted inattention on their part (30.4%). Seven (30.4%) believed that the incident occurred because the horse was in distress or fearful. Poor restraint and inattention by the handler were considered factors in four (17.4%) incidents, and the horse evading capture or poor staffing levels were each believed to contribute to a single incident each. For 12 (52.2%) incidents, respondents indicated that more than one factor might have contributed to the incident (range 0 to 4 factors).

In four (17.4%) incidents, it was the student's perception that no safety precautions other than haltering or tying were being implemented at the time of the event. In the remaining 19 (82.6%) incidents, 15 (65.2%) of the students used protective footwear, three (13.0%) reported the horse was being restrained by an experienced staff member, two (8.7%) reported the horse was being restrained by an inexperienced peer. Additionally, two students used a nose twitch, two used another form of safety precaution outside of the defined categories, one (4.3%) stated the horse was sedated, and another used a helmet.

4. Discussion

Regardless of their experience levels, backgrounds and attitudes towards horses, it is compulsory for all American Veterinary Medical Association accredited university veterinary program students, and many animal science program students, to work with horses for some length of time within their educational programs. In comparison, qualified veterinarians exposed to horses have generally elected to follow this career path. The findings of the current survey support our hypothesis that when compared to reports for practicing veterinarians and others involved in the equine industry from similar cultural backgrounds, injury rates differ [2,4–7,15]. In contrast to equine industry reports, there were no severe injuries among students [3,4,15]. This result, to some degree, reflects favorably on the current procedures, facilities, environment, and staff for on-campus learning, and the selection of host locations for extramural experiences. However, HRI was reported, and the relatively lower rate should not be cause for complacency. An opportunity for comparison of these data with those of other comparable study populations is limited. However, given that attitudes and training with respect to animal safety might reasonably be expected to influence future professional practice, the relationship between the risk of HRI sustained by VS, and that of Australian qualified veterinarians warrants future study [13]. The overall prevalence of all HRI in university students was lower than the prevalence of 16.2% in 2800 qualified Australian veterinarians, but this might be expected as the exposure risk (time) for practicing veterinarians is higher [2]. The real difference in injury rates between the two groups may be greater if the HRAV had accounted for all degrees of injury, not just those that were classified as severe [2]. The design of the questionnaire used for the current study encouraged disclosure of all incidents irrespective of the severity of injuries sustained. The authors contend that this approach may be a more useful one than studies limited to only self-reported severe injuries or hospital admissions, to more accurately reflect the risk of HRI in other settings.

Anecdotally underreporting of incidents is prevalent within the Australian veterinary profession and other equine industries, resulting in underestimation of rates of injury [7]. In agreement with other industry sector reports, formal incident reporting (to the HSO) by the current study population was lacking despite clear policies requiring the documentation of such events [4,7,13]. Injured students also failed to seek independent third party medical assessment. Such behaviors are of concern and may reflect Australian cultural and professional norms related to work practice, lack of safety awareness, and poor self-care [13,15]. It is recommended that greater emphasis be made within educational institutions to develop and monitor a non-punitive and supportive culture of accurate incident and hazard reporting among students and staff [16]. Based upon the data in this report, it is suggested that this process should extend to those incidents that occur during off campus extramural placements. To this end the risk assessment and injury reporting documentation should be provided to hosts and students required or electing to engage in these activities. This would enable a more accurate assessment of the risk of HRI so that evidence based hazard mitigation measures may be identified and taken.

In the current and HRAV study populations the rate of injury in males to females is proportionate to the distribution of gender [2]. Accordingly, a greater net number of females had sustained horse-related injuries in the current student population. In comparison a greater number of males sustain HRI in the qualified veterinary population due to greater proportion of males in equine and mixed practice at the time of the HRAV study [2]. This compares to a study of horse-related injury in Australian riders in which young female riders and older males were found to have the greatest risk of injury [15]. This is likely to be due to the demographics of amateur female riders, and more aged professional equestrians [15,16].

The significant relationship between contact time with horses within the university and the risk of HRI seems intuitive. However, given the limited data available for multivariate analysis, and the relatively uniform amount of equine contact time spent within each cohort, it was not possible to differentiate exposure time from the program-associated risk. Injuries occurred despite the university's current measures to mitigate the risk of HRI. Nevertheless, most incidents occurred during extramural learning activities, and similar safety precautions are less likely to be implemented uniformly. Focused feedback following external placements that addresses these issues is required to evaluate this hypothesis. In lieu of such data, a review of each of the approaches to extramural risk mitigation is necessary to reduce the off campus exposure to HRI [17,18].

Student injuries were predominately sustained as a consequence of being trampled or being kicked by the hind limb, and most commonly resulted in a bruise or open wound. The bodily locations that were most commonly affected were the lower or upper limbs. In comparison, a larger proportion of serious injuries in Australian veterinarians are a result of kicks or strikes (79%), and a smaller proportion from being crushed or trampled (12.3%) [2]. They typically involve the lower extremities (33%); head and neck (26%), and upper extremities (20%), with fractures (27.8%) and bruising (27.6%) described as most common [2]. The findings of the current study are in agreement with a small study of unmounted rider HSI, with half of the riders also sustaining a contusion to the limbs [16]. In contrast facial injuries have also been commonly documented in the riders, and the reasons for this difference are unclear [19,20].

Over half of the HSI occurred at off campus locations for training required by their educational program, including all incidents for two cohorts that occurred during these extramural activities. Insufficient data (incidents) were available to statistically conclude that external placement constitutes an inherently high-risk situation for students. The comparative assessment of risk is confounded by the fact that the nature of the interactions and type of activities with horses on external placements differ from the exposure on-campus premises, as do facilities, and policies with respect to the incident review framework used to formulate the survey. For example, students most commonly reported being situated in a handling yard, stable, or barn at the time of injury. In comparison, qualified veterinarians were most likely to have sustained horse-related injuries in handling yards (37.7%) and stables (15.7%), but additionally in open paddocks (36.6%) [2]. It is possible that such differences arise from the limited or poorly maintained handling yards/crushes on Australian properties often faced by qualified veterinarians, compared to the recently purpose-built facilities at the university campus [7]. The most common safety precautions used by Australian veterinarians are physical restraint of the horse (34%) or hand-held reliance only (9%) [5]. Many large animal veterinarians are required to attend ambulatory visits alone and may not have an adequately skilled person to assist them [7]. In contrast, all on-campus equine-related teaching is supported by qualified staff [5,7]. However, students on external placements may face similar risk conditions as those experienced by practicing veterinarians; these conditions require further study. The authors agree with Jeyaretnam's and others' conclusions that a greater emphasis must be placed on determining the true risk of injuries and thus generating effective strategies to mitigate the hazards [5,21,22].

Currently Adelaide University students are required to wear protective boots during equine activities. However, the data indicated that ~35% of students are not compliant with this policy. Therefore further emphasis on ensuring that utilize sufficient protection against crushing injuries, and compliance with university clothing safety policy is warranted both on and off campus. Consideration should be given to the design and use of coveralls with protective padding over the thighs and knees such as that used for other competitive and recreational pursuits. Stock handling gloves should be worn for horse handling to prevent rope burn injuries. Whilst currently available safety equipment such as chest protection and helmets have been suggested to decrease the likelihood of injury [5] the effectiveness of this equipment for university students, or for personnel on the ground (unmounted) is unsupported by the evidence of injury types presented in the current study and others [13,19,20]. However, the two incidents involving horse riding as part of animal husbandry activities, indicates that their proactive use in this situation is prudent. It is the authors' view that in addition to relying upon protective equipment (where evidence for its use exists), the emphasis placed on educating university students about occupational safety requirements, and specific preventative measures relevant to horse handling and interaction should be maintained [10,13,18].

Overall, these differences between the mechanism, bodily location, and nature of injuries between the populations highlight the markedly different characteristics of injury risk and type for university students, and the need for focused study of the risks to this group [2,17]. Recommendations regarding injury prevention therefore need to be specifically focused on addressing the risks faced by student in university programs. Ideally these should encompass considerations of the specific interactions of students with horses, rather than generally applying those of populations engaged in the equine industry professionally or recreationally [15,16,18]. Not only does this refer to technical risk mitigation, but also socio-cultural mitigation in relation to less tangible risks such as student-specific authority gradients that may discourage students following safety protocol during extramural placements [13,17].

The sample response rate was considered representative of the study population. However caution should be exercised in the broader application of these findings. The survey was retrospective and self-reported, comprising stated attitudes and behaviors within an Australian cultural context, and relied on the ability of the students to recall the details of incidents. Prospective data collection using a standardized approach is recommended to evaluate incidents and near misses, as well as the socio-cultural aspects of accident and safety culture [13], and attitudes among students across programs, institutions and time.

5. Conclusions

Students engaged in university programs involving equine husbandry and veterinary education are at risk of injury, particularly when specific cohorts are engaged in extramural activities. However, injuries are generally not severe in nature, and students report that they normally recovered fully. Based upon injury patterns reported, consideration should be given to the use of protective boots, gloves, and possibly padded coveralls for on-campus and extramural equine-related learning activities. From this report describing the types of HRI in students, it is evident that published data for the practicing veterinarians, equine industry, or equestrian populations should not be relied upon make assumptions about the incidence, nature, and risk of injury to this population. Further study of animal associated risks to students during their education is warranted, and should acknowledge and address the social as well as technical dimensions of safety and risk management [13].

Acknowledgments

The authors would like to acknowledge the assistance Andrew Wright, Health, Safety, and Facilities Officer, and Wendy Hamood, Extramural Studies Coordinator at the University of Adelaide, School of Animal and Veterinary Sciences, Roseworthy Campus.

Author Contributions

Christopher B. Riley, Jessica R. Liddiard and Kirrilly Thompson contributed to the design of the survey, and the preparation of manuscript. Christopher Riley conceived of this project and performed the data analysis. Christopher B. Riley and Jessica R. Liddiard collected and collated the data.

Conflicts of Interest

The authors declare no conflict of interest.

References

1. Fleming, P.R.; Crompton, J.L.; Simpson, D.A. Neuro-ophthalmological sequelae of horse-related accidents. *Clin. Experiment. Ophthalmol.* **2001**, *29*, 208–212. [CrossRef] [PubMed]
2. Lucas, M.; Day, L.; Fritschi, L. Injuries to Australian veterinarians working with horses. *Vet. Rec.* **2009**, *164*, 207–209. [CrossRef] [PubMed]
3. Moss, P.S.; Wan, A.; Whitlock, M.R. A changing pattern of injuries to horse riders. *Emerg. Med. J.* **2002**, *19*, 412–414. [CrossRef] [PubMed]
4. Cowley, S.; Bowman, B.; Lawrence, M. Injuries in the Victorian thoroughbred racing industry. *Br. J. Sports Med.* **2007**, *41*, 639–643. [CrossRef] [PubMed]
5. Lucas, M.; Day, L.; Shirangi, A.; Fritschi, L. Significant injuries in Australian veterinarians and use of safety precautions. *Occup. Med.* **2009**, *59*, 327–333. [CrossRef] [PubMed]
6. Fritschi, L.; Day, L.; Shirangi, A.; Robertson, I.; Lucas, M.; Vizard, A. Injury in Australian veterinarians. *Occup. Med.* **2006**, *56*, 199–203.
7. Fairnie, H.M. Occupational Injury, Disease and Stress in the Veterinary Profession. Ph.D. Thesis, Curtin University of Technology, Perth, November 2005.
8. Loomans, J.B.A.; van Weeren-Bitterling, M.S.; van Weeren, P.R.; Barneveld, A. Occupational disability and job satisfaction in the equine veterinary profession: How sustainable is this "tough job" in a changing world? *Equine Vet. Educ.* **2008**, *11*, 597–607. [CrossRef]
9. BEVA. Survey reveals high risk of injury to equine vets. *Vet. Rec.* **2014**. [CrossRef]
10. Langley, R.L.; Morrow, W.E.M. A survey of personal and occupational health and safety training for US and Canadian veterinary schools. *J. Agromedicine* **1996**, *3*, 23–25. [CrossRef]
11. Wilson, J.H.; Robinson, R.A. Veterinary college safety inspection: A teaching exercise in occupational safety and health and human services. *J. Vet. Med. Educ.* **1999**, *26*, 7–9.
12. Deisch, S.L.; Robinson, R.A.; Hursh, P.W. A health and safety program for students, civil service personnel, and faculty at the College of Veterinary Medicine, University of Minnesota. *J. Am. Vet. Med. Assoc.* **1977**, *171*, 1181–1183.
13. Thompson, K.; McGreevy, P., McManus, P. A critical review of horse-related risk: A research agenda for safe amounts, riders and equestrian cultures. *Animals* **2015**, *5*, 561–575. [CrossRef]
14. Custom Insight. Survey Random Sample Calculator. Available online: http://www.custominsight.com/articles/random-sample-calculator.asp (accessed on 22 October 2012).
15. Cripps, R.A. Horse-related injury in Australia. *Aust. Injury Prev. Bull.* **2000**, *24*, 1–20.
16. Exadaktylos, A.K.; Eggli, S.; Inden, P.; Zimmermann, H. Hoof kick injuries in unmounted equestrians. Improving accident analysis and prevention by introducing an accident and emergency based relational database. *Emerg. Med. J.* **2002**, *19*, 573–575. [CrossRef] [PubMed]
17. Reason, J. Achieving a safe culture: Theory and practice. *Work Stress* **1998**, *12*, 293–306. [CrossRef]

18. Scuffham, A.M.; Firth, E.C.; Stevenson, M.A.; Legg, S.J. Tasks considered by veterinarians to cause them musculoskeletal discomfort, and suggested solutions. *N.Z. Vet. J.* **2010**, *58*, 37–44. [CrossRef] [PubMed]
19. Meredith, L.; Antoun, J.S. Horse-related facial injuries: The perceptions and experiences of riding schools. *Inj Prev* **2011**, *17*, 55–57. [CrossRef] [PubMed]
20. Ueeck, B.A.; Dierks, E.J.; Homer, L.D.; Potter, B. Patterns of maxillofacial injuries related to interaction with horses. *J. Oral Maxill. Surg.* **2004**, *62*, 693–696. [CrossRef]
21. Gabel, C.L.; Gerberich, S.G. Risk factors for injury among veterinarians. *Epidemiology* **2002**, *13*, 80–86. [CrossRef] [PubMed]
22. Jeyaretnam, J.; Jones, H. Physical, chemical and biological hazards in veterinary practice. *Aust. Vet. J.* **2000**, *78*, 751–758. [CrossRef] [PubMed]

Animals **2015**, *5*, 897-909; doi:10.3390/ani5030390

OPEN ACCESS

animals

ISSN 2076-2615

www.mdpi.com/journal/animals

Article

Workplace Injuries in Thoroughbred Racing: An Analysis of Insurance Payments and Injuries amongst Jockeys in Australia from 2002 to 2010

Beverley A. Curry, Peta L. Hitchens, Petr Otahal, Lei Si and Andrew J. Palmer *

Menzies Institute for Medical Research, University of Tasmania, Hobart TAS 7000, Australia; E-Mails: bevcurry@netspace.net.au (B.A.C.); peta.hitchens@slu.se (P.L.H.); petr.otahal@utas.edu.au (P.O.); Lei.Si@utas.edu.au (L.S.)

* Author to whom correspondence should be addressed; E-Mail: andrew.palmer@utas.edu.au; Tel.: +61-3-6226-7729; Fax: +61-3-6226-7704.

Academic Editor: Kirrilly Thompson

Received: 2 June 2015 / Accepted: 28 August 2015 / Published: 8 September 2015

Abstract: *Background*: There is no comprehensive study of the costs of horse-related workplace injuries to Australian Thoroughbred racing jockeys. *Objectives*: To analyse the characteristics of insurance payments and horse-related workplace injuries to Australian jockeys during Thoroughbred racing or training. *Methods*: Insurance payments to Australian jockeys and apprentice jockeys as a result of claims for injury were reviewed. The cause and nature of injuries, and the breakdown of payments associated with claims were described. *Results*: The incidence of claims was 2.1/1000 race rides, with an average cost of AUD 9 million/year. Race-day incidents were associated with 39% of claims, but 52% of the total cost. The mean cost of race-day incidents (AUD 33,756) was higher than non-race day incidents (AUD 20,338). Weekly benefits and medical expenses made up the majority of costs of claims. Fractures were the most common injury (29.5%), but head injuries resulting from a fall from a horse had the highest mean cost/claim (AUD 127,127). *Conclusions*: Costs of workplace injuries to the Australian Thoroughbred racing industry have been greatly underestimated because the focus has historically been on incidents that occur on race-days. These findings add to the evidence base for developing strategies to reduce injuries and their associated costs.

Keywords: injury; jockey; horse; economic; costs; insurance

---

1. Introduction

Thoroughbred racing is a popular sport and major industry, and makes a substantial contribution to the Australian economy. In 2005–2006 the Thoroughbred racing industry provided over 64,000 full-time-equivalent jobs, generated more than 5 billion Australian dollars (AUD) and contributed more than AUD 1 billion in government taxes to the Australian economy [1]. In the 2009–2010 race season, Australia had 374 race clubs that conducted 2694 race meetings and 19,376 races, with 194,736 starters vying for over AUD 427 million in prize money [2].

In Australia, approximately 1000 jockeys are licensed to ride in races annually [2], and for them, it is a dangerous occupation. Jockeys in Australia experience an average of one fall every 240 rides in flat racing, with a third of such falls resulting in injury [3]. A fall can be catastrophic, resulting in the end of the jockeys' career or even death [3–5] An Australian Jockeys' Association survey conducted in 2010, reported that, in the 12 months prior to the survey, at least 50% of jockeys who completed the survey had sustained an injury and 40% had experienced a fall that prevented them from riding [6]. As 41% also reported having no private health insurance and 22% no superannuation, many are dependent on coverage from workers' compensation [6].

As employees of their relevant Principal Racing Authority (PRA), jockeys unable to earn a living because of work-related injury are provided with near-full income replacement for a defined period. Although conditions and entitlements vary between jurisdictions, a lump sum or periodical payment may be provided for permanent impairment, and where injuries result in death, funeral costs and weekly payments for dependents are also provided [7]. The Personal Accident Insurance (PAI) cover for all jockeys and apprentices, introduced in 2009 [8], funded by a 1% levy on the winnings of all Thoroughbred races [9], is now an important safety net for jockeys with low earnings.

Despite the contribution that horse racing makes to the Australian economy, and not withstanding a study of workers' compensation costs from Victoria [10] and studies in Britain [4,5], there has been no national study of the economic impact of injuries to Australian jockeys. This study of Australian workers compensation authority (*WorkCover*) data, on claims for horse-related injuries to licensed Thoroughbred racing jockeys, was undertaken to provide national data on this topic.

2. Methods

2.1. Sources of Data

The Australian Thoroughbred racing industry comprises eight PRAs representing each state or territory of Australia: Racing Victoria Limited; Racing New South Wales; Thoroughbred Racing South Australia Limited; Racing Queensland Limited; Racing and Wagering Western Australia; Tasracing; Thoroughbred Racing Northern Territory and Canberra Racing Club.

With permission from the relevant PRA, a *proforma* spreadsheet, requesting information on all workplace insurance claims by licensed Thoroughbred racing jockeys for the period 1 August 2002 to 31 July 2010, was sent to the *WorkCover* authority in each jurisdiction and the national PAI scheme. Information requested included: the age, sex and experience of the jockey (apprentice/jockey/jumps); the date and a description of the incident; the injury sustained, any absence from work and the (direct) costs associated with the claim.

Introduction of workers' compensation for jockeys was piecemeal across the states and territories of Australia. In Western Australia (WA) and the Northern Territory (NT), workers compensation cover for jockeys was introduced in 2003 and 2004 respectively, while jockeys in Tasmania and South Australia (SA) were not covered until 2007, therefore data for the whole study period were not available for all jurisdictions. In addition, Queensland data for the period 1 July 2004 to 30 June 2005 were not available.

Ethics approval was granted by the Social Sciences Human Research Ethics Committee, University of Tasmania (Reference H0011786).

2.2. Analysis

Descriptive analyses of the costs of claims for workplace injuries (*WorkCover* claims) to licensed jockeys and apprentices were performed. Total costs, mean and standard deviations (SD) are reported. In addition, because a few outliers had a great effect on mean values, median and interquartile ranges (IQR) of distribution are also presented.

As only the cost of claims made to the jockeys' PAI fund were provided for the 2009–2010 season, more detailed analyses of these data were not possible.

The incidence and costs of *WorkCover* claims per race meeting and ride were calculated using denominator data obtained from the Australian Racing Fact Book 2010 [2]. Claim incidence was calculated using the relevant years' denominators only where a full year of claim data were available.

Claims were stratified according to whether the corresponding incident occurred on a race-day or elsewhere (grouped as "race-day" or "other") and whether it was a consequence of a fall from a horse (grouped as "fall" or "non-fall"). The nature and site of injuries were reviewed and in a subset of claims with details available (WA, Queensland, Tasmania and Victoria); the costs associated with each type of claim were reported.

Prior to analyses, payments were adjusted for inflation to 2011 values using the state and territory specific average weekly earnings for full-time adult persons' ordinary time earnings at August 2011 [11].

Differences between groups were determined using Wilcoxon rank sum test (for medians) and tests of the equality of proportions and one sample students' *t*-tests for comparison of means, where appropriate. All analyses were conducted with STATA 12.0 (StataCorp, College Station, TX, USA) with statistical significance at $p < 0.05$.

3. Results

3.1. Overview of Frequency and Cost of Claims

After exclusion of non-horse related claims (n = 43), claims with no cost attached (n = 193) or where misclassification of jockey status was suspected, data from 2817 Australian jockeys' workers compensation claims and 115 PAI claims were available for the period 1 August 2002 to 31 July 2010.

The number, total annual cost, and mean and median cost of *WorkCover* claims in each state per year are presented in Table 1. With the exception of WA, the cost of claims fluctuated considerably between racing seasons. Overall, the costs of *WorkCover* claims for horse-related workplace injuries to jockeys cost the Thoroughbred racing industry at least AUD 72.1 million for the eight year study period, an average of AUD 8.6 million per annum based on the last three years where claims data were available for all states and territories. Furthermore, in the first year of the Jockeys' PAI scheme (2009–2010), 100 claims amounting to AUD 3.3 million were made.

Overall, 39.3% of claims and 51.8% of the total cost of claims were associated with race-day incidents. The mean (AUD 33,756; SD 200,543) and median (AUD 4365; IQR 1104–18,379) costs of race-day incidents were higher than non-race day incidents (AUD 20,338; SD 77,160 and median AUD 3172; IQR 855–12,273).

The incidence of claims according to the number of race meetings and rides in each state are presented in Table 2. Based on the data available for the study period, the overall incidence of claims was around 2.1 per 1000 rides (range 1.43–3.13). The incidence of claims associated with race-day and non-race day incidents was significantly lower than average in Victoria ($p < 0.001$), while SA and WA had a significantly higher than average incidence of total (both $p < 0.001$) and race-day claims ($p < 0.001$ and $p = 0.020$ respectively).

Overall, the mean costs of race-day and non-race day incidents over the years of the study were similar. However, between states there were considerable differences in the costs associated with race-day and other associated claims.

3.2. Causes of Claims

Where the status of the jockey was known, 21.9% (22.1% race-day, 21.8% other) of compensation claims from Victoria and 17.4% (17.1% race-day and 17.5% other) from SA were made by jumps jockeys. Overall the mean (AUD 45,831; SD 134,720) and median (AUD 6291; IQR 2137–25,156) costs of the 128 claims by jumps racing jockeys was significantly higher ($p < 0.001$) than the 2659 claims by flat racing jockeys (mean AUD 24,672; SD 140,321; median AUD 3515; IQR 903–13,887). Such claims accounted for AUD 5.7 million (19.6%; 15.2% of race-day associated and 25.1% of other claims) and AUD 111,974 (11.3%; 26.8% of race-day associated and 4.5% of others) of the total cost of claims for Victoria and SA respectively.

The proportion of claims associated with falls from a horse were similar for racing (76.4%) and other riding activities (80.2%) however only 846 (30%) of 1976 claims associated with a jockey fall from a horse only were race-day falls (Figure 1). The mean costs of a race-day fall (AUD 29,250; SD 92,879) was significantly higher ($p < 0.001$) than falls sustained during other riding activities (AUD 16,519; SD 57,082), but there were no significant differences between the median costs (race-day median AUD 5343 and IQR 1060–22,214; other riding median AUD 2811 and IQR 707–10,841). For incidents that did not result in a fall from a horse, there was no statistically significant difference between the mean (AUD 10,399; SD 22,025) or median (AUD 1968; IQR 494–9032) costs of race-day incidents or those occurring during other horse-related activities (mean AUD 10,074; SD 27,290; median AUD 2014; IQR 470–7773).

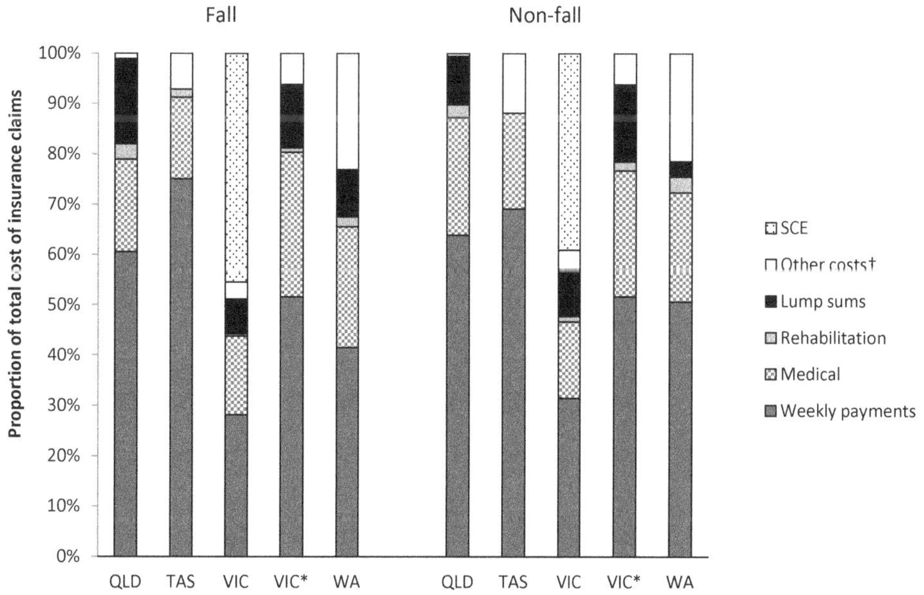

QLD = Queensland; TAS = Tasmania; VIC = Victoria; WA = Western Australia;Other costs include legal fees, investigation fees and fees not otherwise specified; Cost breakdown for SA, NT, NSW and ACT unavailable; SCE = Statistical Case Estimate. SCE was included only in Victorian claims; VIC * Victorian data excluding the SCE contribution to the cost.

Figure 1. Breakdown of total costs of claims, by jurisdiction, according to whether the claim was a consequence of a fall or non-fall.

3.3. Injuries Associated with Claims

Where information was available, there were some differences in the causes of injury between race-day and other incidents. Jockey and or horse falls (74.1% *vs.* 67.9%), hitting the barriers or running rail (11.7% *vs.* 4.5%), being kicked or struck (2.7% *vs.* 9.6%), or trampled (1.2% *vs.* 5.4%) by a horse respectively.

Table 1. Jockeys' workplace compensation claims and costs (AUD 1000s), according to Principal Racing Authority and race season.

State		Racing Season								
		2002–2003	**2003–2004**	**2004–2005**	**2005–2006**	**2006–2007**	**2007–2008**	**2008–2009**	**2009–2010**	**Total**
NSW	N	158	148	167	99	98	73	101	152	996
	TOTAL COST	3634	3534	3143	1966	1794	2098	2769	2637	21,575
	Mean (SD)	23.0 (96.3)	23.9 (120.1)	18.8 (52.6)	19.9 (57.9)	18.3 (48.9)	28.7 (96.2)	27.7 (105.4)	17.4 (34.1)	21.6 (81.1)
	Median (IQR)	2.6 (0.7,9.7)	2.5 (0.8,9.3)	2.9 (0.7,11.8)	2.4 (0.6,12.9)	2.7 (0.7,14.0)	6.1 (0.4,31.2)	3.0 (0.8,12.3)	4.0 (0.6,17.8)	2.7 (0.7,12.7)
NT	N	N/A	N/A	4	7	7	5	9	10	42
	TOTAL COST			6	82	2168	29	490	126	2901
	Mean(SD)			1.6 (1.6)	11.7 (16.6)	309.8 (607.4)	5.7 (7.0)	54.4 (110.5)	12.6 (12.4)	69.1 (262.0)
	Median (IQR)			1.3 (0.4,2.8)	1.2 (0.4,27.5)	0.8 (0.2,515.4)	1.9 (0.5,9.5)	13.1 (6.9,25.6)	10.0 (2.8,18.1)	5.7 (0.8,18.1)
QLD	N	83	69	N/A	97	112	106	95	86	648
	TOTAL COST	1437	862		2798	1532	2335	1535	986	11,485
	Mean(SD)	17.1 (34.4)	12.5 (19.0)		28.8 (60.6)	13.7 (25.2)	22.0 (39.6)	16.2 (28.7)	11.5 (17.6)	17.7 (35.9)
	Median (IQR)	2.6 (0.9,17.4)	6.6 (1.1,13.0)		4.7 (1.7,24.5)	3.1 (0.8,12.3)	4.8 (0.9,22.2)	3.9 (0.7,13.7)	3.3 (0.6,13.6)	4.0 (0.9,17.2)
SA	N	N/A	N/A	N/A	N/A		28	38	32	98
	TOTAL COST						129	670	231	1030
	Mean(SD)						4.6 (7.6)	17.6 (34.5)	7.2 (9.9)	10.5 (23.1)
	Median (IQR)						1.5 (0.2,5.9)	2.3 (0.8,19.9)	1.6 (0.9,10.5)	1.9 (0.5,9.8)
TAS	N	N/A	N/A	N/A	N/A	N/A	7	9	11	27
	TOTAL COST						71	183	47	301
	Mean (SD)						10.1 (7.2)	20.3 (29.6)	4.3(4.7)	11.2(18.4)
	Median (IQR)						5.9 (4.3,16.8)	8.4 (0.6,26.1)	3.0 (0.9,4.2)	4.2 (0.9,12.5)
VIC	N	82	61	51	63	64	86	70	43	520
	TOTAL COST	8842	1223	2327	829	6949	3985	4821	941	29,917
	Mean (SD)	107.8 (607.5)	20.0 (45.2)	45.6 (134.2)	13.2 (22.5)	108.5 (378.6)	46.3 (103.8)	68.9 (155.7)	21.9 (21.6)	57.5 (288.7)
	Median (IQR)	5.5 (1.3,23.8)	5.9 (2.3,19.9)	8.2 (4.1,33.6)	4.3 (2.2,15.1)	10.0 (3.2,37.9)	5.1 (2.0,33.6)	9.0 (3.7,32.2)	16.1 (8.7,23.8)	7.4 (2.7,23.8)
WA	N	N/A	41	62	92	79	70	76	66	486
	TOTAL COST		624	762	1152	706	668	552	458	4922
	Mean(SD)		15.2 (35.6)	12.3 (40.8)	12.5 (43.3)	10.1(29.3)	9.4 (17.6)	7.2 (13.5)	6.9 (10.8)	10.1 (29.3)
	Median (IQR)		3.5 (0.5,11.3)	1.9 (0.7,5.8)	1.6 (0.7,5.6)	1.9 (0.6,7.1)	1.8 (0.5,7.3)	1.6 (0.5,10.1)	1.9 (0.8,6.6)	1.9 (0.6,7.1)

NSW = New South Wales; NT = Northern Territory; QLD = Queensland; SA = South Australia; TAS = Tasmania; VIC = Victoria; WA = Western Australia; N = number; SD = standard deviation; IQR = interquartile range; NSW data includes data for the ACT; N/A = data not available for time period. AUD, Australian dollars expressed in 2011 values.

Table 2. The incidence of workers' compensation claims and the mean cost per ride, stratified by Principal Racing Authority.

State	Total Rides (N)	Denominator Rides (N)	All Claims N (%)	All Claims Claims/1000 rides	All Claims Mean cost/ride (SD)	Race-Day Claims Claims (%)	Race-Day Claims Claims/1000 rides	Race-Day Claims Mean cost/ride (SD)	Other Claims Claims (%)	Other Claims Claims/1000 rides	Other Claims Mean cost/ride (SD)
NSW	438,238	438,238	996 (35.4%)	2.3	49 (12)	304 (30.5%)	0.7	19 (8)	692 (69.5%)	1.6	30 (17)
NT	25,702	16,231	42 (1.5%)	2.6	151 (256)	22 (52.4%)	1.4	144 (269)	20 (47.6%)	1.2	7 (8)
QLD	372,453	324,520	648 (23.0%)	2.0	36 (17)	356 (54.9%)	1.1	23 (11)	292 (45.1%)	0.9	14 (8)
SA	117,874	28,600	98 (3.5%)	3.4	24 (20)	35 (35.7%)	1.2	7 (7)	63 (64.3%)	2.2	17 (13)
TAS	49,514	12,064	27 (1.0%)	2.2	16 (12)	12 (44.4%)	1.0	4 (2)	15 (55.5%)	1.2	12 (13)
VIC	370,145	370,145	520 (18.5%)	1.4	80 (63)	197 (37.5%)	0.5	43 (38)	323 (62.1%)	0.9	37 (26)
WA	176,129	154,959	486 (17.3%)	3.1	32 (12)	180 (37.0%)	1.2	16 (6)	306 (63.0%)	2.0	16 (7)
Total	1,550,055	1,344,757	2817 (100%)	2.1	24 (8)	1106 (39.3%)	0.8	12 (5)	1711 (60.7%)	1.3	12 (4)

NSW = New South Wales, including Australian Capital Territory data; NT = Northern territory; QLD = Queensland; SA = South Australia; TAS = Tasmania; VIC = Victoria; WA = Western Australia; N = number; SD = standard deviation; Total rides = The total number of rides during the study period; Denominator rides = The number of rides during the study period with a full season of claims data.; Claims/1000 rides = Number of claims per 1000 rides.; Mean cost/ride = The mean cost of a claim per ride in AUD, calculated where a full year of claim and ride data were available.; All costs expressed as AUD in 2011 values.

A similar proportion of race-day and other claims were associated with strain injuries (4.4% and 5.8%), being crushed or rolled on by a horse (1.7% and 2.1%), being bitten (0.2% and 0.2%), struck by the horses' head (3.1% and 4.0%) and being dragged by the foot (0.3 and 0.4%), respectively. Where no fall was indicated, three quarters of claims were attributed to jockeys hitting fences or barriers (29.3%), being kicked or struck by a horse (20.1%), strain injuries (20.3%) or being hit by the horses' head (15.0%). The latter was the main contributor to a significantly greater proportion of claims for facial injuries being associated with non-falls compared to falls (15.0 *vs.* 6.8%, $p < 0.001$). Lower limb injuries were also more frequently associated with incidents where no fall was reported (35.7% *vs.* 26.4%, $p < 0.001$), while falls were associated with more intracranial injuries (5.5% *vs.* 1.8%, $p < 0.001$), neck or shoulder injuries (4.9% *vs.* 2.5%, $p = 0.001$), and multiple injuries (8.2% *vs.* 2.9%, $p < 0.001$). Otherwise, the distribution of other injuries associated with falls and non-falls was comparable: back, 8.0% and 9.9%; chest and trunk, 7.6% and 7.5%; and upper limbs, 30.2% and 24.7%, respectively. Overall, fractures were the most common injury. When further investigated, the proportion of claims attributable to fractures was significantly higher in Victoria (37.3%) and the NT (35.7%) than elsewhere (NSW, 28.4%; QLD, 26.5%; SA, 22.7%; TAS, 14.8% and WA 16.7%). In addition, compared to flat racing jockeys, a greater proportion of jumps racing jockeys' claims were for fractures (26.1% *vs.* 44.5%, $p < 0.005$).

With the exception of intracranial injuries, where there was a disproportionate high cost for the number of incidents associated with a fall ($p < 0.001$), the distribution of injuries associated with fall and non-fall claims were similar (Table 3) and the median cost of injury claims were of a similar order of magnitude.

3.4. Indirect Costs of Injury

Three claims were associated with a fatal injury. Amongst those with non-fatal injuries, the mean absence from work was 45 days (SD 119) and this was greater for jumps racing than flat racing jockeys regardless of whether the incident resulted in a fall (78 days, SD 172 *vs.* 56 days, SD 139 $p < 0.001$) or not (112 days, SD 252 *vs.* 36 days, SD 81, $p < 0.001$). Vertebral fractures and intracranial injuries secondary to a fall were associated with the greatest number of days off work (Table 3).

3.5. Breakdown of Direct Costs

Excepting race-day falls in Victoria and WA, which accounted for 44% and 36% of the claims respectively, the majority of the total costs of compensation were for weekly benefits, regardless of the venue or cause of the claim. When documented, around 10% of the total costs of insurance claims were attributed to lump sum payments and 20% to payments for medical services (medical professionals and hospital expenses). Other miscellaneous costs including legal fees, common law payment, investigation costs varied considerably between states. In Victoria, 51% of the total cost of *WorkCover* claims is a Statistical Case Estimate (SCE). The SCE is attributed to each claim to take account of the long-term implications of compensation. After exclusion of SCE from the analysis, the distribution of costs associated with Victorian claims was comparable to other states.

Table 3. The number of non-fatal injuries, and total, mean and median costs (AUD 1000s), associated with workers' compensation claims according to whether or not the injury was sustained as a result of a fall from a horse or where no fall was indicated.

Type of Injury [a]	Incidents Associated with a Fall					No Fall Indicated				
	N (% Total)	Total Cost (% Overall Cost)	Mean Cost (SD)	Median Cost (IQR)	Days Lost [a] (SD)	N (% Total)	Total Costs (% Overall Cost)	Mean Cost (SD)	Median Cost (IQR)	Days Lost [e] (SD)
Fractures [b]	574 (29.1)	20,674 (35.2)	36.0 (76.1)	14.1 (6.0, 33.6)	74 (109)	194 (23.1)	5137 (38.5)	26.5 (111.5)	8.4 (2.6, 20.3)	50 (87)
Sprains, Strains	399 (20.2)	5843 (9.9)	14.6 (41.0)	2.2 (0.7, 9.4)	29 (88)	136 (16.2)	1424 (10.7)	10.5 (24.3)	2.0 (0.4, 10.5)	26 (71)
Contusion & crushing	348 (17.6)	4781 (8.1)	13.7 (92.8)	1.3 (0.4, 4.0)	24 (141)	142 (16.9)	1024 (7.7)	7.2 (26.4)	1.0 (0.3, 3.2)	22 (90)
Muscle/tendon injury	277 (14.0)	4253 (7.2)	15.4 (54.9)	2.6 (0.8, 6.9)	37 (89)	165 (19.6)	2817 (21.1)	17.1 (59.5)	1.9 (0.7, 8.8)	50 (123)
Intracranial injury [c]	105 (5.3)	13,348 (25.0)	127.1 (611.2)	3.0 (1.0, 8.5)	32 (295)	23 (2.7)	105 (0.8)	4.6 (7.1)	1.2 (0.2, 4.4)	14 (33)
Superficial injury	39 (2.0)	281 (0.5)	7.2 (22.9)	0.9 (0.3, 2.8)	17 (54)	19 (2.3)	97 (0.7)	5.1 (8.0)	0.4 (0.2, 12.2)	23 (47)
Vertebral fracture/spinal injury [d]	52 (2.6)	3676 (6.3)	70.7 (117.3)	18.2 (6.8, 86.6)	116 (153)	11 (1.3)	1105 (8.3)	100.5 (192.3)	18.0 (4.1, 178.8)	168 (256)
Open Wound, no amputation	26 (1.3)	154 (0.5)	5.9 (10.6)	3.3 (0.8, 5.5)	21 (57)	43 (5.1)	197 (1.5)	4.6 (12.5)	1.3 (0.5, 4.1)	9 (37)
Dislocation	43 (2.2)	1089 (1.9)	25.3 (36.4)	12.2 (3.0, 31.4)	73 (113)	19 (2.3)	497 (3.7)	26.2 (31.3)	9.8 (8.3, 50.1)	63 (109)
Multiple injuries	31 (1.6)	1042 (1.8)	33.6 (62.4)	14.1 (2.9, 37.8)	74 (118)	10 (1.2)	94 (0.7)	9.4 (7.7)	8.2 (3.0, 12.6)	25 (28)
Other and unspecified injuries	78 (3.9)	2323 (4.0)	29.8 (115.0)	2.1 (0.6, 8.8)	35 (101)	79 (9.4)	835 (6.3)	10.6 (45.0)	2.3 (0.4, 8.2)	9 (21)
Overall	1972 (100)	58,799 (100%)	29.8 (160.4)	4.2 (1.1, 17.1)	49 (128)	841 (100)	13,333 (100%)	15.9 (67.6)	2.7 (0.7, 10.2)	35 (97)

[a] Excludes identified fatalities (3 Intra cranial injuries, total cost AUD 1,335,659 average cost per claim AUD 445,219 SD 390,253) and one claim for replacement of a personal aid; [b] Does not include vertebral fractures; [c] Includes concussion; [d] Includes 2 spinal injuries: 1 associated with a fall, 1 with no fall indicated; [e] Days lost based on 1884 falls and 765 non falls with data available; All costs expressed as AUD × 1000, in 2011 values.

4. Discussion

Between 1 August 2002 and 31 July 2010, the direct cost of workplace injuries to jockeys and apprentice jockeys was at least AUD 9 million per annum. Less than half (41%) of the *WorkCover* claims made by jockeys were the result of an incident at a race meeting.

As individual claims had a large effect on the total cost (particularly in smaller jurisdictions), and because of missing data from some jurisdictions, we were unable to evaluate any trends in the costs of insurance claims over time. Differences observed in the incidence and costs of claims between jurisdictions may be related to the different types of racing in each jurisdiction. Only three states conducted jumps racing during the study period (VIC, SA and TAS), and the overall mean cost of claims by jumps jockeys was higher than for flat racing jockeys. This may reflect the severity of injuries experienced by jumps jockeys, and is apparent in the higher cost per claim in Victoria, where most jumps races are held. These findings are consistent with other studies which have found that, although jumps racing jockeys have a higher incidence of falls, and a lower rate of injury per fall [12–14], their injuries tend to be more severe. A study of insurance payments to injured jockeys in Great Britain reported that compensation for jumps jockeys was almost twice that of flat racing jockeys [5].

Previous studies of the safety of Victorian race tracks [15,16] identified a number of risk factors for injury and made recommendations on how the incidence of workplace incidents might be reduced. The lower incidence of claims in Victoria relative to other states may be a reflection of the increased focus on injury prevention in that state.

The mean costs of claims resulting from a race-day fall or other incident were higher than those that did not occur on race-day. However, as there were a greater number of non-race day incidents, the overall costs of claims from race-day and non-race day incidents were comparable. This indicates that any estimation of the costs of jockeys' compensation claims based solely on race-day claims would underestimate the financial burden to the racing industry of injuries to jockeys. This result is in agreement with two other recent investigations of the Victorian Thoroughbred racing industry where about one third of jockey falls were associated with training activities [10,17]. The higher overall incidence of non-race day related claims compared to other studies [10,17] is a new finding. It is possible that that there may have been some misclassification of employment status in our data (70% of insurance claims made to the Victorian *WorkCover* Authority have previously been identified as being made by track riders and stable assistants) [10]. However, our conclusions were unchanged when the analysis was restricted to jurisdictions where jockey status was well characterised (WA).

As observed by others [3,10,17], regardless of whether the incident occurred on a race-day or involved a fall, the most common sites of injury to jockeys were the lower and upper limbs (>49%). Head injuries were less common but they were associated with a higher mean claim costs and more days off work. For incidents that did not result in a fall, facial injuries were also common. As most epidemiological studies of injuries to jockeys concentrate on falls, facial injuries as a consequence of being hit in the head by the horse have not received much attention to date. However, although the costs associated with them might be low, it may be pertinent to consider incorporating facial protection into helmet design in addition to maximizing protection from the impact of a fall.

Of concern are the results from a questionnaire completed by jockeys that indicated that many experience workplace injuries but do not report or make a *WorkCover* claim because a certain amount of injury is accepted as part of the job [17]. In the same questionnaire jockeys reported that they had at least 5 weeks per year off as a consequence of workplace injury [17]. This is comparable with the average absence observed in these data (6.4 weeks) if we assume that each jockey makes only one claim per year.

One of the main limitations to this study of costs is data quality. Inconsistencies in scheme funding, incident documentation, coding methodology and the breakdown of costs associated with claims, apparent in this study, could result in misclassification of incidents.

In response to previous studies highlighting the scarcity of comprehensive data on jockey accidents in Australia [17], an industry database was developed to facilitate standardised documentation of injuries to workers, and horses, at any horse racing facility and improve ascertainment of injury incidence and to identify potential risk factors. When the web-based Australian Racing Injury Database (ARID) was piloted in Victoria and NSW, 115 ARID incident reports were received for the 2008–2009 season [17]. The proportion of these that were race-day events (73%) was in agreement with the number of jockeys in the AJFD in these states (n = 84) who were declared unfit to ride or were taken to hospital after a fall. Combining information from ARID with insurance claims may provide a clearer and consistent picture of the incidence and costs of horse-related injuries to jockey (and other industry workers) throughout Australia.

5. Conclusions

Considerable interest and emphasis on the human, equine and monetary costs associated with incidents occurring during races highlights the need to improve safety measures in the horse-racing industry. However, in this study, less than half of the compensation claims were associated with race-day injury. The ARID reporting system may help determine the true incidence of workplace injury in this industry, but monitoring costs associated with workplace injury may provide an additional means of assessing the effectiveness of interventions aimed at risk reduction [15,17].

Acknowledgments

The research team would like to thank the Australian Jockeys' Association for provision of Jockeys' Personal Accident Insurance data; the Australian Racing Board; Racing Information Services Australia (RISA) for provision of race data; each Principal Racing Authority for assistance with obtaining insurance and worker's compensation data from their respective WorkCover authorities, and for provision of stewards' reports. This research was funded by WorkCover Tasmania.

Author Contributions

Beverly A. Curry contributed to the study design, statistical analysis and manuscript writing. Peta L. Hitchens contributed to the original idea for the study acquisition of funding, study design and manuscript writing. Petr Otahal contributed to statistical analysis. Lei Si contributed to statistical analysis and manuscript writing. Andrew J. Palmer contributed to the original idea for the study, acquisition of funding, study design and manuscript writing.

Conflicts of Interest

The authors declare no conflict of interest.

References

1. *Economic Impact of Australian Racing*; IER Pty: Melbourne, Victoria, Australia, 2007.
2. Australian Racing Board. *Australian Racing Fact. Book: A Guide to the Racing Industry in Australia 2009–2010*; Australian Racing Board: Sydney, Australia, 2010.
3. Hitchens, P.L.; Blizzard, C.L.; Jones, G.; Day, L.M.; Fell, J. The incidence of race-day jockey falls in Australia, 2002–2006. *Med. J. Aust.* **2009**, *190*, 83–86. [PubMed]
4. Balendra, G.; Turner, M.; McCrory, P. Career-ending injuries to professional jockeys in British horse racing (1991–2005). *Brit. J. Sport Med.* **2008**, *42*, 22–24. [CrossRef] [PubMed]
5. Turner, M.; Balendra, G.; McCrory, P. Payments to injured professional jockeys in British horse racing (1996–2006). *Brit. J. Sport Med.* **2008**, *42*, 763–766. [CrossRef] [PubMed]
6. A Fair Go for Injured Jockeys 2010. Avaiable online: http://www.actu.org.au/images/dynamic/attachments/7129/jockeys_trustcampaign_booklet_print.pdf (accessed on 10 June 2012).
7. Comparison of Workers' Compensation Arrangements in Australia and New Zealand. Avaiable online: http://www.safeworkaustralia.gov.au/sites/SWA/about/ Publications/Documents/173/ComparisonWorkersCompensationArrangementsAustNZ_2010_PDF.pdf (accessed on 10 June 2012).
8. Personal Accident Insurance Cover Achieved for All Australian Jockeys: Australian Racing Board. Avaiable online: http://www.australianracingboard.com.au/Media.personal-accident-insurabce-cover-achieved-for-all-australian-jockeys (accessed on 1 June 2012).
9. Racing for Our Lives: A plan to Protect Australian Jockeys. Avaiable online: http://www.australianracingboard.com.au/Media.personal-accident-insurabce-cover-achieved-for-all-australian-jockeys (accessed on 7 June 2012).
10. Cowley, S.P.; Bowman, B.; Lawrance, M. Injuries in the Victorian thoroughbred racing industry. *Brit. J. Sport Med.* **2007**, *41*, 639–643. [CrossRef] [PubMed]
11. *Average Weekly Earnings*; Australian Bureau of Statistics: Canberra, Australia, 2012.
12. Turner, M.; McCrory, P.; Halley, W. Injuries in professional horse racing in Great Britain and the Republic of Ireland during 1992–2000. *Brit. J. Sport Med.* **2002**, *36*, 403–409. [CrossRef]
13. McCrory, P.; Turner, M.; LeMasson, B.; Bodere, C.; Allemandou, A. An analysis of injuries resulting from professional horse racing in France during 1991–2001: A comparison with injuries resulting from professional horse racing in Great Britain during 1992–2001. *Brit. J. Sport Med.* **2006**, *40*, 614–618. [CrossRef] [PubMed]
14. Oikawa, M. The science of safety helmets—Safety for helmets or safety for jockeys? *Jpn. Racing J.* **2004**, *12*, 5–6.
15. Cowley, S.; Bowman, B.; Lawrance, M. Prevention of Injuries that Result from Working with Horses in the Victorian Thoroughbred Horse Racing Industry: Phase 1 report. Avaiable online: https://www.worksafe.vic.gov.au/__data/assets/pdf_file/0012/9210/SDF_RacingPhase1_Report.pdf (accessed on 1 September 2015).

16. Cowley, S.; Bowman, B.; Lawrance, M. Safety in the Victorian thoroughbred horseracing industry. *J. Occup. Health Safety* **2007**, *23*, 421–431.
17. Foote, C.; McIntosh, A.; V'Landys, P.K.B. *Health and Safety in Australian Horse Racing*; Rural Industries Research & Development Corporation: Canberra, Australia, 2011.

Animals **2015**, *5*, 592-609; doi:10.3390/ani5030374

OPEN ACCESS

animals

ISSN 2076-2615

www.mdpi.com/journal/animals

Article

Inroads into Equestrian Safety: Rider-Reported Factors Contributing to Horse-Related Accidents and Near Misses on Australian Roads

Kirrilly Thompson [1,†,*] and Chelsea Matthews [2,†]

[1] The Appleton Institute, CQUniversity, 44 Greenhill Road, Wayville, SA 5034, Australia

[2] School of Natural and Built Environments, Civil Engineering, University of South Australia, Adelaide, SA 5000, Australia

[†] These authors contributed equally to this work.

* Author to whom correspondence should be addressed; E-Mail: kirrilly.thompson@cqu.edu.au; Tel.: +61-8-8378-4512 (ext. 4512).

Academic Editor: Marina von Keyserlingk

Received: 21 April 2015 / Accepted: 3 June 2015 / Published: 22 July 2015

Simple Summary: Riding horses on roads can be dangerous, but little is known about accidents and near misses. To explore road safety issues amongst Australian equestrians, we conducted an online survey. More than half of all riders (52%) reported having experienced at least one accident or near miss in the 12 months prior to the survey, mostly attributed to speed. Whilst our findings confirmed factors identified overseas, we also identified issues around road rules, hand signals and road rage. This paper suggests strategies for improving the safety of horses, riders and other road users.

Abstract: Horse riding and horse-related interactions are inherently dangerous. When they occur on public roads, the risk profile of equestrian activities is complicated by interactions with other road users. Research has identified speed, proximity, visibility, conspicuity and mutual misunderstanding as factors contributing to accidents and near misses. However, little is known about their significance or incidence in Australia. To explore road safety issues amongst Australian equestrians, we conducted an online survey. More than half of all riders (52%) reported having experienced at least one accident or near miss in the 12 months prior to the survey. Whilst our findings confirm the factors identified overseas, we also identified issues around rider misunderstanding of road rules and driver misunderstanding of rider hand signals. Of particular concern, we also found reports of potentially dangerous rider-directed road rage. We identify several areas for potential safety intervention including (1) identifying equestrians as vulnerable road users and horses as sentient decision-making vehicles; (2) harmonising laws regarding passing horses; (3) mandating personal protective equipment; (4) improving road signage; (5) comprehensive data collection; (6) developing mutual understanding amongst road-users; (7) safer road design and alternative riding spaces; and (8) increasing investment in horse-related safety initiatives.

Keywords: horse-rider; road safety; decision-making vehicle; risk; Australia

1. Introduction

The equine industry is essential to the social and economic wellbeing of Australia. "It is estimated to generate approximately AUS $9 billion each year and employ tens of thousands of people" [1]. About one third of this contribution is attributed to the non-racing sector [2], in which a quarter of a million Australians participate [3]. As well as contributing to the Australian economy, horse ownership and interactions provide numerous benefits for physical, psychological and social health [4]. However, these benefits come at tragic human cost. Approximately 20 Australians die from horse-related accidents per year [5]. "One worker is hospitalised each day in Australia due to a horse related injury. For every worker injured another nine non workers are injured" [6] (p. 2). The repercussions can be tragic: "between July 2000 to June 2012, 98 horse-related deaths occurred" (NCIS [6], p. 25).

Horse riding is undoubtedly a "high risk interspecies sport" [7] and any interaction with horses is dangerous [8]. From the ground, horses can injure humans through biting, kicking and crushing. Even when riding in company, riders can be kicked by other horses. Riders can be transported at speeds of up to 60 km/h with their heads raised three metres above the ground. Upon falling, they may be crushed by their half tonne mount or trampled by other horses (as is common amongst jockeys). It is no surprise that recent calls have been made for a greater understanding of the objective and subjective elements of horse-related risk and equestrian risk-perception [9].

When riders and horses interact on public roads shared with other road users, their vulnerability to injury or death is magnified. Unlike passengers in vehicles, riders are unrestrained. They can be hit by vehicles moving at high speeds, and/or thrown into oncoming traffic, jeopardising the safety of other road users. Although the definition of vulnerable road user (VRU) specifically mentions "pedestrians, pedal cyclists and motorcyclists", a horse/rider could easily be included due to their lack of a "hard metal shell" and their sensitivity to injury in the event of a collision or incident. VRUs are thought to be the "most sensitive to road injury" [10] (p. 1). Five interrelated factors contribute to VRU near misses and accidents: speed, proximity, visibility, conspicuity and mutual misunderstanding.

The sudden or close passing of a vehicle may trigger a horse's dangerous flight response, causing them to bolt blindly forward or veer into traffic [11]. The main causes of 17 horses being killed in the United Kingdom in 2011 following collisions with cars were the "vehicle travelling too close/at speed or the horse becoming spooked" [12]. Sometimes the cause of a near miss or accident arises from a lack of visibility, referring to a drivers' "range of unobstructed vision" [13]. This may be due to a driver rounding a blind corner on the road and quickly coming upon a horse rider. A lack of conspicuity can also be a problem, referring to horse and rider being "clearly discernible" [13]. A driver may not see the horse/rider due to shadows, glare, clothing/coat colour of the horse and rider or other factors. Whilst research in the UK found no significant relationship between the wearing of fluorescent or reflective clothing and the number of near misses experienced by a rider, it did find a significant relationship between riders wearing lights and the incidence of near misses [14]. Riders wearing some form of lights experienced significantly less near-misses than other riders, whether wearing fluorescent/reflective clothing or not. A study from the UK suggests that the root cause of accidents between horse-riders and other road-users may be due to differences in hazard and risk perception, attributed to a lack of empathy. The researchers found that "drivers with horse riding experience or those with family or close friends who rode horses, showed greater ability to consider the scene from both perspectives" [15].

Whilst a UK survey of horse riders found that 60.3% of participants (n = 257) had experienced at least one near miss or accident in the year prior [14], data on the incidence of horse-related road accidents in Australia is limited. One reason for this is variable detail in admissions data from hospitals, although one study estimated that at least 8% of the 20 horse-related deaths each year in Australia occurred on roads (Cripps 2000) [5]; another is the inconsistent recording of large animal rescues by emergency services [16,17], especially those involving extricating horses from vehicle wrecks. Where police records are kept, European research has highlighted massive underreporting of VRU injury in police reports, sometimes as low as 12% [13]. However, the issue of horse-related road accidents is of increasing concern as peri-urban development in Australia could lead to a rise in the frequency of interactions between equestrians and other road users. This is particularly concerning given that data collected by the Victorian Injury Surveillance System two decades ago suggested that 16% of horse riding injuries occurred on public roads [18]. The risk is not only to equestrians. The implications for drivers can also be tragic [19].

One area of potential conflict between equestrians and other road users relates to inconsistencies in the application of national road rules, and legacy issues related to times when horses were more commonly ridden or driven on public roads and eight different sets of road rules prior to the Australian Road Rules being first published in 1999 [20]. Whilst horses are not mentioned specifically in the Australian Road rules (1999), Part 18, Division 2 comprises three "rules for people in charge of animals". Across Australia, horses are allowed on footpaths and nature strips (subject to some conditions), may be ridden two abreast (under some conditions) and must keep left when using a roundabout while giving way to all exiting traffic. The road rules refer to horses indirectly as an animal, with the following two exceptions. The Victorian Road Rules stipulate that any rider under the age of 18 riding a horse on a road must wear a helmet while the Queensland Road Rules require that, on receiving a signal from the person in charge of a restive horse on the road, a driver must keep as far left as practicable, stop the vehicles engine and "not move the vehicle until there is no reasonable likelihood that the noise of the motor, or the movement of the vehicle, will aggravate the restiveness of the horse" [21].

Whilst the piecemeal research overviewed here provides some information on horse-riders as vulnerable road users, little is known about incidence and contributing factors for accidents and near misses to horse riders on Australian roads. Moreover, little is known about rider understanding of road rules. To provide a preliminary overview, horse riders were invited to take part in a pilot survey.

2. Experimental Section

2.1. Materials and Methods

An online survey was conducted in consultation with the Executive Officer of the South Australian Horse Federation who commissioned the research. Surveys were designed to elicit information specifically regarding the roads that the rider frequented with their horse, any near misses or accidents in the preceding 12 months and the cause of the incidents as well as any suggestions for improving safety. The survey was open for 13 days in May 2014 and included the following six open-ended questions:

(1) Do you ride or lead horses on public roads or road related areas?
(2) Please name the roads and area in which you do the majority of your riding.
(3) Have you had any accidents or near misses on these roads in the last 12 months? If so, please describe what happened and include any suggestions for road design or rule improvement that may have helped prevent this from happening.
(4) Have you got any suggestions for updates to the Australian Road Rules you would like to see in relation to the riding or driving of horses? If so, please explain.
(5) Do you have any suggestions for road related infrastructure design or signage that would make roads more horse rider friendly?
(6) Any other suggestions you have to improve general road safety for horse riders on roads?

The survey was digitized using Survey Monkey and the link posted to the "Sa-Horse Federation" Facebook page (approximately 5000 "likes"). It was also made available on the Horse SA website with a link to the page posted in the Horse SA newsletter (approximately 1600 primary recipients).

2.2. Participants

One hundred and forty seven equestrians who ride or lead horses on public roads or road related areas responded to the survey. Almost half (48%) were from South Australia, 28% were from Victoria, 12% were from New South Wales, 4% were from both Queensland and Tasmania, 2% were from Western Australia, 1% were from both the Northern Territory and England while the final 1% did not specify. Participants were not asked to provide their age or gender.

2.3. Analysis

This paper reports on the responses to all questions except Question 2, which is most relevant to a local audience and the results of which could jeopardise the anonymity of respondents. Although questions 4–6 specifically asked for road rule change suggestions, signage change suggestions and any other suggestions in separate questions, respondents did not, on the whole, distinguish between these categories in their responses. These responses were therefore collated for ease of analysis and interpretation.

Whilst respondents were restricted in the amount of open-text responses they could provide, there was sufficient data to apply a modified qualitative data analysis approach based on systematic reading of the data, recording of issues and basic organization of findings around the research questions [22].

Issues identified in open text fields were subject to descriptive statistical analyses for ease of reporting. Where informative, these statistics are illustrated and expanded upon with selected quotations reproduced verbatim from open-text response fields.

3. Results

3.1. Rider-Reported Contributory Factors and Suggestions

Just over half of all riders (52%) reported being involved in at least one near-miss or accident within the 12 months prior to the survey. They were able to specify multiple causes for their accident(s)/near-misses. Speed was the most often cited contributor to near-misses or accidents (72%), operationalized in analysis as a vehicle passing a horse and rider at a greater speed than the rider felt safe/comfortable with. For example,

> *My horse spooked at a ute [utility vehicle] that was flying towards us on a dirt road, even though I was signaling for him to slow down [an arm extended to the left and moved up and down]. He didn't stop until my horse stepped out in front of him*

> *Vehicles move to other side of the road but continue to do same speed (80–100 kph) even when I'm wearing hi vis [high visability clothing] and signaling for [the] driver to slow down.*

Some riders attributed a lack of speed reduction by a driver to their lack of understanding of the unpredictably nature of horses:

> *A horse can spook and put the driver is a precarious position if the horse kicks out or worse jumps on the car. Most drivers have not a clue the danger they put themselves in by passing a horse at speed*

> *There seems to be a lack of understanding by the average driver about the athleticism and unpredictable nature of horses. Simply overtaking on a country road and continuing sometimes at 80+ ks is a serious risk which I experience with some regularity*

> *Most drivers are not aware of how to behave around a horse and assume that if the horse is on the road that it is completely bombproof to traffic.*

Other rider-reported factors contributing to near misses were loud noises (17%), including beeping horns, revving engines, yelling; and "close" proximity (18%), defined as passing a horse and rider at a smaller distance than the rider felt safe/comfortable with. For example,

> *[A] driver tried to pass me, very close, as my horse was baulking at something on the road. Fortunately she was going very slowly and although my horse backed into her car there was no major damage to either party.*

Eleven per cent of riders cited a lack of visibility, where they did not believe they were seen, due to local geography or driver inattention. Fourteen per cent cited a lack of space, where they felt forced to ride on the road due to the lack, or unsuitability, of the nature strip. Eight per cent cited noncompliance with road rules, where they believed drivers were not obeying the road rules (excluding speeding). Four per cent cited bicycles, where a bicycle approached the horse and rider from behind and passed without warning where the horse and rider were unaware of its presence, and five per cent cited the horse spooking at an animal/object not associated with the traffic. One per cent cited "other" without further specification.

As illustrated in the quotation below and throughout many of the quotations in this paper, many rider respondents reported failed attempts to signal to drivers to slow down:

> *When I signal to drivers to slow down they sometimes completely ignore me even if I am clearly having trouble with my horse.*

This quotation demonstrates the ways in which some riders can assume that other road-users can interpret their horse's behavior as not totally under control and be aware of the implications.

Some rider respondents had been explicitly or implicitly made to feel as if they were trespassers on the road:

> *... We ended up down the ditch, at which point he slammed brakes on, and abused me for being on the road on a horse, and both drivers told me horses are not allowed on the road ...*
>
> *Most drivers don't think horses and riders have a right to utilise the road and I have often been abused for being on the road.*
>
> *Had people toot, rev engines, yell out windows causing my horse to react. Never been hit but I always try to stay well off the road edge if cars are coming.*

Seventeen per cent of rider respondents involved in near-misses or accidents reported being abused by the driver of a car. Abuse of a horse and rider was defined as the rider being yelled or cursed at, intentionally chased, having objects thrown at them or being "beeped" repeatedly.

> *3X the same guy has driven at me deliberately & pulled out at the last second. 1/2 full beer bottles thrown at me & the horse, while crossing the bridge, along with having a bicycle pass me on the inside, a car sit no more than 15cms behind the horses back legs, horns tooted, abuse yelled to get off the road, all while crossing the bridge. A learner driver pass me missing my right foot by mere cms, cars speed up when over taking & or drop down a gear for more power & revved engines. So it goes on.*

One responded recalled receiving abuse from a motorcycle rider:

> *Have had deliberate attempts to frighten my horse (Harley Davidsons revving their motors while stationary beside me! Passengers waving and yelling as they drive past)*

Another noted the difficulty of reporting offenders to police:

> *If there is an incident/accident the rider is usually trying to control their horse or is on the ground after falling and there is no opportunity to get rego [registration] numbers of drivers. No point reporting the incident to the police as the driver can't be identified. Could be worthwhile to have a study on where these incidents tend to happen though—riders might report at least locations if they knew someone/police were collating data for a study.*

3.2. Rider Understanding of Road Rules

Not all rider respondents knew the rules that applied to them when riding on the road. Such riders either incorrectly stated that horses have right of way or described the Queensland law where a driver must pull over and turn off their engine when a horse becomes restive. For example:

> *I believe a very old but valid law exists. i.e., if a motorist sees a horse fractious/frightened he should pull over and turn off his engine (South Australian rider)*

As noted above, this law exists only in Queensland. Even so, some variation of it was specifically mentioned by nine other non-Queenslander respondents who ride horses on roads:

> *Drivers must give way and slow down and even stop if requested. The problem is our ability, or lack thereof, to enforce these rules (Victorian rider)*
>
> *... need better publishing of the existing laws—most drivers don't know they are required to slow down or stop if signaled (South Australian rider)*

A Tasmanian respondent was under the false impression that:

> *a driver only has to do what a rider says if a horse becomes "Restive" meaning unsettled. By then it's too late!!!! The law needs to state: keep a minimum of 5 m away from the back of a horse & a minimum of 2 m away from a horse when overtaking & a speed limit of 30 km maximum when passing a horse ... the law needs to be the same in all states of Australia.*

As none of the questions specifically asked respondents if they believed horses had some form of right of way on roads or if some variation of the Queensland law existed in their state, it could be surmised that at least six per cent of rider respondents not residing in Queensland or overseas demonstrated an incorrect perception about how the road rules relate to horses.

3.3. Rider Suggestions for Improving Safety

The survey returned a total of 295 suggestions for making roads safer for horses and riders. They were allocated to one of the following seven categories derived inductively from the data: education (cited by 50% of respondents), better/different signage (45%), more room for horses off roads (34%), road rule changes (27%) helmets or high visibility equipment (22%), awareness (15%) and miscellaneous other (8%).

Suggestions attributed to the category "education" related to having more information in the official drivers' handbook, more training and "better" information more readily available. The category "awareness" refers more to raising awareness in the general population that horses are allowed to ride on roads as well as scare campaigns. These responses were clearly aimed at television, perhaps similar to the "Ride to Live" Campaign as mentioned by one respondent.

With regard to suggesting that riders should be encouraged to wear, or have to wear by law, high visibility and safety clothing of some kind, some riders believed that their experience riding on the road was more positive while they were wearing high visibility clothing.

I believe if riders ride on the road they should wear a high vision (sic) vest. I have recently started doing this every time I leave my property gates and I feel it makes a HUGE difference to my safety.

Possibly consider riders wearing safety vest and/or hi viz for their horses make them more visible. I have done this a few times now and found drivers acknowledge the risk and slow down.

Regarding better/different signage, riders reinforced a perceived lack of understanding amongst drivers about the nature and behaviour of horses. That is, they considered the current signage inefficient and meaningless (see Figure 1).

Figure 1. Image of sign taken at the intersection of Dressage Avenue and Pimpala Road in South Australia. Note the absence of a helmet from the visual image and a lack of any informative text. Photo Credit: Chelsea Matthews, 2015.

Some riders recommended the addition of specific instructions to signage:

I think a "slow when passing" or something similar in addition to the picture would be helpful.

I think it could help to include "Pass Slow & Wide" with the pictured horse & rider sign commonly used. Just telling people there are horses about doesn't help them know what to do.

Maybe the yellow signs with horse pics [pictures] on them could include lower speed limit or state "reduce speed".

One respondent provided a passionate rationale for why any changes or improvements would be fortified by a public education campaign (caps in original):

> *... I think the best way to achieve this is a major ... road safety campaign that is put in newspapers and on the television, something particularly graphic that highlights that many of these people out there riding are young girls, mothers, daughters, brothers, sisters, uncles etc. They are PEOPLE. And they can and will die from people doing stupid things past them. There are numerous road safety campaigns for horses but they are all online and will only be seen by those people looking for them, which is mainly other horse riders. This campaign needs to reach the people who aren't looking for it, hence suggesting mass media. A recent Facebook post regarding a tabard with a built in camera has attracted numerous comments from people saying they get annoyed by horse riders and deliberately speed up to teach them a lesson and hopefully they won't ride on the roads anymore. They need to learn that this is ILLEGAL and could not only kill the rider but kill them too. A nice image of a driver being taken to hospital and dying while a horses legs stick out his windscreen might go some way to doing the trick.*

This quotation demonstrates a riders' awareness of the interdependence of rider, horse and driver safety. Whilst riders and drivers may seem at odds when sharing roads, they have a shared responsibility for each other's safety.

4. Discussion

4.1. Incidents and Contributing Factors

The data suggest an incidence rate similar to the 60% reported for riders in the UK [15]. More than half (52%) of riders had experienced at least one accident or near miss within the 12 months prior to the survey. Based on this figure, and using a conservative estimate of 50% of the 226,100 Australians participating in horse-riding (ABS, 2000) [3], this would suggest that around one hundred thousand horse riders and handlers are at risk of an accident or near-miss whilst riding their horse on an Australian road.

Our findings also confirm the factors contributing to near misses and accidents identified in other countries [15], namely speed, proximity, low visibility, low conspicuity and mutual misunderstanding. Speed was cited as a contributing factor in just over half of all rider-reported near misses and accidents. In addition to these five factors, we also found evidence of rider misunderstanding of road rules and driver misunderstanding of rider hand signals. The latter is unsurprising given that hand signals used by riders to request a driver to take an action are not included in the South Australian Drivers Handbook [23] (p. 33) or the road rules of any Australian state or territory.

Of particular concern were the experiences of road rage reported by riders. Although it may be a result of mutual misunderstanding or rider misunderstanding of road rules, it is illegal. The South Australian Drivers Handbook specifically mentions that drivers should not accelerate or rev their engine near a horse, sound their horn or make unnecessary noise and not throw objects or shout at a horse or rider [23] (p. 33). Directing road rage at horse riders can also have fatal results. By frightening horses and riders, abuse itself could contribute to an (additional) accident or near miss. For example, one incident of driver road rage directed at a horse-rider resulted in the death of a horse and the injury of a rider in Florida, America [24].

The contributing factors of speed, proximity, visibility, conspicuity and mutual misunderstanding are discussed in this paper from the perspective of VRUs. However, it should be noted that riders, other road users and horses can all contribute to, or mitigate, risks. For example, when riders ride at speed, they restrict a driver's ability to safely reduce speed when passing. Similarly, when there is "mutual misunderstanding" between rider and horse, they are less predictable to other road users. Following research considering risk and safety as emergent properties of socio-technical networks [25], further research should consider the ways in which risk factors are distributed across the horse-rider-road-driver-car network. It is also important to recognize the perspective of each actor in the network. For horses, the visibility and conspicuity of other road users is particularly aural. This raises important concerns for electric vehicles and bicycles that are less audible and may "spook" horses more easily.

4.2. Research Limitations

Findings in this paper are biased towards self-selection and the experiences of South Australian riders. Almost half (48%) of respondents resided/rode in South Australia while data from the Australian Bureau of Statistics 2000 suggests that South Australia had one of the lowest participation rates in horse activities of all states and territories in Australia at 1.4%. The Australian Capital Territory had the highest participation rate of 2.4% followed by Queensland and New South Wales at 2%. Victoria and Western Australia had participation rates of 1.8% followed by Tasmania at 1.5% while the Northern Territory was equal with South Australia at 1.4% [3]. Despite the Horse Federation of South Australia being active nationally, it is unsurprising that the majority of participants were located in South Australia. This is consistent with other national surveys promoted primarily by the Horse Federation of South Australia [26,27]. To validate the findings in this exploratory research, and determine if there are any statewide differences, further research should recruit participants from across Australia.

Further research will need to distinguish between a near miss and an actual accident, and collect demographic data including age, gender, riding experience. As these are all important factors [28], this data should be collected in future research with a more representative sample size.

4.3. Potential Safety Interventions

Based on the findings from the survey and following suggestions made by riders themselves, we have identified several areas for potential safety intervention

4.3.1. Identifying Equestrians on Roads as VRUs and their Horses as Sentient Decision-Making Vehicles

Terminology matters. Although riders are "vulnerable road users", the term is not defined in Australian law, nor are VRUs referred to in any legislation with respect to road or traffic laws. The legal definition of horse riders, horse drivers (*i.e.*, cars, carriages and racing harnesses) or horse handlers at a national level as vulnerable road users could change driver perceptions, promote research and provide a rationale for funding.

Terminology also has implications for animals, including the ways in which they are valued and prioritized relative to other lives and things [29]. This has been noted by White in relation to natural disaster response where he refers to a "strict legal categorisation of companion animals as personal property, or things, rather than legal persons" [30] (p. 381). Emergency services personnel risk their lives to save human life, property and the environment—in that order. Animals such as horses may be considered "property" or "environment". The choice affects where attention is placed in preventing horse-related road accidents and prioritizing resources when attending callouts or making triage decisions for people and animals. From a perspective of psychological first aid, considering the horse "just a vehicle" may trivialize the trauma suffered by those at the scene and even compound the distress reported by some responders attending accidents involving horses [17]. "Vehicalising" horses also ignores the extent to which horses are not just "things" to their humans. Consistent with Belk's use of the term "extended self" [31,32], horses are extensions of their riders' selves. As such, the behaviour and controllability of each individual horse is at least as diverse as that of every individual rider, driver and handler. From an actor-network theory perspective, horse-riders are sentient assemblages generated by the interactions of human, technology and animal being, who often share roads with other assemblages, such as driver-cars [33].

Terminology is particularly important in incident analysis and accident reporting. In determining causal factors, investigators may consider road, vehicle or human factors. However, the interaction of human and vehicle factors is more straightforward when the vehicle is a car or motorbike, and much less so when the vehicle is a horse (or a donkey, mule, bullock, yak or camel). This has been demonstrated in research on horse-rider relations where subjective experiences of trust have been found to have objective consequences for riders' risk-taking behaviours [7] and the degrees of "control" sought over their horses [34].

At worst, horses can be viewed as decision-making vehicles. They are like vehicles in that they transport humans and goods and are subject to road rules, but they are unlike vehicles in that they are sentient creatures capable of making their own decisions and subject to their own instincts and training. Horses (at least, well trained ones) have even been proposed as a useful metaphor for designing automated vehicles [35]. Defining horses in road rules as something more specific and sentient than just a vehicle may strengthen efforts to educate other road users about the "risky" nature and behaviour of horses—especially in relation to how horses respond to stressors. The essential differences between horse-riders, cyclists and motor vehicles could be emphasized in the development of specific road rule terminology for horses other than "vehicle" or "animal". As there are multiple alternatives (if not just "horse"), each with different implications for the perception of horse-riders by other road-users, discursive and empirical research is required to identify and evaluate the terminology most likely to reduce horse-related near misses and accidents on public roads.

4.3.2. Harmonising Laws Regarding Passing Horses

The implementation of a law that stipulates the manner in which a driver must pass a horse being ridden, driven or led on a road may make roads safer for horse riders, drivers and handlers. One possible remedy would be to implement the Queensland law that drivers must pull over and turn off their engine if a horse becomes restive. Another option could be a law requiring drivers to pull over and turn off their engine. Although the respondent quoted above was incorrect about the existence of such a law in Tasmania, they raise a valid point; By the time a horse becomes "restive", both the horse and rider have already been placed in danger. Her suggestions for keeping a 5 m buffer behind a horse and 2 m to the side with a speed limit of 30 km/h for overtaking could prove quite effective for reducing the instances of near misses due to horses being frightened or spooked by fast passing or close vehicles but may prove impractical in some cases. However, reducing speed may not always be a possible or safe option, such as when a vehicle on a blind corner sees a horse after entering a turn at speed, or on the rare occasion that there are horses on both sides of a road.

One practical solution with respect to law changes could be to require that drivers slow down when passing a horse and give the horse a lateral buffer where possible when passing. Further research is needed to determine a buffer distance and safe and practical speed. They may differ for large vehicles and trucks that can scare horses if they pass too slowly and/or if air brakes are applied. Harmonising road rules across Australia's states and territories may also reduce general misunderstanding or confusion about the legal specifications for the interactions of horses and other road users. Riders could be kept informed of road rules through a test similar to the British Horse Society's "Riding and Road Safety Test".

4.3.3. Mandating Personal Protective Equipment

Horse riding helmets are one form of "shell" that horse riders drivers and handlers can wear as a proven form of protection whilst riding, driving or leading horses. A 1995 draft of the Proposed Australian Road Rules included "requirements that all horse riders wear a helmet, reflectors when riding at night and be allowed to use footpaths and nature strips" [36]. However, with the exception of riders 18 years and under on Victorian roads, wearing helmets whilst riding on private property or public roads is still voluntary in Australia. Whilst wearing appropriate footwear when riding is a widespread practice, the low use of helmets when riding is alarming. For instance, a study of fall related injuries in Australian agriculture reported that 79% of hospital presentations "did not report use of a safety device at the time of injury" and "only 18% of injury presentations resulting from a height-related fall reported wearing head protection" [37]. Twenty years ago, researchers suspected that "[t]he promotion of equestrian helmets in Australia is likely to need a similar approach to that used to promote and effectively introduce mandatory helmet wearing" [36]. Given the multiple psychological, social and cultural barriers to increasing the voluntary use of helmets amongst horse riders [38], compulsory use and appropriate enforcement of wearing helmets on public roads should be seriously considered in Australia.

Despite research finding no relationship between the use of high visibility clothing and reduced risk [14], some respondents in this Australian survey recalled more positive road riding experiences while wearing high visibility clothing. Their perception requires further investigation; especially as those who wear high visibility clothing may be more predisposed to "safer" riding practices in general. The use of "GoPro" style camcorders should also be considered as a form of personal protective equipment. Research is required to determine their impact on discouraging drivers from dangerous behavior such as road-rage, as well as their usefulness in recording number plates of dangerous drivers.

4.3.4. Improving Road Signage

Forty seven per cent of respondents recommended more or improved road signage. Road signage in Australia with respect to horse riders is fairly limited and usually consists simply of a yellow sign with a black horse and rider silhouette (Figure 1, above). Reviewing the current signage for horses on roads and making it more informative for drivers could be a very positive step forward for horse and rider safety. By providing some indication of how the horse needs to be treated when approaching and passing, drivers should be better able to react appropriately when they see a horse. One very simple addition to current horse signage could be the phrase "pass wide and slow", as is currently available on some specialty high visibility rider clothing. Similarly, a diagram showing the dimensions of a buffer zone around the horse for passing drivers could inform drivers about how to pass horses more safely.

The location of signage also needs to be reviewed as many respondents reported that there were not enough signs in their area or that they had asked for some/more signage and it had not been provided. Councils could be encouraged to provide signage for riders, particularly in areas with a higher concentration of horse traffic.

4.3.5. Data Collection

There is no comprehensive data on horse-related injury in Australia, let alone that occurring on roads. More rigorous data collection is required to:

(a) Determine the distribution, frequency and consequence of horse-related near-misses and accidents
(b) Identify at-risk rider and driver groups, or high risk locations
(c) Justify and prioritise interventions
(d) Evaluate interventions

The implementation of a mapping system over a number of years could be of assistance, similar to that used by the British Horse Society. In 2010, they launched a website dedicated solely to equestrian safety where riders can report incidents including the location and the type of incident as well as find further advice and information on road safety. Data obtained from this reporting system is intended to "*lobby those in power to make the changes that are required to ensure riding is safer for all*" [39].

4.3.6. Developing Mutual Understanding amongst Road-Users

As noted above, drivers often misperceive the amount of control that riders have over their horses [15], and riders assume that other road users can interpret horse behaviour. The present study reinforces the existence of a general mutual misunderstanding between riders and drivers that in some instances leads to road rage. We also identified a lack of rider understanding of road rules. Although clarified or additional laws and signage changes may assist with reducing horse-related road incidents, these legal and technical interventions would be more successful in combination with social, educational and behaviour change interventions. Formalising and effectively communicating protocols for safely interacting with horses on roads will require engagement with all relevant stakeholders and end-users. Focus group research similar to that conducted in the UK [15], may enable the identification of important barriers and enablers to mutual understanding and the adoption of safer behaviours by all road users, for the benefit of all road users. Findings could be used to design fact sheets, infographics or other information-based safety interventions.

Interventions could draw from and enhance social connectedness rather than reinforce division and competition amongst road users. Thompson proposed the "pet as protective factor" principle to engage pet owners in activities that can increase their chances of surviving a natural disaster [40]. The principle aims to reconfigure pet ownership from a risk factor for survival to a protective factor by leveraging from people's desire to save their pets. As the preparations required to improve an animal's chance of surviving a natural disaster are almost always the same ones that improve human survival, this provides an opportunity for engaging with those members of the population who wouldn't make natural disaster preparations for themselves. And for those pet guardians who wouldn't deliberately seek information on natural disaster preparedness, animal-related social networks may provide a conduit for the dissemination of that information [41]. More recently, the pet as protective factor has been refined through "who depends on you?" messaging and extended from a premise of "save your pets and you might save yourself too" to one of "save yourself so you can take care of your pets now and after a disaster or emergency" [42].

Research on the risk perception of riders in high-risk equestrian sports found that riders were more likely to worry about their horses being injured than themselves [7]. This suggests that horse riders, handlers, drivers and guardians could be motivated to engage in personal injury prevention measures on public roads for the benefit of their horses. Due to the taken for granted riding relationship in human-horse relationships [8], the "pets as protective factor" principle and "who depends on you?" message [42] take on multiple meanings. First, riders could be more likely to engage in precautionary behaviours if the benefits to their horses are emphasized. Second, precautionary and "personal protective behaviour" could include improved horse training and education to increase rider control and/or reduce horse unpredictability [43]. In this regard, a well trained and understood horse could in certain circumstances be a literal form of rider protections. Third, horses depend on riders to keep them safe on the roads, and they depend on riders keeping themselves safe so that they can continue to care for their horses. Finally, road users depend on one another. Emphasizing the ways in which horses and their riders depend on drivers of cars, and vice versa, could create a platform of care rather than conflict.

4.3.7. Safer Road Design and Alternative Riding Spaces

Increasing population growth, urban densification and peri-urban development could increase interactions between multiple road users. In addition to improving the safety of those interactions on public roads, there is a need to consider the provision of alternative spaces for horses. These could range from specially marked laneways or "bridleways" for horses in high horse-traffic areas to open spaces including ovals, trails, stock routes, parks and forests. Where open spaces are shared with other users such as cyclists, motorised trail bikes, walkers and people with dogs, there is still a need to promote mutual understanding, respect and rules of engagement. In a previous survey of horse owners in South Australia, rider access to national parks was raised as a safety control to keep riders away from roads [44]. However, sharing parks with off-road motorbike users was also noted as risky for horse riders, suggesting the same need for protocols around multiple user interactions on trails and in parks as identified for public roadways. Furthermore, alternative riding spaces require horse-friendly design, regular maintenance and possible biosecurity controls where horse hooves and manure may introduce germs or weeds (as do walking shoes, bicycle tyres and dog faeces). The issue of horse access to parks and other public spaces should also be seen as a safety issue.

4.3.8. Increasing Investment in Horse-Related Safety Initiatives

All of the aforementioned measures require resources. Despite an average of 20 deaths per year in Australia's equestrian sector being a figure that would be completely unacceptable in any other industry or activity, there is currently little financial support of horse-related safety research or interventions. By contrast, an average of 1.7 people die each year in Australia from shark-related injury [45], yet the State Government of Western Australia invested AUS $22 million in its Shark Hazard Mitigation Program [46]. To secure much-needed funding, the Australian equestrian industry must draw attention to the scale, severity and consequence of horse-related accident and injury. Whilst effective technical interventions such as helmets, back protectors and high visibility clothing are freely available and affordable, they lack equally effective legal and behaviour change interventions to mandate or increase their adoption. One avenue for acquiring funds to safeguard the people sustaining Australia's $9 billion equestrian industry [1] could be the introduction of a national horse registration system. This could also help to address the lack of important data noted above.

5. Conclusions

Insufficient action is being taken to improve the safety of around a quarter of a million horse people in Australia in general, let alone their safety whilst sharing public roads. This places other road users and pedestrians at risk of serious injury or worse. By reporting findings from an exploratory study of 147 equestrians who ride or lead their horses on public roads or road-related areas, we were able to identify that just over half of those riders had experienced a near miss or accident in the 12 months preceding the survey, and that speed was identified as a contributing factor in almost three quarters of those cases. As road incidents involving horses can put the lives of all road users at risk, there is an urgent need for more research to mitigate horse-related road accidents and near misses. This paper identified several areas for potential risk management spanning technical, social, legal, informational, behavioural and environmental interventions. The safety of Australia's horse people and those around them would be well supported by the establishment of a national task force on equestrian safety with a working group on road safety. To address the complexity of the issue collaborations should be established between safety scientists, horse behaviourists, behaviour change experts, social marketers, civil engineers, town planners, police and emergency services, large animal rescuers and relevant motor accident authorities, not to mention horse riders, drivers and handlers themselves.

Acknowledgments

The authors would like to thank Julie Fiedler, Executive Officer of Horse SA for initiating this project. This project would not have been possible without the voluntary contribution of the respondents who gave their time to provide at times very detailed free-text responses in the online survey. Thanks also to Holly Bowen for her assistance with formatting the manuscript. This paper has been greatly improved by the insightful comments of two reviewers to whom we are grateful.

Author Contributions

The survey was drafted, analysed and reported by Matthews. The framing, interpretation and discussion of results were undertaken by Thompson and Matthews.

Conflicts of Interest

The authors declare no conflict of interest.

References

1. Howell, G. Equine influenza-horsing about with a quarantine crisis. *Austr. J. Emerg. Manag.* **2010**, *25*, 23.
2. Gordon, J. *The Horse Industry—Contributing to the Australian Economy*; Centre for International Economics: Canberra, Australia, 2001.
3. *Participation in Sport and Physical Activities*; Australian Bureau of Statistics: Canberra, Australia, 2000.
4. University of Brighton and Plumpton College. The Health Benefits of Horse Riding in the Uk. Available online: http://www.bhs.org.uk/~/media/BHS/Files/PDF Documents/Health Benefits of Riding in the UK Full Report.ashx (accessed on 21 January 2015).
5. Cripps, R.A. *Horse-Related Injury in Australia*; Flinders University: Adelaide, Australia, 2000.
6. *Guide to Managing Risks When New and Inexperienced Persons Interact with Horses*; Safe Work Australia: Canberra, Australia, 2014.
7. Thompson, K.; Nesci, C. Over-riding concerns: Developing safe relations in the high-risk interspecies sport of eventing. *Int. Rev. Sociol. Sport* **2013**. [CrossRef]
8. Thompson, K. Theorising rider-horse relations: An ethnographic illustration of the centaur metaphor in the spanish bullfigh. In *Theorising Animals*; Taylor, N., Signal, T., Eds.; Brill: Leiden, the Netherlands; Boston, MA, USA, 2011; pp. 221–253.
9. Thompson, K.; Adelman, M. Epilogue: A research agenda for putting gender through its paces. In *Gender and Equestrian Sport*; Adelman, M., Knijnik, J., Eds.; Springer: New York, NY, USA, 2013; pp. 195–211.
10. Standing Committee on Planning Environment and Territory and Municipal Services. *Inquiry into Vulnerable Road Users*; ACT Legislative Committee: Canberra, Australia, 2014.
11. NZ Horse Network. Horse Sense on the Roads. Available online: http://www.horsesenseontheroads.info/Home/about-horses (accessed on 10 January 2015).
12. Road Safety Observatory. Horse Riders. Available online: http://www.roadsafetyobservatory.com/Review/10092 (accessed on 15 January 2015).
13. Constant, A.; Lagarde, E. Protecting vulnerable road users from injury. *PLoS Med.* **2010**, *7*, e1000228. [CrossRef] [PubMed]
14. Scofield, R.M.; Savin, H.; Randle, H. Road safety: Is there a relationship between 'near misses' and the use of rider and horse reflective/fluorescent equipment. In Proceedings of the 9th International Equitation Science Conference, University of Delaware Printing Services, Newark, DE, USA & University of Pennsylvania, New Bolton Center, Kennett Square, PA, USA, 17–20 July 2013; Heleski, C., Wickens, C., Eds.; p. 30.
15. Chapman, C.; Musselwhite, C.B.A. Equine road user safety: Public attitudes, understandings and beliefs from a qualitative study in the united kingdom. *Accident Anal. Prev.* **2011**, *43*, 2173–2181. [CrossRef] [PubMed]

16. Smith, B.; Thompson, K.; Taylor, M. What's the big deal? Responder experiences of large animal rescue in australia. *PLOS Curr. Dis.* **2015**, *1*. [CrossRef]
17. Thompson, K.; Leighton, M.; Riley, C. Helping hands, hurting hooves: The need for a multidisciplinary paradigm of large animal rescue. *Autr. J. Emerg. Manag.* **2015**, *30*, 53–58.
18. Williams, F.; Ashby, K. *Horse Related Injuries*; Monash University Accident Research Centre: Victorian Injury Surveillance System: North Melbourne, Australia, 1995.
19. Ramp, D.; Roger, E. Frequency of animal-vehicle collisions in NSW. In *Too Close for Comfort: Contentious Issues in Human-Wildlife Encounters*; Lunney, D., Munn, A., Meikle, W., Eds.; Royal Zoological Society of New South Wales: Mosman, Australia, 2008; pp. 118–126.
20. Australian Transport Council. *Australian Road Rules*; National Road Transport Commission: Sydney, Australia, 1999.
21. Queensland Consolidated Regulations. Giving Way to Restive Horses. Available online: http://www5.austlii.edu.au/au/legis/qld/consol_reg/toumrr2009629/s303a.html (accessed on 27 March 2015).
22. Thompson, K. Qualitative research rules—Using qualitative and ethnographic methods to access the human dimensions of technology. In *A Practical Guide to Evaluating the Human Factors Issues of New Rail Technologies*; Bearman, C., Naweed, A., Dorrian, J., Rose, J., Dawson, D., Eds.; Ashgate: Aldershot, UK, 2013; pp. 75–110.
23. Department of Planning Transport and Infrastructure. *Driver's Handbook*; Government of South Australia: Adelaide, Australia, 2000.
24. Graaf, M.D. Driver 'deliberately slammed his pickup into horse'—Killing it and injuring the rider—after road rage dispute. *DailyMail* **2015**.
25. Wetmore, J.M. Redefining risks and redistributing responsibilities: Building networks to increase automobile safety. *Sci. Technol. Human Values* **2004**, *29*, 377–405. [CrossRef]
26. Clarkson, L.; Thompson, K. *Australian Horse Welfare & Well-Being Survey: Analysis*; CQUni Appleton Institute: Wayville, Australia, 2013.
27. Thompson, K.; Clarkson, L.; Matthews, R.; Kosmadopoulos, S. *The Future of Horse Research in Australia: Strategic Directions Survey Results*; The Appleton Institute, Central Queensland University: Wayville, Australia, 2012.
28. Wolframm, I. *The Science of Equestrian Sports: Theory, Practice and Performance of the Equestrian Rider*; Routledge: London, UK; New York, NY, USA, 2013.
29. Gupta, A.F. Foxes, hounds, and horses: Who or which? *Soc. Anim.* **2006**, *14*, 107–128. [CrossRef]
30. White, S. Companion animals, natural disasters and the law: An australian perspective. *Animals* **2012**, *2*, 380–394. [CrossRef]
31. Belk, R.W. Metaphoric relationships with pets. *Soc. Anim.* **1996**, *4*, 121–145. [CrossRef]
32. Belk, R.W. Possessions and the extended self. *J. Consumer Res.* **1988**, *15*, 139–168. [CrossRef]
33. Dant, T. The driver-car. *Theory Culture Soc.* **2004**, *21*, 61–79. [CrossRef]
34. Thompson, K.; Birke, L. The horse has got to want to help: Human-animal habituses and networks in amateur show jumping. In *Sport, Animals, and Society*; Gillett, J., Gilbert, M., Eds.; Routledge: New York, NY, USA, 2014; pp. 69–84.

35. Flemisch, F.O.; Adams, C.A.; Conway, S.R.; Goodrich, K.H.; Palmer, M.T.; Schutte, P.C. *The H-Metaphor as a Guideline for Vehicle Automation and Interaction*; National Aeronautics and Space Administration, Langley Research Center: Hampton, VA, USA, 2003.
36. Williams, F.; Ashby, K. *Horse Related Injuries*; Monash University Accident Research Centre, Victorian Injury Surveillance System: Clayton, Australia, 1995.
37. *Fall Related Injuries in Australian Agriculture*; Australian Centre for Agricultural Health and Safety and Rural Industries Research Development Corporation: Canberra, Australia, 2004.
38. Haigh, L.; Thompson, K. Helmet use amongst equestrians: Harnessing social and attitudinal factors revealed in online forums. *Animals* **2015**, *5*, 576–591.
39. British Horse Society. Horse Accidents. Available online: http://www.bhs.org.uk/safety-and-accidents/horse-accidents (accessed on 15 January 2015).
40. Thompson, K. Save me, save my dog: Increasing natural disaster preparedness and survival by addressing human-animal relationships. *Austr. J. Commun.* **2013**, *40*, 123–136.
41. Thompson, K.; Every, D.; Rainbird, S.; Cornell, V.; Smith, B.; Trigg, J. No pet or their person left behind: Increasing the disaster resilience of vulnerable groups through animal attachment, activities and networks. *Animals* **2014**, *4*, 214–240. [CrossRef]
42. Thompson, K. For pets' sake, save yourself: Motivating emergency and disaster preparedness through relations of animal guardianship. *Autr. J. Emerg. Manag.* **2015**, *30*, 43–46.
43. Thompson, K.; McGreevy, P.; McManus, P. A critical review of horse-related risk: A research agenda for safer mounts, riders and equestrian cultures. *Animals* **2015**, *5*, 561–575.
44. Thompson, K.; Clarkson, L.; Fiedler, J.; Rebbeck, M. *Climate Change for Horse Owners: Horse Keeping in Our Community Online Survey Report*; Horse Federation of South Australia, Australian Horse Industry Council: South Australia, Australia, 2012.
45. Taronga Conservation Society Australia. Australian Shark Attack File. Available online: http://taronga.org.au/conservation/conservation-science-research/australian-shark-attack-file (accessed on 23 January 2015).
46. Government of Western Australia. Figures support shark mitigation strategy. Available online: http://sharksmart.com.au/news/figures-support-shark-mitigation/ (accessed on 15 January 2015).

Animals **2015**, *5*, 576-591; doi:10.3390/ani5030373

OPEN ACCESS

animals

ISSN 2076-2615

www.mdpi.com/journal/animals

Article

Helmet Use Amongst Equestrians: Harnessing Social and Attitudinal Factors Revealed in Online Forums

Laura Haigh † **and Kirrilly Thompson** *,†

The Appleton Institute, Central Queensland University, 44 Greenhill Road, Wayville, SA 5034, Australia; E-Mail: kirrilly.thompson@cqu.edu.au

† These authors contributed equally to this work.

* Author to whom correspondence should be addressed; E-Mail: kirrilly.thompson@cqu.edu.au; Tel.: +61-8-8378-4512.

Academic Editor: Marina von Keyserlingk

Received: 20 March 2015 / Accepted: 6 July 2015 / Published: 17 July 2015

Simple Summary: Epidemiological research details a high rate of horse-related injury, despite technical countermeasures being widely available and largely affordable. Whilst barriers to engaging in preventative behavior such as helmet-use have been identified, less attention has been given to enabling factors. These factors could contribute to the design of more effective injury prevention interventions. To identify barriers as well as enablers in an Australian context, we explored how riders discussed helmet use amongst one another in an online setting. Our analysis revealed that social relations heavily influenced safety behavior. In particular, we identified three attitudes that affected helmet use: "I Can Control Risk", "It Does Not Feel Right" and "Accidents Happen".

Abstract: Equestrian activities pose significant head injury risks to participants. Yet, helmet use is not mandatory in Australia outside of selected competitions. Awareness of technical countermeasures and the dangers of equestrian activities has not resulted in widespread adoption of simple precautionary behaviors like helmet use. Until the use of helmets whilst riding horses is legislated in Australia, there is an urgent need to improve voluntary use. To design effective injury prevention interventions, the factors affecting helmet use must first be understood. To add to current understandings of these factors, we examined the ways horse riders discussed helmet use by analyzing 103 posts on two helmet use related threads from two different Australian equestrian forums. We found evidence of social influence on helmet use behaviors as well as three attitudes that contributed towards stated helmet use that we termed: "I Can Control Risk", "It Does Not Feel Right" and "Accidents Happen". Whilst we confirm barriers identified in previous literature, we also identify their ability to support helmet use. This suggests challenging but potentially useful complexity in the relationship between risk perception, protective knowledge, attitudes, decision-making and behavior. Whilst this complexity is largely due to the involvement of interspecies relationships through which safety, risk and trust are distributed; our findings about harnessing the potential of barriers could be extended to other high risk activities.

Keywords: equestrian; horse; injury; helmet; safety; risk; online forum; barriers; enablers; behavior change; injury prevention

---

1. Introduction

Involvement in sport and physical recreation has a positive impact on health and quality of life, providing socio-economic benefits to wider public health [1]. Despite this, the potential risk and occurrence of injury associated with physical activity can act as a barrier to continued participation [1]. This has prompted researchers to advocate for the incorporation of injury prevention strategies into sport promotion [1–4]. Horse riding and related activities are particularly dangerous [5–8]. Equestrian activities pose injury risks to an estimated quarter of a million Australians [9]. "The rider is unrestrained and is travelling on a largely unpredictable animal capable of speeds of up to 70 kph and of kicking with a force of up to 1 ton" [5] (p. 5). Not only are horses large, fast and strong, they are decision-making animals of prey with a heightened flight response [10].

There is a sophisticated body of work documenting injury statistics, identifying horse-related injury risk factors and determining high-risk groups [5,8,11–18]. Both Australian and International research into horse related injury highlight the importance of encouraging the use of safety standard helmets to be consistently adopted by riders [19–22]. This is further demonstrated by research that charts the positive impact of safety helmet use [23,24]. Indeed, Safe Work Australia [25] advocate the use of Australian standard safety helmets alongside the development of skills and confidence in horse interaction. Yet, despite awareness of the critical importance of helmet use in preventing head injury, adoption by equestrians still lacks consistency [11,19,26–28]. Until the use of helmets whilst riding horses is legislated in Australia, there is an urgent need to improve voluntary use.

Recommendations to increase helmet use and improve overall safety typically include training, education, and awareness-raising activities such as wider dissemination of injury statistics [12,19,25,27]. However, increasing the levels of risk literacy amongst equestrians may not necessarily lead to safer equestrian practices [29]. This is largely because the ways in which people perceive and respond to risk are not straightforward. They are subject to psychological and cultural contextualization [30], as are the values, attitudes, beliefs and practices of equestrians towards issues of horse-related risk and safety. As such, equestrian injury statistics and risk factors can be considered a product of individual, social and cultural attitudes and behaviors.

Unlike helmet use amongst cyclists and motorcyclists on Australian public roads [31,32], helmet use is not mandatory for horse riders. Helmet use amongst equestrians therefore provides an illuminating case study of voluntary helmet use. Whilst voluntary helmet use has been discussed in detail in relation to cycling [33,34], motorcycling [35,36] and alpine sports [37–39], those activities do not involve human interaction with another sentient, decision-making being such as the horse [10]. Indeed, previous research on human and horse relations demonstrates the ways in which riders' risk and safety decisions depend on how they perceive the quality of their human–horse relationship [40,41].

Moreover, there is a particular need to understand and address the ways in which equestrians perceive, respond to and talk about risk and safety amongst themselves. These can be inferred from "everyday talk" [42] in discussions and conversations where beliefs and behaviors are often advocated, justified and challenged. For example previous research [28,29] demonstrates the impact that negative attitudes, altered risk perception and physical discomfort has had on the inconsistent uptake of helmet use. As that research was biased towards barriers, there is a need to identify enablers and consider their value in behavior change campaigns.

2. Method

To identify attitudes and beliefs surrounding helmet use, we analyzed two public-access forums on Australian equestrian websites. This provided an unobtrusive means of exploring the "everyday talk" [42] of equestrians, driven by their own concerns. The construction of forum-as-field-site has been successfully utilized by health researchers looking into discourses surrounding topics such as breastfeeding, early onset dementia and disordered eating habits [42–44].

A sampling strategy was devised to select equestrian-based discussion forums, following Callaghan and Lazard [42]. An Internet search was conducted using the terms "Australia", "Forum" and "Horse" or "Equestrian". The inclusion criterion for forum selection was: high variety of topics; high level of posting activity; ability to search the forums; publically accessible content and, containing a ".au" identifier. The two forums selected were accessed publicly, without the need for payment or membership. They contained a broad variety of topics (from horse health to training) but also had differing target audiences (one was noted to be a "junior rider" forum while the other was a general equestrian discussion forum). The forums were initially searched using terms: "Helmets" and "Helmet use". Inclusion criteria for threads included: relevance of the topic to the research aims; high level of response/posting activity and variance of posters. Both threads commenced with a forum member enquiring about, or commenting on, the helmet wearing behaviors of fellow forum members. Thread 1 contained 50 posts from 37 members, with discussion taking place within the time span of a month in 2013. Thread 2 contained 53 posts from 43 members, taking place over a weeklong period in 2004. Neither of the authors were involved in thread instigation or participation.

Data collection followed a protocol used by Rodriquez [43], who only analyzed posts from publically published and accessible forum posts, and devised pseudonyms for the already generally anonymous posters' nicknames. This study was approved by the CQUni Human Research Ethics Committee.

The method can be summarized in the flowchart below:

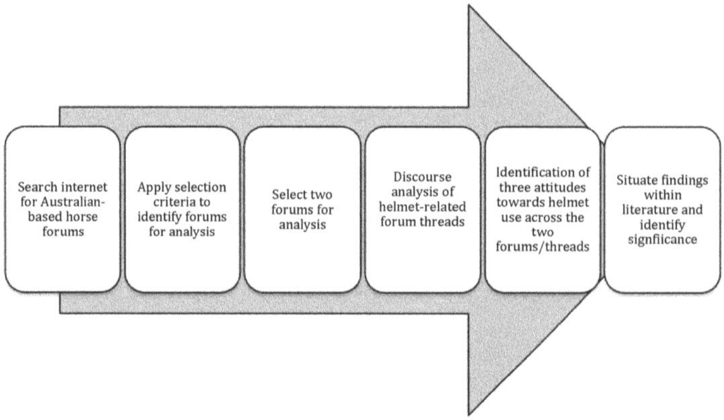

Figure 1. Six steps comprising the research methodology.

Discourse analysis enabled the identification of factors affecting helmet use. The 103 separate posts were analyzed. First, they were read and re-read before being inductively analyzed as whole statements. Each post, or "statement", was broken down and analyzed sentence by sentence with particular focus on discursive attitudes expressed towards helmet use. Following Green and colleagues [45], relationships between these codes were explored. Patterns between factors, risk perception and different stated safety behaviors were documented, as were differing values associated with various equestrian activities within the posts. To identify attitudes or factors that enabled helmet use, we focused attention on discourse surrounding stated consistent helmet use or stated opposition to non-use. Analysis was conducted primarily by Haigh in conjunction with Thompson whose personal involvement in equestrian activities allowed for a nuanced interpretation of horse-specific terminology and concepts.

3. Results

In the "everyday talk" [42] of forum users, we identified numerous social and attitudinal factors significantly affecting helmet use, risk perception and injury mitigation behaviors. Overall, they were deeply entrenched in the social relations of their "eque-cultures" [46]. Three attitudes that affected helmet use have been selected for discussion in this paper, based on their overarching representation across both forums:

(1) "I Can Control Risk";
(2) "It Does Not Feel Right";
(3) "Accidents Happen".

3.1. The Impact of Social Relations on Helmet Use

In forum threads, what was often perceived as a "personal" choice to wear or not wear a helmet was socially influenced. Indeed, a large number of forum posters justified their helmet (non)use through stories of peer behaviors. Social influence was seen to act negatively on some forumites, who both explicitly and implicitly linked their helmet use to the riding culture or social environment they were within. For example:

- One rider openly claimed to be inconsistent with helmets, asserted this as a personal choice, but justified their actions with the claim that: "*I know people that don't even own one*".
- A decision to not wear a helmet was stated as supported by family.
- The influence of experienced riding peers altered the helmet use of one rider, who noted that "*I spent a few months training with a well-known showjumper in NSW, basically nobody wears a helmet there and I just got used to it*". This same rider made an exception and noted that they would wear a helmet "*If kids are watching*".

This last example highlights that while social influence and peer behavior operated on occasion as a negative force on helmet use, it also had the potential to enable and support helmet use. This can be seen in the following examples:

- One rider justified their helmet use by suggesting that non-helmeted riders "*are being a bit selfish. Not because they are doing a sport where there are safety measures that may limit the damage done but because they aren't considering their families*".
- Another rider stated that being injured while not wearing a helmet "*is not fair to those left behind*".
- Helmet use was positioned as something you do for "*those who love you*".
- One rider linked their helmet use to being raised in a family involved in racing (Racing Australia [47] mandates helmet use at any competition).

In these examples, attitudes and behaviors towards helmet use were overtly influenced by perceived social responsibility and "care" for some riders, and more subtly formed by social influences and norms for other riders. This gave rise to a judgmental rhetoric. Alongside this pervasive social dimension, three attitudes were found to influence helmet use: "I Can Control Risk", "It Does Not Feel Right" and "Accidents Happen". These attitudes are outlined below.

3.2. Attitudes Influencing Helmet Use

3.2.1. "I Can Control Risk"

Most riders were generally positive about helmet use. Despite this, when an attitude of "I Can Control Risk" was present, the necessity of helmet use was affected by varying risk mitigation strategies. The extent of risk was perceived as depending on the type of riding activity, the age and training of the horse and importantly, the relationship between horse and rider. Risk mitigation strategies available to these riders included: the rider's own skill set and experience; knowledge of horse behavior; strength of horse and human relationship, and helmet use. Interplay between these factors affected helmet use even for those who stated to be generally consistent in wearing. For example:

- One rider noted: *I very rarely ride without one. In fact the only time I ride without one, is when I bring them up from the paddock or I ride* (my own horse-name removed) *tackless* (without a saddle and bridle) *for a few minutes'.*
- Another discussed riding helmetless on a few occasions while bringing the horse they trust "110%" up from their own paddock. This rider also noted that they would only ever walk their horse up (travelling at a slower and safer pace) and would not risk it if their horse were having a "*hyper*" day.

- Another suggested that the "*1%*" of the time they would not ride with a helmet would only be when "*jumping on*" their old horse to bring them up to their gate.
- The attitude that "safer" situations could potentially negate the need to wear a helmet was even held by one member who claimed quite strongly that riding without a helmet "*made zero sense*", as it was later noted that the reason why they did not ride without a helmet was they knew their horse to be "*unpredictable*".

These situations suggest that trusting and being able to read equine behavior are factors favored above and potentially *instead of* wearing a safety helmet, although a rider's ability to read behavior was taken for granted as ethologically accurate. The attitude that risk could be avoided by other mitigation strategies, above and potentially instead of helmet use, was particularly evident in comments by those who openly declared non-consistent helmet use:

- One rider explained that, "*I don't usually ride with a helmet at home doing flat work but if I go out, ride an un-educated horse or new horse or do any jumping at all I'll pop one on*". Only situations self-categorized as unsafe and unpredictable required a helmet, and this rider later noted that they saw helmet use as dependent on knowing a horse and the level of rider experience.
- Similarly, another rider noted that they usually rode without a helmet, unless riding a "*new or young*" horse.
- One rider stated that they did not wear a helmet while training for dressage but would do so if they were riding a young horse, jumping or riding on the roads.
- While listing a number of situations in which they did wear a helmet, one rider mentioned being *a* "*culprit for not wearing a helmet when schooling dressage at home*", but only while riding their or their mothers "*very experienced*" horses as they "*know damn well they won't do anything*".
- Surprisingly, one rider attributed their head injury not to failure to wear a helmet, but a failure to correctly interpret the horse's behavior.

When risk was present as uncontrollable and outside of the horse/human dyad, then helmet use was considered necessary:

- An openly helmet resistant rider stated that horse riding is "*dangerous as hell*" but that they were not consistent in wearing a helmet. Despite this, they stressed that "*I will however, always wear one on the road!*"

In these examples, the risks involved in equestrian activity were perceived and understood in varying ways by forum members. Helmet use was seen as a tool to mitigate some risks, but other risk limitation strategies, such as trusting and knowing horses, familiarity with riding situations and locations, relationship between the horse and human and levels of rider and horse experience/expertise were often valued over, and often *instead of,* the use of a safety helmet.

3.2.2. "It Does Not Feel Right ... "

While some riders found helmets uncomfortable, others found not wearing a helmet more uncomfortable. In this, the attitude that "It Does Not Feel Right" had an unpredictable and bi-directional affect on helmet use behavior. Further, in some cases the strength of social influence meant that the perceived physical discomfort of helmets did not always prevent use.

"... to wear a helmet"

A few riders commented that wearing a helmet did not feel right. In some cases, this attitude resulted in non-use, but in others helmets were still worn:

- Justifying their general lack of helmet use, one rider noted "*My usual place of residence is in the tropics and I prefer to wear something that shades my face etc. and allows for airflow*".
- Similarly, another exclaimed "*All I can say is once you try riding without a helmet it is really hard to go back!*" After training in an environment that non-wearing was encouraged they stated to "*much more able to concentrate without that heat, sweat and constriction*".

In some cases, the discomfort of wearing a helmet was not a barrier to use, as other factors that encouraged the use of helmets had more behavioral impact:

- One rider commented that they did not grow up wearing helmets, "*hates*" them as they are hot and uncomfortable, but started to wear once their kids started riding as you "*can't really expect them to wear one if your not*". For this rider, being a positive role model for their family was valued more than the perceived negative sensation of helmet use.
- Another rider stated simply that they "*always have and always will wear a helmet*" despite hating the red lines they leave on their forehead.
- Despite the negative sensation associated with helmet use, another associated the decision to wear with wanting to be around to see their grandchildren.
- Another stated that they never wore a helmet when younger but had got used to wearing one after the rules changed at their agistment/livery to mandate helmet use and despite still hating them they would feel odd getting onto their horse without one. Though, they noted that they would still wear an Akubra (an iconic Australian cowboy style hat) on long rides for sun protection.

"... to not wear a helmet"

While some equestrians viewed the helmet as hot and uncomfortable, presenting a barrier to use (although, as seen, not always), our findings suggest that the physicality of the helmet was not always seen as a negative:

- Wearing a helmet was likened on a number of occasions to putting on clothes everyday, with one rider stating they would feel "*naked*" without their helmet, and another noting that "*not wearing a helmet to me feels like not wearing a bra*".
- For three riders, wearing a helmet was such a habit that they joked they felt weird not wearing it in other "risky" situations such as driving a car.
- One rider noted wearing their helmet added to their comfort levels as it warmed their head on cold days.
- Three riders simply stated that wearing a helmet was "*second nature*" and not wearing feels "*odd*".

These findings suggest that helmet use was not always considered uncomfortable, that discomfort was not always a barrier to use and helmet wearing had the potential to become a physically comforting habit, enabling use.

3.2.3. "Accidents Happen"

Similar to the attitude "I Can Control Risk", those expressing the attitude that "Accidents Happen" perceived risk on a scale, and valued a range of risk mitigation strategies. However, for a majority of those who held this attitude, nothing negated the need to wear a safety helmet. Forum members with this attitude stressed that "accidents happened", even to the most experienced riders, and that a helmet provides an easy method of injury prevention in these random circumstances:

- One rider noted that in their experience, the only injury incidents they had experienced were from "*accidents*" such as the horse tripping.
- Another stated that wearing a helmet is a "*very basic*" precaution that could change the severity of injury in an accident event.
- Another noted wearing a helmet on their "*quiet horse*" during equestrian sports as it meant they could focus on their/their horse's performance instead of worrying that the "*unexpected*" might happen.
- One stressed that they always wore a helmet, as "*even 'bomb proof' horses can trip. A friend of mine broke her neck riding a bomb proof school horse that tripped in the paddock at home (she is a professional rider)*".
- The view that even quiet and trusted horses can fault and experienced riders can have accidents was also held by another rider who noted a few occasions where a helmet had saved them in an accident with a quiet, known and trusted horse.
- One rider told the story of a professional dressage rider who had sustained an injury to stress that "*ANYONE can come off, no one is invincible*".
- Another shared a similar story of a fatal accident in which a very experienced rider fell off, stating, "*experienced people do fall off too*".

The attitude of "Accidents Happen" provided protection for some riders against the association of experience with a reduced need to wear a helmet. It allowed these riders to present themselves as experienced equestrians while still choosing to wear a helmet.

The attitude that experience negates the need for helmet use is challenged by (self-stated) experienced riders through tales of "accidents that really did happen". Here, understanding the importance of wearing a helmet had come about through the personal experience of being involved in, or knowing of someone involved in, an accident:

- While admitting not liking helmets when younger one rider stated that they would never "*dream of getting on a horse without one these days*", sharing the story of being involved in an accident where they had luckily decided to wear a helmet which had lessened the severity of injury.
- Similarly, another highlighted that after witnessing a horrific accident of a non-helmeted rider, they "*would not consider wearing a helmet these days*" despite previously being a bit "*hit and miss*".
- Another equestrian wrote that they have never ridden without a helmet since being in an accident as "*I shudder when I think back that I had tossed up whether I should wear the helmet before I got on him that fateful day*".

Riders such as those represented above demonstrated a positive attitude towards helmet use and injury mitigation; they valued risk limitation strategies such as knowing and trusting horses, rider experience and knowing how to read equine behavior *on top of* the added safety measure of safety helmet use. For this group, safety helmets were stated as consistently used because, despite best intentions, "Accidents Happen".

4. Discussion

Through discourse analysis of discussions on two Australian based public-access online forums we found that helmet use was intrinsically linked to the social worlds of equestrians, despite often being viewed as a "personal choice". Further, we found two attitudes that demonstrated a bidirectional affect on helmet use: "I Can Control Risk" and "It Does Not Feel Right . . . " and one that enabled helmet use: "Accidents Happen". We have summarized these attitudes in the table below.

Table 1. Barriers and enablers to helmet usage.

Attitude	Encourages Helmet Use When	Discourages Helmet Use When
"I Can Control Risk"	Helmet use seen as part of risk control	Helmet use seen as extemporaneous to other controls such as being a good rider, having a good horse, and having a good relationship with that horse
"It Does Not Feel Right . . . "	Wearing a helmet becomes a habitual sensation of riding, even if that sensation is discomfort	Wearing a helmet is considered intolerable
Accidents Happen	Accidents accepted as beyond control, and unrelated to rider skill, horse temperament or the quality of their human–horse relationship	The rider has a fatalistic risk perception

The social worlds of the equestrian cohort had a significant influence on helmet use attitudes and behaviors. Attitudes and behaviors towards helmet use were overtly influenced by perceived social responsibility and "care" as well as social regulation in particular riding contexts for some riders, and more subtly formed through social influence and behavioral norms for others. The latter point is consistent with previous research that has demonstrated the negative influence of peer pressure and cultural norms of particular riding styles on helmet use [27–29]. The influence of riding peers as a barrier to helmet use was exemplified by one rider who noted that "*I spent a few months training with a well-known show jumper in NSW, basically nobody wears a helmet there and I just got used to it*". Yet, peer behavior and perceived social responsibility also supported helmet use. For example, one rider noted that wearing a helmet is something one does for "*those who love you*" and another wore a helmet particularly to be a good role for their children, as you "*can't really expect them to wear one if you're not*". This strength could be targeted in injury prevention campaigns. Indeed, research specifically into the adoption of other types of safety helmets has highlighted the significant positive influence peer behavior and social norms could have on consistent use. In an Austrian study, it was found that the positive effect on youth ski helmet adoption that mandatory youth ski helmet laws had in one province was mirrored in another province without mandatory laws [48]. This was attributed to the high ski helmet use by adults and parental figures, which acted to model appropriate and safe behavior [48]. The potential of adults acting as behavioral role models towards helmet use was seen as well in an observational study of bicycle riders in the USA [49], which found that helmet use was higher in pairs of cyclists than in solo riders, and particularly child adult pairs where the adult wore a helmet [49], supporting the strength of companionship as a potential enabler in the promotion of public health interventions. The suggestion of using role modeling to promote desired safety behaviors is stressed in a US study of factors affecting children's bicycle helmet use which found that regardless of the presence or absence of mandatory laws, use of helmet was associated with beliefs of social consequences and peer use [50]. The researchers suggest communication targeting the social desirability of helmets alongside helmet use by parents to model appropriate behavior for children [50].

The adoption of safety helmets by those in equestrian "role model" positions could also act to break down the association between helmet use and inexperience. Moves by prominent horse media magazines or websites (such as *Dressage Today*) to only publish images of helmeted riders are excellent initiatives in this regard. In this, we suggest investigating and trialing avenues for the promotion of helmet use as a normalized and social activity, no matter what the level of perceived risk, or ability to mitigate risks through skill is held by the rider. This might include petitioning for widespread horse media adoption of helmet only image publication policy as well as injury prevention campaigns that leverage off the helmet use enabler presented by those riders who saw helmet wearing as an action of responsible horse owners and family/community members.

The attitude "I can control risk" demonstrated that the risks involved in equestrian activities were perceived and understood in varying ways. Positively, helmet use was seen as a tool to mitigate risks, but worryingly other risk mitigation strategies were often valued over and above the use of safety helmets. This finding is consistent with that of Reed and colleagues [28] and Condie and colleagues [29], who identified differing risk perceptions in different equestrian contexts as potential barriers to the use of helmets. Moreover, the relationship between risk perception and prevention is not straightforward. Some riders may (more or less consciously) not want to acknowledge risk by taking precaution.

Further, the construction of horse riding, and horses in general as dangerous and unpredictable animals could unintentionally reduce the energy that riders invest in building a "strong" relationship with their horses [40,51,52]. Challenging this relationship might prove difficult, especially as riders rarely described their own horses as dangerous or unpredictable. Knowing a horse, and being able to read equine behavior was favored strongly as a risk mitigation strategy as part of the attitude that "I can control risk". This was exemplified in the comment by one equestrian that they ride their family's "*very experienced*" horses without helmets, as they "*know damn well they won't do anything*".

In a similar light, our findings demonstrated some of the factors contributing to risk (mis)perception were possibly inflated self-perceptions of the equine knowledge, skills and experience that are preventative factors for equestrian injuries and have positive consequences for equine welfare [53–56]. There is a particular challenge in removing any stigma attached to helmet use that riders may interpret as an indicator of a rider's poor equestrian skills, a horse's bad temperament or their weak horse-rider relationship. However, our research suggests risk-reduction value in injury prevention campaigns that leverage off the attitude "Accidents Happen" and also normalize helmet use by increasing usage amongst equestrian role models. There may also be value in creating an association between helmet use and being a "good carer" for the horses that depend on riders [57], and an equestrian sector that could be jeopardized by increasing injury statistics.

The finding that physical sensation, comfort and habit influenced helmet use behavior is consistent with Condie and colleagues' [29] study that found that the perception of helmets as uncomfortable presented a barrier to use. This was also highlighted in studies of English-style riders [58], farm youth [28] and roughstock rodeo athletes [27]. Yet, while the perception that a helmet is uncomfortable and "It Does Not Feel Right" was similarly observed in this data, some riders expressed an opposing positive attitude of "It Does Not Feel Right ... to not wear my helmet". This suggests that the sensation of helmet wearing can be viewed positively and has been effectively normalized by some riders. Further, even when wearing helmets was seen as uncomfortable, this did not present a barrier for some riders, as other values and attitudes exerted more influence on the decision to wear, Particularly strong was the desiring to be a good role model or wanting to "*be around*" for family. This shows again the advantage of utilizing and promoting valued social roles to encourage use and as a result normalize the helmet wearing sensation. The view of equestrians who noted that wearing a helmet felt normal demonstrates there is potential for supporting the idea that helmets can become "*second nature*" and that not wearing can also become an uncomfortable sensation. Indeed, several horse riders likened not wearing their helmet to being uncomfortably "*naked*". While the physical sensation of helmets has previously been identified as a barrier to use, our analysis suggests that this is not always the case. There is need for current phenomenological and ergonomic research into the physicality of helmet wearing. Promotion of values that override the negative side of the attitude of "It Does Not Feel Right", particularly those focused on the social would help contribute towards consistent helmet use. Supporting and promoting the attitude that "It Does Not Feel Right ... to not wear my helmet" could further work towards normalizing, for a larger number of equestrians, the sensation of helmet wearing.

Interestingly, some forum members displayed attitudes of "Accidents Happen" *and* "I Can Control Risk". Whilst most saw the use of helmets as not negotiable, there were a few for whom the attitude "Accidents Happen" was interpreted fatalistically to justify not wearing a helmet. Overall, the "Accidents Happen" attitude was linked with the use of safety helmets. It also provided protection for riders against the association between helmet use and inexperienced suggested in the attitude of "I Can Control Risk". By maintaining the attitude that "Accidents Happen", riders could present the idea that their experience did not detract from the need (or "choice") to wear a helmet. Supporting and promoting this attitude could help to alter the inconsistent helmet use of those who predominantly believe "I Can Control Risk" but without threatening the positive injury mitigation strategies that this attitude also contains. Harnessing this attitude could encourage holistic equestrian safety habits that understand the importance of injury countermeasures such as the development of skills and confidence in equine interaction but *alongside* the adoption of an Australian safety standard helmet.

In conclusion, our findings suggest that there is unharnessed potential in injury prevention approaches that strengthen the enablers of equestrian helmet use. In this regard our research extends previous studies of voluntary use of safety equipment in activities such as cycling [31,33,34,49,59], motorcycling [35,36] and winter sports [38,39,48]. Previous studies on equestrian helmet use provide a valuable illustration of the range of attitudes, beliefs and factors that result in helmet non-use [27–29,58]. Yet, through accessing naturalistic helmet use discussions between equestrians, we found that many of these factors can also support helmet use. We also highlighted the added sensitivity required in the promotion of equestrian safety equipment, due to the complexity of the horse and human relationship in not only creating [25], but also mitigating, risks [40]. We have illustrated an unexplored and unrealized inroad for equestrian injury prevention interventions by shifting the predominant research focus away from barriers to use, towards working with already existing strengths and values held by "horse people" [60] that support and promote voluntary helmet use [61]. Importantly, we identified an enabling attitude that "Accidents Happen". This attitude contained a holistic approach to injury prevention which valued increasing equestrian skills, experience, confidence and positive horse/human interaction *alongside* the consistent use of safety helmets because, despite best intentions, "Accidents Happen". The findings suggest that values important to equestrians already have the potential to assist mitigate risk. These included the importance of a "strong" human and horse relationship, the influence of local "eque-cultural" communities, the importance of acquiring knowledge of equine behavior as well as the impact of comfort and physical sensation of riding. Injury prevention interventions might benefit from maintaining these important values, and strengthening their influence on the desired behavior of helmet use.

5. Limitations and Further Research

The qualitative depth of this study exempts the findings from generalizability, but offers a unique glimpse into the "everyday talk" [42] of equestrians about helmet use, providing a range of avenues for further research. Whilst not immediately evident in the analytic phase, the data may have been affected by differences in the target audiences of each forum (*i.e.*, junior and adult, modern and traditional riding styles, 'English' and 'Western' styles). Due to the online sampling, the lack of demographics available about the authors of forum comments restricted larger conclusions that could link types of riders, form of riding discipline, level of experience or population subgroups to particular helmet use attitudes. Given that a rider's ability to interpret horse behavior and body language could increase the predictability of horses [62], it could also be beneficial to compare self-reported skills with objectively evaluated skills. Such research should also determine the impact of high correlations on helmet use, as an ability to correctly read and predict horse behavior is more likely to reduce the likelihood of an injury more than its consequence. As with any piece of research, opinions expressed in the forums may have subsequently changed. However, there have been no major environmental or legal changes in Australia since 2004 that are considered to have significantly impacted helmet use or attitudes.

To determine the generalizability of these attitudes arising from a particular group of people using a particular forum, further research could utilize surveys, interviews or focus groups with equestrians. Doing so would provide the added bonus of gathering demographic information of riders, which would add to a more detailed analysis of helmet use attitudes. Further research could also involve quantitative research incorporating validated psychometric surveys of sensation seeking [63], as well as evaluating self-efficacy or the physical and psychological impact of previous injuries on helmet use. Focus groups could be particularly beneficial in relation to developing behavior change initiatives and trialing public health messaging and communications. They would be particularly suited to evaluating interventions leveraging from the attitude that "Accidents Happen" and the social implications for individual helmet use decisions on the horses under riders' care, their families and the broader equestrian community. Similarly, surveys and/or observational research methods could also be useful to extend our understanding of the social influences on helmet use by considering using role models to promote voluntary helmet use. Importantly, our research has shown that shifting the research focus away from helmet use barriers has allowed the identification of already existing helmet use enablers. As a result, behavior change may be acheivable not by challenging long-held and stubborn individual beliefs, but by redirecting them into precautionary behaviors that benefit the whole community.

Acknowledgments

This research was undertaken whilst the first authored was funded by a summer scholarship provided by CQUniversity. This paper has been greatly improved by the insightful comments of two reviewers to whom we are very grateful.

Author Contributions

The study was conceived and managed by Thompson. Data collection and analysis were undertaken by Haigh. Both authors contributed to the presentation and interpretation of findings.

Conflicts of Interest

The authors declare no conflict of interest.

References

1. Finch, C.F.; Owen, N. Injury prevention and the promotion of physical activity: What is the nexus? *J. Sci. Med. Sport* **2001**, *4*, 77–87. [CrossRef]
2. Finch, C.; Cassell, E. The public health impact of injury during sport and active recreation. *J. Sci. Med. Sport* **2006**, *9*, 490–497. [CrossRef] [PubMed]
3. Boufous, S.; Dennis, R.; Finch, C. *A Profile of Hospitalisations and Deaths due to Sport and Leisure Injuries in New South Wales, 2000–2004*; NSW Injury Risk Management Research Centre: Sydney, Australia, 2006.
4. Andrew, N.E.; Gabbe, B.J.; Wolfe, R.; Cameron, P.A. Trends in sport and active recreation injuries resulting in major trauma or death in adults in victoria, australia, 2001–2007. *Injury* **2012**, *43*, 1527–1533. [CrossRef] [PubMed]
5. Hasler, R.M.; Gyssler, L.; Benneker, L.; Martinolli, L.; Schötzau, A.; Zimmermann, H.; Exadaktylos, A.K. Protective and risk factors in amateur equestrians and description of injury patterns: A retrospective data analysis and a case-control survey. *J. Trauma Manag. Outcomes* **2011**, *5*. [CrossRef] [PubMed]
6. Thompson, K.; Adelman, M. Epilogue: A research agenda for putting gender through its paces. In *Gender and Equestrian Sport*; Adelman, M., Knijnik, J., Eds.; Springer: New York, NY, USA, 2013; pp. 195–211.
7. Wolframm, I. *The Science of Equestrian Sports: Theory, Practice and Performance of the Equestrian Rider*; Routledge: London, UK; New York, NY, USA, 2013.
8. Petridou, E.; Kedikoglou, S.; Belechri, M.; Ntouvelis, E.; Dessypris, N.; Trichopoulos, D. The mosaic of equestrian-related injuries in greece. *J. Trauma Acute Care Surg.* **2004**, *56*, 643–647.

9. *Participation in Sport and Physical Activities*; Australian Bureau of Statistics: Canberra, Australia, 2015.
10. Matthews, C.; Thompson, K. Inroads into equestrian safety: Rider-reported factors contributing to horse-related accidents and near misses on australian roads. *Animals* **2015**, *5*, 592–609.
11. Havlik, H.S. Equestrian sport-related injuries: A review of current literature. *Curr. Sports Med. Rep.* **2010**, *9*, 299–302. [CrossRef] [PubMed]
12. Loder, R.T. The demographics of equestrian-related injuries in the united states: Injury patterns, orthopedic specific injuries, and avenues for injury prevention. *J. Trauma Acute Care Surg.* **2008**, *65*, 447–460. [CrossRef] [PubMed]
13. Ball, C.G.; Ball, J.E.; Kirkpatrick, A.W.; Mulloy, R.H. Equestrian injuries: Incidence, injury patterns, and risk factors for 10 years of major traumatic injuries. *Am. J. Surg.* **2007**, *193*, 636–640.
14. Carmichael, S.P.; Davenport, D.L.; Kearney, P.A.; Bernard, A.C. On and off the horse: Mechanisms and patterns of injury in mounted and unmounted equestrians. *Injury* **2014**, *45*, 1479–1483.
15. Hitchens, P.L.; Hill, A.E.; Stover, S.M. Jockey falls, injuries, and fatalities associated with thoroughbred and quarter horse racing in california, 2007–2011. *Orthop. J. Sports Med.* **2013**, *1*. [CrossRef]
16. Clarke, C.N.; Tsuei, B.J.; Butler, K.L. Equine-related injury: A retrospective analysis of outcomes over a 10-year period. *Am. J. Surg.* **2008**, *195*, 702–704. [CrossRef] [PubMed]
17. Mayberry, J.C.; Pearson, T.E.; Wiger, K.J.; Diggs, B.S.; Mullins, R.J. Equestrian injury prevention efforts need more attention to novice riders. *J. Trauma Acute Care Surg.* **2007**, *62*, 735–739.
18. Smartt, P.; Chalmers, D. A new look at horse-related sport and recreational injury in new zealand. *J. Sci. Med. Sport* **2009**, *12*, 376–382. [CrossRef] [PubMed]
19. Papachristos, A.; Edwards, E.; Dowrick, A.; Gosling, C. A description of the severity of equestrian-related injuries (eris) using clinical parameters and patient-reported outcomes. *Injury* **2014**, *45*, 1484–1487. [CrossRef] [PubMed]
20. Lim, J.; Puttaswamy, V.; Gizzi, M.; Christie, L.; Croker, W.; Crowe, P. Pattern of equestrian injuries presenting to a sydney teaching hospital. *ANZ J. Surg.* **2003**, *73*, 567–571. [CrossRef] [PubMed]
21. Abu-Zidan, F.M.; Rao, S. Factors affecting the severity of horse-related injuries. *Injury* **2003**, *34*, 897–900. [CrossRef]
22. Cripps, R.A. *Horse-Related Injury in Australia*; National Injury Surveillance Unit: Canberra, Australia, 2000.
23. Bond, G.R.; Christoph, R.A.; Rodgers, B.M. Pediatric equestrian injuries: Assessing the impact of helmet use. *Pediatrics* **1995**, *95*, 487–489. [PubMed]
24. Chitnavis, J.; Gibbons, C.; Hirigoyen, M.; Parry, J.L.; Simpson, A. Accidents with horses: What has changed in 20 years? *Injury* **1996**, *27*, 103–105. [CrossRef]
25. Safe Work Australia. *Guide to Managing Risks When New and Inexperienced Persons Interact with Horses*; Safe Work Australia: Canberra, Australia, 2014.
26. Finch, C.F.; Watt, G.M. *Locking the Stable Door: Preventing Equestrian Injuries*; Monash University Accident Research Centre: Melbourne, Australia, 1996.
27. Ross, D.S.; Ferguson, A.; Bosha, P.; Cassas, K. Factors that prevent roughstock rodeo athletes from wearing protective equipment. *Curr. Sports Med. Rep.* **2010**, *9*, 342–346. [CrossRef] [PubMed]

28. Reed, D.B.; Novak, S.P.; Heath, R.L. Farm youth and horse-related injuries: A case for safety helmets. *J. Agromed.* **1998**, *5*, 45–57. [CrossRef]
29. Condie, C.; Rivara, F.P.; Bergman, A.B. Strategies of a successful campaign to promote the use of equestrian helmets. *Public Health Rep.* **1993**, *108*, 121–126. [PubMed]
30. Lupton, D. *Risk and Sociocultural Theory: New Directions and Perspectives*; Cambridge University Press: New York, NY, USA, 1999.
31. Curnow, W.J. The efficacy of bicycle helmets against brain injury. *Accid. Anal. Prev.* **2003**, *35*, 287–292. [CrossRef]
32. *Australian Road Rules*; The National Road Transport Commission: Melbourne, Australia, 2012.
33. Thompson, N.J.; Sleet, D.; Sacks, J.J. Increasing the use of bicycle helmets: Lessons from behavioral science. *Patient Educ. Couns.* **2002**, *46*, 191–197. [CrossRef]
34. Villamor, E.; Hammer, S.; Martinez-Olaizola, A. Barriers to bicycle helmet use among dutch paediatricians. *Child Care Health Dev.* **2008**, *34*, 743–747. [CrossRef] [PubMed]
35. Hung, D.V.; Stevenson, M.R.; Ivers, R.Q. Barriers to, and factors associated, with observed motorcycle helmet use in vietnam. *Accid. Anal. Prev.* **2008**, *40*, 1627–1633. [CrossRef] [PubMed]
36. Papadakaki, M.; Tzamalouka, G.; Orsi, C.; Kritikos, A.; Morandi, A.; Gnardellis, C.; Chliaoutakis, J. Barriers and facilitators of helmet use in a greek sample of motorcycle riders: Which evidence? *Trans. Res. Part F Psychol. Behav.* **2013**, *18*, 189–198. [CrossRef]
37. Peterson, A.R.; Brooks, M.A. Pilot study of adolescent attitudes regarding ski or snowboard helmet use. *WMJ Off. Publ. State Med. Soc. Wis.* **2010**, *109*, 28–30.
38. Evans, B.; Gervais, J.T.; Heard, K.; Valley, M.; Lowenstein, S.R. Ski patrollers: Reluctant role models for helmet use. *Int. J. Injury Control Saf. Promot.* **2009**, *16*, 9–14. [CrossRef] [PubMed]
39. Cundy, T.P.; Systermans, B.J.; Cundy, W.J.; Cundy, P.J.; Briggs, N.E.; Robinson, J.B. Helmets for snow sports: Prevalence, trends, predictors and attitudes to use. *J. Trauma Acute Care Surg.* **2010**, *69*, 1486–1490. [CrossRef] [PubMed]
40. Thompson, K.; Nesci, C. Over-riding concerns: Developing safe relations in the high-risk interspecies sport of eventing. *Int. Rev. Sociol. Sport* **2013**. [CrossRef]
41. Thompson, K.; Birke, L. The horse has got to want to help: Human-animal habituses and networks in amateur show jumping. In *Animals, Sport and Society*; Routledge: London, UK, 2013.
42. Callaghan, J.E.; Lazard, L. "Please don't put the whole dang thing out there!": A discursive analysis of internet discussions around infant feeding. *Psychol. Health* **2012**, *27*, 938–955. [CrossRef] [PubMed]
43. Rodriquez, J. Narrating dementia self and community in an online forum. *Qual. Health Res.* **2013**, *23*, 1215–1227. [CrossRef] [PubMed]
44. Flynn, M.A.; Stana, A. Social support in a men's online eating disorder forum. *Int. J. Men's Health* **2012**, *11*, 150–169. [CrossRef]
45. Green, J.; Willis, K.; Hughes, E.; Small, R.; Welch, N.; Gibbs, L.; Daly, J. Generating best evidence from qualitative research: The role of data analysis. *Aust. N. Zeal. J. Public Health* **2007**, *31*, 545–550. [CrossRef] [PubMed]
46. McManus, P.; Albrecht, G.; Graham, R. *The Global Horseracing Industry: Social, Economic, Environmental and Ethical Perspectives*; Routledge: London, UK, 2012.

47. Racing Australia. Australian Rules of Racing. Available online: http://www.australianracingboard.com.au/racing_rules.aspx (accessed on 23 February 2015).
48. Ruedl, G.; Brunner, F.; Kopp, M.; Burtscher, M. Impact of a ski helmet mandatory on helmet use on austrian ski slopes. *J. Trauma Acute Care Surg.* **2011**, *71*, 1085–1087. [CrossRef] [PubMed]
49. Dannenberg, A.L.; Gielen, A.C.; Beilenson, P.L.; Wilson, M.H.; Joffe, A. Bicycle helmet laws and educational campaigns: An evaluation of strategies to increase children's helmet use. *Am. J. Public Health* **1993**, *83*, 667–674. [CrossRef] [PubMed]
50. Gielen, A.C.; Joffe, A.; Dannenberg, A.L.; Wilson, M.E.; Beilenson, P.L.; DeBoer, M. Psychosocial factors associated with the use of bicycle helmets among children in counties with and without helmet use laws. *J. Pediatr.* **1994**, *124*, 204–210. [CrossRef]
51. Thompson, K. *Theorising Rider-Horse Relations: An Ethnographic Illustration of the Centaur Metaphor in the Spanish Bullfight*; Brill: Leiden, The Netherlands, 2011.
52. Schuurman, N. Blogging situated emotions in human-horse relationships. *Emot. Space Soc.* **2014**, *13*, 1–8. [CrossRef]
53. Hawson, L.A.; McLean, A.N.; McGreevy, P.D. The roles of equine ethology and applied learning theory in horse-related human injuries. *J. Vet. Behav. Clin. Appl. Res.* **2010**, *5*, 324–338.
54. Hausberger, M.; Roche, H.; Henry, S.; Visser, E.K. A review of the human-horse relationship. *Appl. Anim. Behav. Sci.* **2008**, *109*, 1–24. [CrossRef]
55. McGreevy, P.D. The advent of equitation science. *Vet. J.* **2007**, *174*, 492–500. [CrossRef] [PubMed]
56. Ladewig, J. Human safety and horse welfare-two sides of the same coin. *J. Vet. Behav. Clin. Appl. Res.* **2011**, *6*, 292–293.
57. Thompson, K. For pets' sake, save yourself: Motivating emergency and disaster preparedness through relations of animal guardianship. *Aust. J. Emerg. Manag.* **2015**, *30*, 43–46.
58. Nelson, D.E.; Rivara, F.P.; Condie, C. Helmets and horseback riders. *Am. J. Prev. Med.* **1994**, *10*, 15–19. [PubMed]
59. Ritter, N.; Vance, C. The determinants of bicycle helmet use: Evidence from germany. *Accid. Anal. Prev.* **2011**, *43*, 95–100. [CrossRef] [PubMed]
60. Cassidy, R.L. *Horse People: Thoroughbred Culture in Lexington and Newmarket*; JHU Press: Baltimore, MD, USA, 2010.
61. Krumeich, A.; Weijts, W.; Reddy, P.; Meijer-Weitz, A. The benefits of anthropological approaches for health promotion research and practice. *Health Educ. Res.* **2001**, *16*, 121–130. [CrossRef] [PubMed]
62. Thompson, K.; McGreevy, P.; McManus, P. A critical review of horse-related risk: A research agenda for safer mounts, riders and equestrian cultures. *Animals* **2015**, *5*, 561–575.
63. Zuckerman, M. The sensation seeking scale v (SSS-V): Still reliable and valid. *Pers. Individ. Differ.* **2007**, *43*, 1303–1305. [CrossRef]

Animals **2015**, *5*, 561-575; doi:10.3390/ani5030372

OPEN ACCESS

animals

ISSN 2076-2615

www.mdpi.com/journal/animals

Article

A Critical Review of Horse-Related Risk: A Research Agenda for Safer Mounts, Riders and Equestrian Cultures

Kirrilly Thompson [1,†,*], Paul McGreevy [2,†] and Phil McManus [3,†]

[1] The Appleton Institute, Central Queensland University, 44 Greenhill Road, Wayville, SA 5034, Australia

[2] Faculty of Veterinary Science, University of Sydney, Sydney, NSW 2006, Australia; E-Mail: paul.mcgreevy@sydney.edu.au

[3] School of Geosciences, University of Sydney, Sydney, NSW 2006, Australia; E-Mail: phil.mcmanus@sydney.edu.au

† These authors contributed equally to this work.

* Author to whom correspondence should be addressed; E-Mail: kirrilly.thompson@cqu.edu.au; Tel.: +61-08-8378-4512 (ext. 4512).

Academic Editor: Marina von Keyserlingk

Received: 24 June 2015 / Accepted: 7 July 2015 / Published: 17 July 2015

Abstract: While the importance of improving horse-related safety seems self-evident, no comprehensive study into understanding or reducing horse-related risk has been undertaken. In this paper, we discuss four dimensions of horse-related risk: the risk itself, the horse, the rider and the culture in which equestrian activities takes place. We identify how the ways in which risk is constructed in each dimension affects the applicability of four basic risk management options of avoidance, transference, mitigation and acceptance. We find the acceptance and avoidance of horse-related risk is generally high, most likely due to a common construction of horses as irrevocably unpredictable, fearful and dangerous. The transference of risk management is also high, especially in the use of protective technologies such as helmets. Of concern, the strategy least utilised is risk mitigation. We highlight the potential benefit in developing mitigation strategies directed at: (a) improving the predictability of horses (to and by humans), and (b) improving riders' competence in the physical skills that make them more resilient to injury and falls. We conclude with the presentation of a multidisciplinary agenda for research that could reduce accident, injury and death to horse-riders around the world.

Keywords: horse-riding; risk; mitigation; culture; research; review; safety; behavior change; eque-culture; motivator

1. Background

Horse-riding is a hazardous activity. Each year, horse riders are injured, hospitalized or killed as a result of horse-related accidents and injuries. Despite technological advancements in equestrian safety equipment [1], horse riding continues to be found more dangerous than motorcycling, skiing, football, and rugby [2,3]. Whilst injury can occur simply from handling horses [4–7], falling from a horse constitutes a dangerous fall from height, possibly at speed. A rider's head can be elevated up to 3 m from the ground and horses can travel at speeds around 50 km/h. one study of children under 15 years of age found that a 'mean Modified Injury Severity Scale score of injured riders was exceeded only by that of pedestrians struck by a car' [8] (p. 487).

Improving safety for the millions of horse-riders around the globe is significant for moral, economic, socio-cultural, and public health reasons. While the importance of improving horse-related safety seems self-evident, no rigorous study into understanding or reducing horse-related risk has been undertaken internationally in the academic literature. This may be due to a historical legacy of horse-riding being a pre-modern sport with a robust culture that accepts the dangers of riding. It may also be due to the difficulty of analyzing and mitigating the risks inherent in, and generated by, a complex socio-technical network and interspecies interaction that is historically, socially and culturally constructed. Nonetheless, complexity must be addressed to enhance the safety of millions of equestrians around the globe.

In this paper, we discuss four inter-related dimensions of horse-related risk: the risk itself, the horse, the rider and the cultural context through which equestrian activities become meaningful. For each dimension, we present an overview of current knowledge and offer a list of important questions that remain. Moreover, we identify the repercussions of how risk has been constructed for each dimension. Risk management strategies are commonly presented as four options: avoidance, transference (of responsibility to a third party), mitigation and acceptance [9]. As argued throughout this paper, the acceptance and avoidance of horse-related risk is high. Horses are frequently constructed as irrevocably unpredictable and dangerous. As a result, most horse-related safety information concentrates on avoiding injury by keeping out of horse-related danger. Risk transference is also high, in relation to a reliance on insurance companies, professional trainers to manage difficult horses or the use of protective technologies such as helmets, body protectors or even sedative substances. Lowest of all is mitigation, particularly mitigation directed at: (a) improving the predictability of horses (through making horses more predictable and making riders more able to predict their behavior); and (b) improving riders' competence in the physical skills with the explicit intention of making them more resilient to injury and falls.

To address the questions raised throughout the paper, we conclude with a list of recommended methods for future data collection. Together, these questions and methods comprise a multidisciplinary agenda for research that could reduce accident, injury and death to millions of horse-riders around the world.

2. Current Knowledge and Critical Questions

2.1. The Risk

As noted above, horses are dangerous and interacting with them is more or less risky. Horses are often pushed to their physical and physiological limits in equestrian pursuits [10], resulting in risk to both horses and riders. Research on horse-related risk is biased towards cross-sectional epidemiological studies of injury type and severity [7], or the efficacy of technical interventions such as frangible pins on jumping obstacles, helmets and back protectors [8,11–13]. With very little exception [13–15], a focus on the causes and consequences of horse-related injury rather than prevention, could lead to researchers being accused of 'shutting the gate after the horse has bolted'.

Whilst risk analysis is an important aspect of risk research, especially for ranking risk, prioritizing intervention and evaluating campaigns, risk-reduction requires changing the attitudes and behaviours of participants in risky activities. Sandman's model of risk communication is particularly useful in this regard [16]. According to Sandman, 'risk' is most usefully understood not as the standard arithmetical product of likelihood and consequence of an incident occurring (ISO 31000), but as the cultural sum of 'hazard + outrage'. Sandman discusses combinations of high/low risk/outrage on a 2X2 matrix and their implications for crisis management. For example, the risk of autism being induced by childhood immunization is a 'low risk/high outrage' concern. The cultural response of high outrage outweighs and overshadows the objective risk but has had a significant impact on rates of failure to immunize.

Horse-related risk can be considered 'high risk/low outrage', suggesting a level of wide-spread recognition of risk but a high degree of acceptance, complacency or inaction. The recommended strategy for encouraging behaviours that reduce horse-related risk would then be to reduce the risks (see sections on the rider and the horse below), whilst also increasing outrage (see 'culture', below). In particular, outrage over horse-related injury needs to be increased amongst riders as well as the governments that have the ability to financially support or legislate for change. As increasing outrage when there is no available risk mitigation can lead to paralyzing fear and perpetuate complacency, there is a need to fortify existing technical risk controls with behavioural, physical and cultural controls.

In relation to the risk dimension of horse-related risk, we therefore need to ask: How can horse-related injuries be prevented?

(a) What contributes to complacency around horse-related risk?
(b) How can outrage over horse-related risk be usefully increased?
(c) What behavioural, physical and cultural controls can supplement existing technical controls?

2.2. The Horse

The risk of horse-related accidents and injury are well known and widely accepted. Risk is frequently attributed to the 'nature' of horses as irrevocably unpredictable, fearful and flight-wired [17]. As noted by Lawrence, even the most highly trained horses can be treated with suspicion, lest they go 'feral' and return to a state of unruly wildness [18]. This essentialist view of horses as more or less 'unchangeable' could explain why horse-related risk intervention is largely technical, and risk management is avoidance-based (as discussed in more detail below).

In equestrianism, there has been a strong focus on using personal protective equipment (e.g., helmet, boots, gloves and, more recently, body protectors) and maintaining horse equipment to safe standards (e.g., checking stitching on saddlery). Whilst technical intervention has an undeniable role in reducing the likelihood and consequence of an adverse equestrian event, the voluntary use of basic protective equipment such as helmets use is low to inconsistent [11,12,15]. Technical intervention should therefore not overshadow attention to horse behavior, or rider skill (discussed in more detail below).

Overall, there seems to have been more discussion about why horses are dangerous (e.g., unpredictable, large and flighty), rather than if, how and to what extent those risk factors can be mitigated or even controlled. In fact, research suggests that the unpredictability of horses can be reduced through behavioural interventions and approaches [19]. This is particularly encouraging, given that a taken-for-granted assumption that horses are irrevocably unpredictable may contribute to a lack of outrage about current levels of horse-related accident, injury and death. Even if horses are irrevocably unpredictable based on the innate fears or a prey animal, there has been no consideration of the extent to which their own need to feel safe could be reconfigured to reduce risk to humans. As noted by McGreevy and colleagues, 'the value of safety to animals is often overlooked by trainers and handlers' [20]. This is a firm reminder that feelings of risk, fear and safety are at least as important to horses as well as people. As these equine desires may contribute to risks where their importance goes unrecognized, the corollary is that they could be used to reduce risk where their importance is accommodated and addressed.

The unpredictability of horses suggests that risks may be higher amongst people with low levels of knowledge of horse ethology and behaviour, since this can help to anticipate natural but undesirable equine behaviours and responses. Evidence of breed differences in personality [21] and vision [22] has been revealed, but there is little guidance on how to use this information to reduce the risks associated with handling and riding certain horses. One exception is the recent 'Guide to managing risks when new and inexperienced persons interact with horses' [23], that singles out Thoroughbred horses as unsuitable for beginners. This seminal guide highlights the potential risk reduction benefits of applying WHS principles to equestrian cultures. In workplaces, workers are given guidance and decision-making support tools to safely operate dangerous plant, machinery or equipment through the identification and control of hazards. However, professional, amateur or leisure riders are rarely given decision-making tools for riding a given horse on a given occasion (*i.e.*, the equivalent of a pre-flight safety check). Although evidence of its adoption is lacking, the following seven point pre-ride equestrian checklist has been suggested by Guyton *et al.* [24]:

(1) Am I wearing adequate protective gear?
(2) Is the tack durable and well-fit to the horse?
(3) Are the environmental conditions (weather, ground footing) safe for riding?
(4) Does the arena contain unfamiliar objects, animals, or people that may alarm the horse?
(5) Is the horse healthy and prepared for riding? Is the mood or behavior of the horse uncharacteristic or concerning?
(6) Am I healthy and prepared for riding? Do I have an emotional or physical condition that may impair my ability to safely ride this horse?
(7) Do the horse and I have a healthy relationship? Do I have concerns about my ability to assert myself with this horse? [24]

The list is entirely relevant, but can be significantly improved. The first two checks relate to technology and are under direct control of the human. Checks Three and Four relate to the environment, over which riders have less control, especially when riding in open spaces. Check Six relates to the rider but 'emotional or physical' condition is rather broad, especially around the degree to which safety is jeopardized. Checks Five and Seven relate to the horse, but require familiarity with individual horses or a perceptive eye informed by knowledge of equine behavior. The idea of a 'healthy relationship' and being 'assertive' are both open to interpretation and relative to different equestrian disciplines. Moreover, being able to interpret and predict equine behaviour may be a more useful risk-management tool than being assertive. Check Six relates to the rider, but 'emotional or physical' condition is rather broad, especially around the degree to which safety is jeopardized by particular conditions. Furthermore, Check Six does not address specific physical skills essential for riding, such as those discussed in more detail below.

Overall, whilst the checklist comprises questions that are undoubtedly important to ask, most if not all rely on a level of knowledge and expertise in their interpretation and assessment. There is particular scope to incorporate objective checks of safety, quality of the human-horse relationship and resilience to external or environmental stimuli. These can be undertaken first from the ground and then under saddle in a confined area through testing basic responses such as deceleration, acceleration (sometimes necessary to avoid hazards such as vehicles or other horses), reversing and turning.

Moreover, riders are not routinely given explicit advice on what to do if they assess a horse as unsafe, *i.e.*, additional controls, alternative (less dangerous) activities, strategies to improve the safety of a mount (*i.e.*, install robust deceleration responses and habituate the horses to common hazards). Whilst riders may be given instruction that has the effect of improving safety, it may not be understood or framed explicitly as a risk-management strategy. Rather, it may be seen as taken for granted tradition or simply 'just the way things are done', as has been found in relation to risk management by volunteer firefighters [25]. Riders too may already be engaging in risk-reduction practices that are tacit. The explicit identification of practices that improve safety as risk-reduction strategies may encourage riders not only to value, improve and maintain high standards of those practices and the horse's response to them, but to communicate those practices to other riders to create safer equestrian cultures.

It is tempting to construct the horse as an independent source of risk. Certainly, without a horse there is no risk. However, from a safety systems perspective, there is a need to recognize the ways in which risk is generated in, by and through socio-technical networks. Despite the importance of matching riders with appropriate horses being acknowledged [24] and largely carried out on a basis of experienced/inexperienced horse/rider [26], there is no widely accepted or validated means of assessing experience for the purpose of determining safe horse-rider combinations. As environmental conditions and stressors can impact the behavior of horses and riders, assessment needs to be undertaken in general as well as on specific occasions, such as during competition, on return to work after an injury or in unfamiliar surroundings.

In relation to the equine dimension of horse-related risk, we therefore need to ask:

(a) How well do riders understand horse behavior?
(b) How can increasing levels of knowledge about horse behavior improve safety?
(c) Which horses are most likely to be unpredictable for riders (e.g., age, breed, level of experience, level of horse education, early preparation and history of unwelcome behaviour)

(d) What physical and behavioural conditioning increases a horse's predictability or make a horse reasonably safe to ride?
(e) How might a horse's preference for 'safety' be used to reduce risk?
(f) What benefits may be taken from OHS approaches to safety?
(g) How can a rider assess, train, maintain and improve a horse's level of risk?
    i From the ground
    ii From the saddle
(h) What tacit practices do riders currently engage in without being aware of their risk-reduction benefits, which could be emphasized?
(i) What tools could enable safer matching of horse and rider?

2.3. The Rider

Most—if not all—riders are aware of the risk of equestrian sports and mindful of their own safety [27]. Pony club manuals are replete with references to safe practices around horses [28–30] and the pony club movement itself has a 'safety first' attitude. However, from a risk management perspective, the most common strategies for improving safety around horses can be characterized as avoidance strategies—keeping out of harm's way. Aside from the safe use of safe equipment, being 'horse safe' usually refers to how to avoid being kicked, bitten, trampled or crushed. Once again, the reliance on avoidance rather than control could be a repercussion of the construction of horses as irrevocably unpredictable. As discussed above, there is very little overt discussion about how to manage or reduce 'equinogenic' risk.

Similarly, there is little overt discussion of the physical and postural skills that can increase a rider's resilience to injury or falls. Riders communicate to horses through 'natural aids'; their hands (rein tension increases for 'stop' and 'turn'), legs (leg pressure increases for 'go'), seat (classically conditioned weight and balance cues for 'stop' or 'go'), voice and various 'artificial aids' such as whips and spurs. The self-awareness, timing and accuracy of the application of aids have been related to rider safety, especially (1) rein pressure [tension and release], (2) leg stability and (3) balance and trunk stability [31]. When these stimuli are applied incorrectly, inadvertently or simultaneously, they can cause miscommunication, confusion, discomfort or pain for the horse. The horse may respond with evasive behaviour, resistance or flight behaviour and anti-predator responses such as rearing, bucking or bolting, potentially leading to rider injury or even death. Riders' cues must be applied with awareness, accuracy, timing and consistency to: (a) avoid or reduce the likelihood of a horse-related incident occurring; and (b) provide the rider with the ability to safely respond to an untoward incident, thereby reducing the significance of its consequences [32].

The mensuration of cues from riders' hands and legs has commenced as part of equitation science [33]. There may also be benefits in measuring levels of cognitive and somatic anxiety in riders, as anxiety can impact horse-rider communication by affecting fine motor control and decision-making [34,35]. High self-efficacy may reduce anxiety [36], although little is known about riders' subjective confidence in their proficiency in the application of those cues, based on self-report. From a psychological perspective, high self-efficacy may mislead riders into over-estimating their safety and/or taking increased risks [37,38].

Furthermore, there is a need to know if and how riders associate their physical skills with safety, especially in some of the more artistic equestrian disciplines such as dressage where a rider's skills have significant aesthetic value. There are several well established approaches to improving riding position such as 'centred riding' [39] and 'riding with your mind' [40] that focus on riders developing better feel, relating better with the horse, and becoming more effective as well as feeling and looking more aesthetically pleasing. The simultaneous effect of these approaches on improving safety is largely taken for granted and remains to be determined, as does the usefulness of 'improved safety' as a means to motivate riders to adopt such approaches.

In relation to the rider dimension of horse-related risk, we therefore need to ask:

(a) How riders acquire and objectively perform skills related to horse safety?
(b) How riders subjectively self-evaluate the skills essential to horse safety?
(c) If and how riders and trainers associate physical ability with safety?
(d) What is the impact of existing rider position programmes on safety?

2.4. The Culture

As acknowledged by Sandman's hazard/outrage model descried above, subjective perceptions of risk do not always align with objective risk calculations [41], and not all strategies that reduce fear actually reduce risk. Perceptions of risk can have a significant impact on the uptake of safety behaviours and protective equipment. 'Safety culture' is a contested term that has been used in studies of organizational culture to describe the ways in which workplaces do (or do not) value safety [42,43] or perform on measures thought to indicate an organisation's 'safety climate' [44]. The concept can be used to consider the ways in which various equestrian cultures or 'eque-cultures' [45] impact on the safety of equestrians, through cultural variation in risk perceptions, attitudes, beliefs and behaviours. In relation eque-culture in general, safety is often poorly evaluated especially at the point of sale [14,46]. This is despite a growing appreciation of the horse's need to feel safe [20] and the range of responses a horse offers when threatened [47]. In relation to racing eque-cultures, the common rehoming of racehorses to novice riders is one example of a contemporary eque-cultural practice that presents risks to horses and humans. Whilst the aim is often to reduce the risk of horses being unnecessarily euthanized, this practice may increase the risk of a horse-related risk or injury to a human, and the potential mistreatment of a horse that is subsequently labeled 'dangerous'.

From a more socio-historical standpoint, the romantic cultural construction of horse riding as an art can also conflict with practical considerations of risk. Since the Classical Greek period and especially in the renaissance in Europe, equitation has been constructed as an art [48–50]. As with any artistic endeavor, there is reverence for those who seem blessed with a natural affinity or 'feel' that eludes explanation. In equestrian culture, there has always been an acceptance that some equestrians have a natural ability to communicate seemingly telepathically with horses such that the two become one [51,52]. These riders are said to have natural 'feel' [26]. However, equitation science is helping to demystify many of the qualities that distinguish such individuals [32,53–55]. For example, communication from rider to horse always relies upon at least some form of pressure cues, however subtle [55]. These cues appear imperceptible in the best cases having been reduced from larger operant cues [19] by a process of classical conditioning [56]. Different eque-cultures have different tolerances for the 'volume' of cues to the horse and their visibility to bystanders.

Whilst equitation science has the potential to significantly reduce horse-related injury and death, it would be disingenuous to present it as a panacea, at least uncritically. In any given population, there will be varying levels of scientific literacy and diverse attitudes towards 'scientific' modes of thought, ranging from supporters to skeptics. For some riders, the elusive experience of harmony is precisely what attracts them to horse-riding [51]. For others, instruction from elite competitors is highly valued [26,27]. Some resist a scientific framework of horse training, preferring leadership narratives that place them above their horses in a perceived hierarchy [57]. Moreover, research has identified that the desire to achieve a 'good' human-horse relationship (or at least to 'perform' the achievement of a 'good' relationship) can be a barrier to precautionary behaviour when such riders consider their 'good' relationship as evidence of reduced risk and therefore a reduced need to wear protective equipment such as helmets [15].

This heterogeneity of styles of engagement with equestrianism, or 'equestrian dispositions', suggests that safety information needs to be communicated in forms that resonate with riders of a multitude of dispositions including scientific and artistic approaches to equitation. Understanding the cultural specificity and generalizability of eque-cultures and equestrianism more broadly will be crucial to effectively applying or adapting longstanding behavior change theories and models from health psychology [58–61]; as well as theories of decision-making and threat and error management from human factors and safety science [62]. Whilst different equestrian dispositions have been operationalized from researcher experience [57] and through factor analysis [63], there remains a need to evaluate their resonance with equestrians themselves. There is a particular need to understand what motivates the behaviour of different equestrians to identify useful 'irrelevant motivators' [64] that could encourage the adoption of protective behaviours in the short term and cultivate safety values in the medium to long term. For example, signaling being fashionable or professional may be a powerful motivator for some riders to wear helmets [65]. Others might be more motivated to wear helmets if they become a symbol of the 'toughness' of their discipline, especially in traditional cowboy equestrian disciplines associated with masculinity, bravery and resilience [66]. Regardless of the motivation, many riders will be influenced by their peers, and could be engaged to replicate desired precautionary behaviours through 'participant modeling' [36] by their role models.

In relation to the cultural dimension of horse-related risk, we therefore need to ask:

(a) How do riders perceive equestrianism in general and equitation science in particular?
(b) What are the different styles of engagement with equestrianism?
(c) What are the most powerful motivators for equestrians?
(d) How do riders consume safety-related information?
(e) How are the causes of horse-related injuries and deaths understood by riders?
(f) How is the use of horse safety equipment represented in the horse community?
(g) What are the socio-cultural dimensions and determinants of risk and safety?
(h) What are the socio-cultural barriers and enablers to improve rider safety?
(i) What are the elements of effective behavior change campaigns and programmes that increase rider safety?

3. Discussion

In this paper we critically reviewed four dimensions important to horse-related risks to horse riders. Whilst they were presented discretely, they are mutually inclusive within a complex socio-technical network and interspecies interaction that is historically, socially and culturally constructed. In relation to the dimensions of risk, horses, riders and culture, we found that research on risk explains the intricacies of what can go wrong, how often and what the consequences are. Research on horses is also problem focused, arguing why horses are a source of risk. These biases favour risk-management options of acceptance and avoidance. To take full advantage of risk-management strategies of mitigation, further research is required on the extent to and ways in which the behavior of horses can be made more predictable, and riders can be made more capable of predicting their behaviour. There is also a need for research evaluating rider proficiency at performing and self-assessing the physical skills that increase resilience to horse-related injury whilst riding. Finally, we considered the unavoidable and omnipotent cultural context that affects riders' behaviours, values, attitudes and beliefs regarding risk and safety. Inconsistent levels of voluntary helmet use suggest that increasing the safety of eque-cultures and equestrianism in general will require external legislation and internal transformation. Overall, we found enormous potential for reducing horse-related risk through the risk management strategy of mitigation.

For each element of horse-related risk, we presented a list of research questions. Together, they comprise a multidisciplinary agenda for further research that could significantly reduce accident, injury and death to millions of horse-riders around the world. As these questions are multi-disciplinary, so too do they require data from various sources, including but not limited to:

- Surveys and questionnaires—to obtain a wider perspective on a range of issues, and to generate quantitative results for policy development and advocacy around horse and rider safety [67,68].
- Interviews and focus groups—to explore controversial subjects that pertain to risk and safety of both the horse and rider [45,69].
- Ethnographic research—to study the actual practices of equestrians; and identify risk management strategies consistent with the motivations, beliefs and values of eque-cultures [26,70–73].
- Media analysis—to identify how particular incidents and risks are reported (or not reported) and relationships to the values, beliefs and practices of equestrians [15,67,74,75].
- Physiometry—to measure rider position and identify physical attributes positively correlated with safety or resistance to being unseated [33,76–78].
- Psychometric research into fear [79], risk-taking propensity [80,81] and sensation seeking [82] amongst riders and equestrian discipline—to identify target groups and tailor behavior change interventions [38,83].
- Analysis of accident and injury reports (*i.e.*, from inquests, insurance records and hospital admission data)—to enable triangulation of objective and self-report data, especially around risk [6,84,85].
- Inferential modeling - to determine predictors of risk and safety, animal attachment and target group archetyping [86,87].

To ensure that data are translated into effective safety intervention tools that can reduce numbers of horse-related injury and death, researchers should focus on developing initiatives that (a) increase outrage about preventable horse-related injury and death, and (b) reduce horse-related risk, such as:

- Ethical techniques and behavioral interventions to increase the predictability of horses
- Interventions to improve riders' ability to predict horse behaviour
- Horse safety assessment and decision-making support tools
- Rider safety skills assessment tools
- Validated measure of horse training/riding style
- Behavior change for safe equestrian cultures

4. Conclusion: Horse/Human-Related Risk/Safety

As demonstrated throughout this paper, horse-related risk is generated through a complex socio-technical network of risk, horses, humans and culture. Whilst these dimensions have been recognized and in some cases researched, horse-related risk has typically been constructed anthropocentrically; it originates in horses and it impacts humans. Moreover, safety is largely seen as a concern for humans only, despite a desire for safety being a powerful driver of 'unpredictable' behavior in horses (and an excellent reward for behavioural interventions). To overcome these biases, there is apparent advantage in the widespread adoption of a more anthrozoological approach to horse-related risk that includes human-related risk as well as human/horse-related safety. With human-horse safety as the ultimate goal, this paper has identified unrealised opportunity to mitigate horse-related risk with behavioural, physical and socio-cultural interventions that could make horses safer mounts, humans safer riders, and equestrianism a safer culture.

Acknowledgments

The authors would like to thank Holly Bowen for her assistance in the preparation and proofreading of this manuscript.

Author Contributions

Thompson conceived this article. All authors contributed to the writing and approval of the article's content.

Conflicts of Interest

The authors declare no conflicts of interest.

References

1. Foote, C.; Gibson, T.; McGuaran, P. *Health and Safety in Australian Racing-Evaluation of Safety Vests*; Rural Industries Research and Development Corporation: Canberra, Australia, 2014.
2. Ball, C.G.; Ball, J.E.; Kirkpatrick, A.W.; Mulloy, R.H. Equestrian injuries: Incidence, injury patterns, and risk factors for 10 years of major traumatic injuries. *Amer. J. Surg.* **2007**, *193*, 636–640. [CrossRef] [PubMed]

3. Paix, B.R. Rider injury rates and emergency medical services at equestrian events. *Brit. J. Sport. Med.* **1999**, *33*, 46–48. [CrossRef]
4. Nunnink, L.; Abu-Zidan, F. Accidental carotid artery injury caused by a horse rope. *Brit. J. Sports Med.* **2003**, *37*, 460–461. [CrossRef]
5. Exadaktylos, A.K.; Eggli, S.; Inden, P.; Zimmermann, H. Hoof kick injuries in unmounted equestrians. Improving accident analysis and prevention by introducing an accident and emergency based relational database. *Emerg. Med. J.* **2002**, *19*, 573–575. [CrossRef] [PubMed]
6. Thomas, K.E.; Annest, J.L.; Gilchrist, J.; Bixby-Hammett, D.M. Non-fatal horse related injuries treated in emergency departments in the united states, 2001–2003. *Brit. J. Sports Med.* **2006**, *40*, 619–626. [CrossRef] [PubMed]
7. Carmichael, S.P., II; Davenport, D.L.; Kearney, P.A.; Bernard, A.C. On and off the horse: Mechanisms and patterns of injury in mounted and unmounted equestrians. *Injury* **2014**, *45*, 1479–1483.
8. Bond, G.R.; Christoph, R.A.; Rodgers, B.M. Pediatric equestrian injuries: Assessing the impact of helmet use. *Pediatrics* **1995**, *95*, 487–489. [PubMed]
9. Hillson, D. Extending the risk process to manage opportunities. *Int. J. Project Manag.* **2002**, *20*, 235–240. [CrossRef]
10. McLean, A.N.; McGreevy, P.D. Ethical equitation: Capping the price horses pay for human glory. *J. Vet. Behav. Clin. Appl. Res.* **2010**, *5*, 203–209. [CrossRef]
11. Nelson, D.E.; Rivara, F.P.; Condie, C. Helmets and horseback riders. *Am. J. Prev. Med.* **1994**, *10*, 15–19. [PubMed]
12. Finch, C.; Watt, G. *Locking the Stable Door: Preventing Equestrian Injuries*; Monash University Accident Research Centre: Clayton, Australia, 1996.
13. Condie, C.; Rivara, F.P.; Bergman, A.B. Strategies of a successful campaign to promote the use of equestrian helmets. *Public Health Rep.* **1993**, *108*, 121–126. [PubMed]
14. Hawson, L.A.; Oddie, C.; McLean, A.N.; McGreevy, P.D. Is safety valued in the australian pony market? *J. Vet. Behav. Clin. Appl. Res.* **2010**, *6*, 254–260. [CrossRef]
15. Haigh, L.; Thompson, K. Helmet use amongst equestrians: Harnessing social and attitudinal factors revealed in online forums. *Animals* **2015**, *5*, 576–591.
16. Sandman, P.M. *Responding to Community Outrage: Strategies for Effective Risk Communication*; The American Industrial Hygiene Association: Fairfax, VA, USA, 1993.
17. LeGuin, E. Man and horse in harmony. In *The Culture of the Horse: Status, Discipline, and Identity in the Early Modern World*; Raber, K., Tucker, T.J., Eds.; Palgrave Macmillan: New York, NY, USA, 2005; pp. 175–196.
18. Lawrence, E.A. *Rodeo: An anthropologist looks at the wild and the tame*; University of Chicago Press: Chicago, IL, USA, 1982.
19. Hawson, L.A.; McLean, A.N.; McGreevy, P.D. The roles of equine ethology and applied learning theory in horse-related human injuries. *J. Vet. Behav. Clin. Appl. Res.* **2010**, *5*, 324–338. [CrossRef]

20. McGreevy, P.D.; Henshall, C.; Starling, M.J.; McLean, A.N.; Boakes, R.A. The importance of safety signals in animal handling and training. *J. Vet. Behav. Clin. Appl. Res.* **2014**, *9*, 382–387. [CrossRef]
21. Lloyd, A.S.; Martin, J.E.; Bornett-Gauci, H.L.I.; Wilkinson, R.G. Horse personality: Variation between breeds. *Appl. Anim. Behav. Sci.* **2008**, *112*, 369–383. [CrossRef]
22. Evans, K.; McGreevy, P. The distribution of ganglion cells in the equine retina and its relationship to skull morphology. *Anat. Histol. Embryol.* **2007**, *36*, 151–156. [CrossRef] [PubMed]
23. Safe Work Australia. *Guide to Managing Risks When New and Inexperienced Persons Interact with Horses*; Safe Work Australia: Canberra, Australia, 2014.
24. Guyton, K.; Houchen-Wise, E.; Peck, E.; Mayberry, J. Equestrian injury is costly, disabling, and frequently preventable: The imperative for improved safety awareness. *Amer. Surg.* **2013**, *79*, 76–83.
25. Dawson, D.; Mayger, K.; Thomas, M.J.; Thompson, K. Fatigue risk management by volunteer fire-fighters: Use of informal strategies to augment formal policy. *Accid. Anal. Prevent.* **2015**. in press.
26. Thompson, K.; Birke, L. The horse has got to want to help: Human-animal habituses and networks in amateur show jumping. In *Sport, Animals, and Society*; Gillett, J., Gilbert, M., Eds.; Routledge: New York, NY, USA, 2014; pp. 69–84.
27. Thompson, K.; Nesci, C. Over-riding concerns: Developing safe relations in the high-risk interspecies sport of eventing. *Int. Rev. Sociol. Sport 2013,* **2013**. [CrossRef]
28. Knox-Thompson, E.; Dickens, S. *Pony Club Manual No. 1: D and C Certificates*; Ray Richards Publisher: Auckland, New Zealand, 1994.
29. Knox-Thompson, E.; Dickens, S. *Pony Club Manual No. 2: C+, B, A And H Certificates*; Ray Richards Publisher: Auckland, New Zealand, 1994.
30. Knox-Thompson, E.; Dickens, S. *Pony Club Manual No. 3: For Instructors*; Ray Richards Publisher: Auckland, New Zealand, 1993.
31. Wolframm, I. *The Science of Equestrian Sports: Theory, Practice and Performance of the Equestrian Rider*; Routledge: London, UK; New York, NY, USA, 2013.
32. McGreevy, P.D.; Murphy, J. Equitation science offers new horizons in the understanding of equine performance and horse-human relationships. *Vet. J.* **2009**, *181*, 1–5. [CrossRef] [PubMed]
33. Pierard, M.; Hall, C.; von Borstel, U.K.; Averis, A.; Hawson, L.; McLean, A.; Nevison, C.; Visser, K.; McGreevy, P. Evolving protocols for research in equitation science. *J. Vet. Behav. Clin. Appl. Res.* **2015**, *10*, 255–266. [CrossRef]
34. Miu, A.C.; Heilman, R.M.; Houser, D. Anxiety impairs decision-making: Psychophysiological evidence from an iowa gambling task. *Biol. Psychol.* **2008**, *77*, 353–358. [CrossRef] [PubMed]
35. Paus, T. Primate anterior cingulate cortex: Where motor control, drive and cognition interface. *Nat. Rev. Neurosci.* **2001**, *2*, 417–424. [CrossRef] [PubMed]
36. Bandura, A. Self-efficacy: Toward a unifying theory of behavioral change. *Psychol. Rev.* **1977**, *84*, 191. [CrossRef] [PubMed]
37. Rimal, R.N.; Real, K. Perceived risk and efficacy beliefs as motivators of change. *Hum. Commun. Res.* **2003**, *29*, 370–399. [CrossRef]

38. Beauchamp, M.R.; Whinton, L.C. Self-efficacy and other-efficacy in dyadic performance: Riding as one in equestrian eventing. *J. Sport Exercise Psychol.* **2005**, *27*, 245–252.
39. Swift, S. *Centred Riding*; William Heinemann Ltd.: London, UK, 1985.
40. Wanless, M. *Ride with Your Mind: A Right Brain Approach to Riding*; Methuen Publishing Ltd.: York, UK, 1987.
41. Lupton, D. *Risk*; Routledge: London, UK, 1999.
42. Hopkins, A. Studying organisational cultures and their effects on safety. *Saf. Sci.* **2006**, *44*, 875–889. [CrossRef]
43. Haukelid, K. Theories of (safety) culture revisited-an anthropological approach. *Saf. Sci.* **2008**, *46*, 413–426. [CrossRef]
44. Cooper, M.D.; Phillips, R.A. Exploratory analysis of the safety climate and safety behavior relationship. *J. Safety Res.* **2004**, *35*, 497–512. [CrossRef] [PubMed]
45. McManus, P.; Albrecht, G.; Graham, R. *The Global Horseracing Industry: Social, Economic, Environmental and Ethical Perspectives*; Routledge: London, UK, 2013.
46. McGreevy, P.D.; Oddie, C.F.; Hawson, L.A.; McLean, A.N.; Evans, D.L. Do vendors value safety in thoroughbred horses in the australian recreational riding horse market? *J. Vet. Behav. Clin. Appl. Res.* **2014**, *10*, 153–157. [CrossRef]
47. McGreevy, P.D.; Oddie, C.; Burton, F.L.; McLean, A.N. The horse-human dyad: Can we align horse training and handling activities with the equid social ethogram? *Vet. J.* **2009**, *181*, 12–18. [CrossRef] [PubMed]
48. Raber, K.; Tucker, T.J. *The Culture of the Horse: Status, Discipline, and Identity in the Early Modern World*; Palgrave Macmillan: New York, NY, USA, 2005.
49. van Orden, K. From gens d'armes to gentilshommes: Dressage, civility, and the ballet à cheval. In *The Culture of the Horse: Status, Discipline, and Identity in the Early Modern World*; Raber, K., Tucker, T.J., Eds.; Palgrave Macmillan: New York, NY, USA, 2005; pp. 197–222.
50. Xenophon. *The Art of Horsemanship*; J.A. Allen & Co.: London, UK, 1962.
51. Thompson, K. Theorising rider-horse relations: An ethnographic illustration of the centaur metaphor in the spanish bullfight. In *Theorising Animals*; Taylor, N., Signal, T., Eds.; Brill: Leiden, the Netherlands; Boston, MA, USA, 2011; pp. 221–253.
52. Game, A. Riding: Embodying the centaur. *Body Soc.* **2001**, *7*, 1–12. [CrossRef]
53. Goodwin, D.; McGreevy, P.; Waran, N.; McLean, A. How equitation science can elucidate and refine horsemanship techniques. *Vet. J.* **2009**, *181*, 5–11. [CrossRef] [PubMed]
54. Henshall, C.; McGreevy, P.D. The role of ethology in round pen horse training-a review. *Appl. Anim. Behav. Sci.* **2014**, *155*, 1–11. [CrossRef]
55. Boot, M.; McGreevy, P.D. The x files: Xenophon re-examined through the lens of equitation science. *J. Vet. Behav. Clin. Appl. Res.* **2015**, *8*, 367–375. [CrossRef]
56. McGreevy, P.; Boakes, R.; Boakes, R.A. *Carrots and Sticks: Principles of Animal Training*; Darlington Press: Sydney, Australia, 2011.
57. DeAraugo, J.; McLean, A.; McLaren, S.; Caspar, G.; McLean, M.; McGreevy, P. Training methodologies differ with the attachment of humans to horses. *J. Vet. Behav. Clin. Appl. Res.* **2014**, *9*, 235–241. [CrossRef]

58. Gielen, A.C.; Sleet, D. Application of behavior-change theories and methods to injury prevention. *Epidemiol. Rev.* **2003**, *25*, 65–76. [CrossRef] [PubMed]
59. Ajzen, I. From intentions to actions: A theory of planned behavior. In *Action Control: From Cognition to Behavior*; Kuhl, J., Beckman, J., Eds.; Springer: Heidelberg, Germany, 1985; pp. 11–39.
60. Prochaska, J.O.; DiClemente, C.C.; Norcross, J.C. In search of how people change: Applications to addictive behaviors. *Amer. Psychol.* **1992**, *47*, 1102–1114. [CrossRef]
61. Fishbein, M. A theory of reasoned action: Some applications and implications. *Nebr. Symp. Motiv.* **1980**, *27*, 65–116. [PubMed]
62. Helmreich, R.L.; Davies, J.M. Culture, threat, and error: Lessons from aviation. *Can. J. Anesth.* **2004**, *51*, R1–R4. [CrossRef]
63. Visser, E.K.; Van Wijk-Jansen, E.E.C. Diversity in horse enthusiasts with respect to horse welfare: An explorative study. *J. Vet. Behav. Clin. Appl. Res.* **2012**, *7*, 295–304. [CrossRef]
64. Sandman, P.M. *Media Campaigns. Environmental Education & Communication for a Sustainable World*; Academy for Educational Development: Washington, DC, USA, 2000; pp. 79–84.
65. Veblen, T. *The Theory of a Leisure Class*; Penguin Books: New York, NY, USA, 1899.
66. Busick, J. Getting workers to wear ppe: Communication is key. Safety Compliance Letter, 2005, Available online: http://www.psandman.com/articles/Busick.pdf(accessed on 14 July 2015).
67. McManus, P.; Graham, R. Horse racing and gambling: Comparing attitudes and preferences of racetrack patrons and residents of sydney, australia. *Leisure Stud.* **2014**, *33*, 400–417. [CrossRef]
68. Thompson, K.; Matthews, C. Inroads into equestrian safety: Rider-reported factors contributing to horse–related accidents and near misses on Australian roads. *Animals* **2015**, *5*, 592–609.
69. Chapman, C.; Musselwhite, C.B.A. Equine road user safety: Public attitudes, understandings and beliefs from a qualitative study in the united kingdom. *Accid. Anal. Prevent.* **2011**, *43*, 2173–2181. [CrossRef] [PubMed]
70. Cassidy, R.L. *Horse People: Thoroughbred Culture in Lexington and Newmarket*; John Hopkins University Press: Baltimore, MD, USA, 2010.
71. Krumeich, A.; Weijts, W.; Reddy, P.; Meijer-Weitz, A. The benefits of anthropological approaches for health promotion research and practice. *Health Educ. Res.* **2001**, *16*, 121–130. [CrossRef] [PubMed]
72. Thompson, K. For pets' sake, save yourself: Motivating emergency and disaster preparedness through relations of animal guardianship. *Autr. J. Emerg. Manag.* **2015**, *30*, 43–46.
73. Brandt, K. A language of their own: An interactionist approach to human-horse communication. *Soc. Anim.* **2004**, *12*, 299–316. [CrossRef]
74. Montoya, D.; McManus, P.; Albrecht, G. Jumping to conclusions? Media coverage of jumps racing debates in australia. *Soc. Anim.* **2012**, *20*, 273–293. [CrossRef]
75. McManus, P.; Montoya, D. Toward new understandings of human-animal relationships in sport: A study of australian jumps racing. *Soc. Cult. Geogr.* **2012**, *13*, 399–420. [CrossRef]
76. Eckardt, F.; Münz, A.; Witte, K. Application of a full body inertial measurement system in dressage riding. *J. Equine. Vet. Sci.* **2014**, *34*, 1294–1299. [CrossRef]

77. Janura, M.; Peham, C.; Dvorakova, T.; Elfmark, M. An assessment of the pressure distribution exerted by a rider on the back of a horse during hippotherapy. *Hum. Movement Sci.* **2009**, *28*, 387–393. [CrossRef] [PubMed]
78. Peham, C.; Kotschwar, A.B.; Borkenhagen, B.; Kuhnke, S.; Molsner, J.; Baltacis, A. A comparison of forces acting on the horse's back and the stability of the rider's seat in different positions at the trot. *Vet. J.* **2010**, *184*, 56–59. [CrossRef] [PubMed]
79. Tsohantaridis, V.A. Development of the Jewel Equestrian Scale. Unpublished Thesis, George Fox University, Portland, OR, USA, 2004.
80. Lyng, S. Edgework: A social psychological analysis of voluntary risk taking. *Amer. J. Sociol.* **1990**, *95*, 851–886. [CrossRef]
81. Giulianotti, R. Risk and sport: An analysis of sociological theories and research agendas. *Sociol. Sport J.* **2009**, *26*, 540–556.
82. Zuckerman, M.; Neeb, M. *Sensation Seeking and Psychopathology*; Hogrefe & Huber Publishers: Washington, DC, USA, 1979.
83. Wolframm, I.A.; Shearman, J.; Micklewright, D. A preliminary investigation into pre–competitive mood states of advanced and novice equestrian dressage riders. *J. Appl. Sport Psychol.* **2010**, *22*, 333–342. [CrossRef]
84. McManus, P.; Graham, R.; Ruse, K. The construction of human-animal relations. National jumps day 2013 at Te Rapa, Hamilton, New Zealand. *N. Z. Geographer* **2014**, *70*, 190–200. [CrossRef]
85. Lawrence, K.; McManus, P. Towards household sustainability in sydney? Impacts of two sustainable lifestyle workshop programs on water consumption in existing homes. *Geogr. Res.* **2008**, *46*, 314–332. [CrossRef]
86. Trigg, J.; Smith, B.; Thompson, K. Does emotional closeness to pets motivate their inclusion in bushfire survival plans? Implications for emergency communicators. *Aust. J. Emerg. Manag.* **2015**, *30*, 24–30.
87. Trigg, J.; Thompson, K.; Smith, B.; Bennett, P. Engaging pet owners in disaster risk and preparedness communications: Simplifying complex human–animal relations with archetypes. *Environ. Hazards* **2015**. [CrossRef]

MDPI
St. Alban-Anlage 66
4052 Basel
Switzerland
Tel. +41 61 683 77 34
Fax +41 61 302 89 18
www.mdpi.com

Animals Editorial Office
E-mail: animals@mdpi.com
www.mdpi.com/journal/animals

www.ingramcontent.com/pod-product-compliance
Ingram Content Group UK Ltd.
Pitfield, Milton Keynes, MK11 3LW, UK
UKIIW060124010826
14089UKWH00006B/133

9 783036 550466